AF358699

Fungal Biotechnology and Application 2.0

Fungal Biotechnology and Application 2.0

Editor

Baojun Xu

Basel • Beijing • Wuhan • Barcelona • Belgrade • Novi Sad • Cluj • Manchester

Editor
Baojun Xu
Beijing Normal
University-Hong Kong
Baptist University United
International College
Zhuhai
China

Editorial Office
MDPI
St. Alban-Anlage 66
4052 Basel, Switzerland

This is a reprint of articles from the Special Issue published online in the open access journal *Journal of Fungi* (ISSN 2309-608X) (available at: www.mdpi.com/journal/jof/special_issues/685O0VFCHP).

For citation purposes, cite each article independently as indicated on the article page online and as indicated below:

Lastname, A.A.; Lastname, B.B. Article Title. *Journal Name* **Year**, *Volume Number*, Page Range.

ISBN 978-3-7258-0670-6 (Hbk)
ISBN 978-3-7258-0669-0 (PDF)
doi.org/10.3390/books978-3-7258-0669-0

Contents

About the Editor

Baojun Xu

Dr. Xu is a Chair Professor in Beijing Normal University-Hong Kong Baptist University United International College (UIC, a full English teaching college in China), Fellow of the Royal Society of Chemistry, Zhuhai Scholar Distinguished Professor, Department Head of Department of Life Sciences, Program Director of Food Science and Technology Program, and author of over 330 peer-reviewed papers. Dr. Xu received his Ph.D. in Food Science from Chungnam National University, South Korea. He conducted postdoctoral research work in North Dakota State University (NDSU), Purdue University, and Gerald P. Murphy Cancer Foundation in USA during 2005-2009. He conducted short-term visiting research in NDSU in 2012 and University Georgia in 2014, followed by visiting research during his sabbatical leave (7 months) in Pennsylvania State University in USA in 2016. Dr. Xu is serving as the Associate Editor-in-Chief of *Food Science and Human Wellness*, Associate Editor of *Food Research International*, Associate Editor of *Food Frontiers*, and an Editorial Board Member of around 10 international journals. He received the inaugural President's Award for Outstanding Research of UIC in 2016, and the President's Award for Outstanding Service of UIC in 2020. Dr. Xu was listed in the world's top 2% scientists by Stanford University in 2020, 2021, 2022, and 2023 and was listed in the Best Scientist in the World in the field of Biology and Biochemistry at Research.com in 2023.

Journal of
Fungi

Review

Unlocking the Power: New Insights into the Anti-Aging Properties of Mushrooms

Jing Luo [1,2,†], Kumar Ganesan [1,3,†] and Baojun Xu [1,*]

[1]   Food Science and Technology Programme, Department of Life Sciences, BNU-HKBU United International College, Zhuhai 519087, China; jing.luo@tum-create.edu.sg (J.L.); kumarg@hku.hk (K.G.)
[2]   TUMCREATE, 1 Create Way, Singapore 138602, Singapore
[3]   School of Chinese Medicine, Li Ka Shing Faculty of Medicine, The University of Hong Kong, Hong Kong, China
[*]   Correspondence: baojunxu@uic.edu.cn; Tel.: +86-756-3620636
[†]   These authors contributed equally to this work.

Abstract: Aging is a complex biological process that is influenced by both intrinsic and extrinsic factors. Recently, it has been discovered that reactive oxygen species can accelerate the aging process, leading to an increased incidence of age-related diseases that are characteristic of aging. This review aims to discuss the potential of mushrooms as a dietary intervention for anti-aging, focusing on their nutritional perspective. Mushrooms contain various bioactive compounds, including carbohydrates, bioactive proteins, fungal lipids, and phenolic compounds. These compounds have shown promising effectiveness in combating skin aging and age-related diseases. In vitro and in vivo studies have demonstrated that treatments with mushrooms or their extracts can significantly extend lifespan and improve health span. Furthermore, studies have aimed to elucidate the precise cellular and molecular mechanisms of action and the structure–activity relationship of mushroom bioactive compounds. These findings provide a strong basis for further research, including human clinical trials and nutritional investigations, to explore the potential benefits of mushrooms in real-life anti-aging practices. By exploring the anti-aging effects of mushrooms, this review aims to provide valuable insights that can contribute to the development of broader strategies for healthy aging.

Keywords: Mushrooms; anti-aging; age-related disease; cellular mechanisms; bioactive compounds

Citation: Luo, J.; Ganesan, K.; Xu, B. Unlocking the Power: New Insights into the Anti-Aging Properties of Mushrooms. *J. Fungi* **2024**, *10*, 215. https://doi.org/10.3390/jof10030215

Academic Editor: Mingwen Zhao

Received: 22 February 2024
Revised: 10 March 2024
Accepted: 12 March 2024
Published: 14 March 2024

1. Introduction

The global population is currently experiencing a significant expansion of aging populations compared to previous years. This trend is reflected in the increase in average life expectancy at birth, which has risen by 6.2 years from 65.3 years in 1990, to 71.5 years in 2013. Additionally, individuals who reach the age of 60 can now expect to live for another 22 years on average [1]. By the year 2040, it is projected that the average life expectancy will increase by 4.4 years for both men and women. Men can expect to live an average of 74.3 years, while women can expect to live an average of 79.7 years. However, these numbers may vary depending on individual health conditions [2]. As the population ages, there has been a noticeable increase in the prevalence of chronic degenerative diseases such as neurodegenerative and cardiovascular diseases, diabetes, and cancer. These diseases contribute to up to 70% of global mortality each year, including premature deaths occurring between the ages of 30 and 70 [1]. It is important to note that, while aging is often accompanied by deteriorative changes and an increased risk of functional declines or diseases, aging itself is not considered a disease. The focus of anti-aging strategies is not to reverse or halt the aging process, but rather to promote healthy aging and reduce the incidence of age-related diseases. The World Health Organization recommends adopting healthy dietary habits, engaging in regular physical activity, and controlling tobacco use as effective measures to alleviate or prevent the incidence of chronic diseases. By following these guidelines, the risk of developing age-related diseases can be reduced [3].

There is growing evidence to suggest that healthy aging can be promoted by consuming nutraceuticals and following various dietary patterns, such as caloric restriction, intermittent fasting, a Mediterranean diet, an Okinawan diet, and a Nordic diet. These dietary patterns have been evaluated for their negative correlation with aging and age-related conditions and diseases [4,5], which has led to a search for anti-aging components from food sources and an investigation of the underlying mechanisms of anti-aging pathways. Bioactive compounds derived from plant sources, including fruits and vegetables, roots, seeds, and edible flowers, have been suggested to exert anti-aging effects. These compounds include certain polysaccharides, phenolic compounds, and peptides [6,7]. In recent years, mushrooms—filamentous fungi with fruiting bodies—have also been shown to possess enormous pharmacological attributes that are valuable for healthy aging. These attributes include anti-oxidant, immunomodulatory, neuroprotective, anti-inflammatory, and anti-cancer properties [8–11].

Mushrooms are nutritious foods that are rich in carbohydrates and proteins, with a lower content of lipids [12]. In addition to their nutritional value, mushrooms contain various bioactive compounds, such as β-glucans, lectins, and linolenic acids, which can be isolated through different extraction methods. These compounds confer a variety of pharmacological activities and may enhance the immune system and strengthen the biological function of the body [13]. Regular intake of mushrooms or their extracts may help alleviate age-related diseases. This review focuses on the anti-aging properties of mushrooms from the perspective of aging and age-related diseases, with a brief introduction of the major bioactive compounds found in edible and medicinal mushrooms.

2. Aging

2.1. Aging and Age-Related Diseases

Aging is a complex process that involves the time-dependent accumulation of diverse deleterious changes in cells, tissues, organs, or systems that increase vulnerability to chronic illness and death [14,15]. Nine candidate hallmarks of aging have been identified and classified, including primary hallmarks (genomic instability, telomere attrition, epigenetic alterations, and loss of proteostasis), antagonistic hallmarks (deregulated nutrient sensing, mitochondrial dysfunction, and cellular senescence), and integrative hallmarks (stem cell exhaustion and altered intercellular communication), all of which are correlated with each other [16]. The antagonistic hallmarks exert positive effects at low levels but negatively affect the organism at high levels [16]. For example, reactive oxygen species (ROS) are important signaling molecules that play a role in regulating cellular functions, but excessive levels can lead to oxidative damage and contribute to aging. The primary hallmarks are the contributors to molecular damage during aging, while the integrative hallmarks are signs of failure of cellular homeostasis and metabolism mechanisms to ameliorate the damage. These hallmarks are interconnected with each other and could serve as a guidance to decipher the mechanistic molecular basis for prolonging health span and development of strategies for longevity, such as stem-cell-based therapies, epigenetic drugs, anti-inflammatory drugs, and dietary restrictions [16].

The free radical theory of aging, proposed in 1956 by Denham Harman [17], is a widely accepted theory of aging. The theory postulates that the aging process is triggered by the initiation of free radical reactions, leading to increased generation of free radicals by damaged mitochondria with increasing age [18]. Major sources of free radical reactions in mammals include non-enzymatic reaction of oxygen, ionizing radiation, cytochrome P-450 system, respiratory chain, phagocytosis, and prostaglandin synthesis, which lead to the accumulation of oxidative damage and may shorten the lifespan. Several defenses that alleviate the damage of the reactions include DNA repair mechanisms, superoxide dismutase, glutathione peroxidase, and anti-oxidants (e.g., carotenes and vitamin E) [15,19].

ROS are byproducts of oxidative metabolism that can induce cellular defense mechanisms against oxidative invasion at low doses, potentially prolonging health span and lifespan. However, long-term excessive exposure to ROS can lead to the oxidation of nucleic acids, proteins, and lipids, causing damage to macromolecules and mitochondrial dysfunction. This can disrupt cell homeostasis and result in cellular death [20]. ROS production is driven by progressive mitochondrial dysfunction with increasing age, creating a positive feedback loop of ROS generation and oxidative damage accumulation [18]. Concurrently, oxidative stress arises due to excessive ROS levels and limited anti-oxidant defense capability, leading to cellular senescence and a shortened lifespan. The accumulation of oxidative damage to macromolecules and mitochondria contributes to detrimental consequences, such as pathophysiological changes, functional decline, and accelerated aging, which are associated with age-related conditions such as inflammation, cardiovascular diseases, neurodegenerative diseases, autoimmune diseases, and cancer [21].

It is important to note that aging itself is not a disease. Age-related diseases can be considered "symptoms" of aging, initiated by minor disturbances that are intensified via vicious positive feedback loops, destabilizing the physiology of an organism and potentially leading to destruction (i.e., mortality) if no negative feedback loops are in place [22]. For example, low-grade inflammation can intensify in chronic inflammation, leading to decreased muscle mass, decreased physical activity, and excess fat deposition. This can further contribute to obesity, diabetes, and cardiovascular problems. Eventually, cardiovascular diseases can arise and worsen the physiological status of an individual by triggering chronic inflammation. To minimize cumulative damage to different organs and maintain cell function for healthy aging, interventions that can interrupt or break the vicious cycles of age-related diseases can be implemented, including medications, lifestyle adjustments, and dietary management (Figure 1).

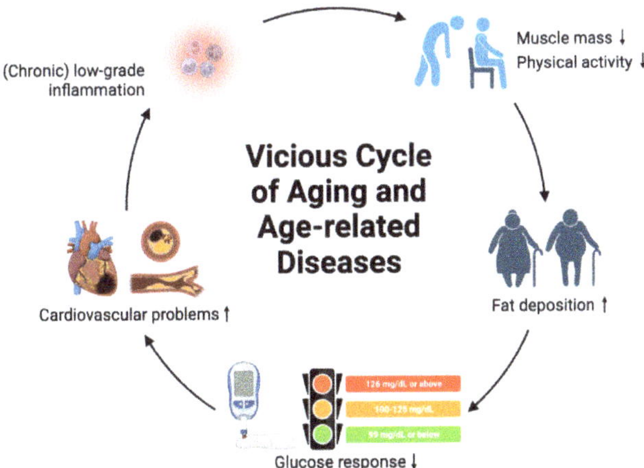

Figure 1. Concept of vicious cycle of aging and age-related diseases. Symbol ↑ denotes increase; symbol ↓ denotes decrease.

2.2. Aging and Dietary Intervention

The lifestyle of an individual is closely linked to their health span and lifespan. One of the main ways to modify lifestyle for better health maintenance and to reduce the incidence of age-related diseases is through dietary management. Unhealthy dietary habits and lifestyle can accelerate the aging process by causing molecular and cellular damage. For example, a sedentary lifestyle, combined with a "Western diet", that is high in energy but lacking in nutrition, has been associated with reduced lifespan and increased occurrence of age-related conditions such as obesity, type 2 diabetes, and cancer [23]. On the other hand, caloric restriction (CR) has been shown to slow down the rate of aging and extend health span. CR involves reducing total energy intake by 20% to 40% while ensuring optimal nutrition, compared to an ad libitum diet. This approach has been demonstrated to extend lifespan and health span in various experimental models, including yeast, fruit flies, mice, nonhuman primates, and even humans [24–26].

According to the theory of aging, CR enhances longevity by reducing oxidative damage and increasing resistance to oxidative stress through specific signaling pathways. The stress caused by CR, such as nutrient deprivation, activates defense mechanisms against oxidative damage, thereby slowing down the aging process [27]. CR also affects physiological pathways that may mediate anti-aging effects, such as the insulin-like growth factor-1 and insulin signaling pathways, the mammalian target of rapamycin (mTOR) pathway, and the sirtuins pathway [24,28,29]. Previous studies have demonstrated the potential of implementing CR as an anti-aging regimen, as adherence to this dietary management reduces biomarkers associated with the development of age-related diseases, including cardiovascular diseases, autoimmune disorders, neurodegenerative diseases, diabetes, and cancer [29–31]. Therefore, CR can be considered as the mechanistic foundation for healthy aging strategies involving dietary intervention, which can prolong lifespan and maintain physiological function for an extended health span.

Despite the potential benefits of CR, it can be challenging for individuals to adhere to it in the long term due to various pitfalls and health concerns, such as hypotension, osteoporosis, slower wound healing, depression, and irritability [32]. As a result, scientists have explored alternative diet regimens and studied different dietary patterns that may offer similar benefits to CR but are more feasible for humans to sustain. One such approach is intermittent fasting, which shares the same concept as CR. Intermittent fasting activates cellular pathways that enhance the body's intrinsic defense against oxidative stress, promotes the removal of damaged molecules, and facilitates tissue repair and growth. It also helps to suppress inflammation and improve stress resistance [33,34].

In addition to dietary modifications, researchers have developed anti-aging drugs that mimic the effects of CR. Examples include rapamycin and metformin, which have shown promising effects in various model organisms and clinical trials. Rapamycin delays aging by inhibiting mTOR, thereby maintaining the normal functioning of mitochondria and stem cells. Metformin, on the other hand, affects telomere length, reduces oxidative damage to DNA, and modulates the synthesis and degradation of age-related proteins [35,36]. However, it is important to note that there are concerns and side effects associated with the use of these drugs. For instance, rapamycin may lead to nephrotoxicity and thrombocytopenia, while metformin may cause vitamin B12 deficiency and lactic acid accumulation [37,38]. Therefore, there is a need to explore naturally occurring compounds that have significant anti-aging effects with minimal side effects.

Nutraceuticals and dietary supplements are also viable alternatives for anti-aging and extending health span. Examples include curcumin, quercetin, ginseng, and medicinal mushrooms, which exhibit anti-inflammatory, immunomodulatory, and antioxidative effects [39–41]. A diet rich in fruits and vegetables, which provide a significant number of nutraceuticals and phytochemicals, is crucial for maintaining overall health. Interestingly, mushrooms, although not classified as animals or plants but as part of the fungal kingdom, are often considered as vegetables. They are low in calories, sodium, and fat, while being a valuable source of fiber, phenolic compounds, β-glucans, selenium, glutathione, B vitamins, and vitamin D. These components serve as protective agents against oxidative damage, which accelerates aging [12]. Medicinal mushrooms have also been used for centuries in traditional therapies, like Chinese medicine and Indian Ayurveda medicine, to alleviate symptoms of various diseases [42]. The bioactive compounds found in mushrooms may contribute to their anti-aging effects through various physiological pathways involved in aging and age-related diseases.

2.3. Ageing, Mental Health and Gender

Gender and mental health can significantly impact ageing experiences. Gender influences ageing in various ways, including health outcomes, social roles and expectations, and economic status. Women are more likely to experience depression, anxiety, and stress due to factors such as caregiving responsibilities, hormonal changes, and discrimination [43]. Women also tend to report higher levels of loneliness and social isolation in later life. In contrast, men may experience social isolation and mental health issues due to societal expectations of masculinity, which can lead to reluctance in seeking help for mental health problems [44].

Gender differences in health outcomes are well-documented, with women living longer but experiencing more chronic health conditions than men. Women are more likely to experience osteoporosis, urinary incontinence, and depression than men. Women also experience menopause, which can lead to physical and psychological symptoms [45]. Men, on the other hand, are more likely to experience heart disease, stroke, and certain types of cancer. Biological factors such as sex hormones, genetics, and lifestyle factors like diet, exercise, and smoking influence gender differences in health outcomes [46].

Gender roles and expectations can influence ageing experiences [47]. Women are often expected to take on caregiving roles for children, spouses, or ageing parents, which can lead to stress and impact their own health and well-being. Women may also face ageism and discrimination in the workplace, leading to financial insecurity in later life. Men, on the other hand, may experience pressure to maintain their independence and financial stability, leading to social isolation and mental health issues [43,44,48].

Gender differences in economic status can also impact ageing experiences. Women often earn less than men over their lifetimes, leading to lower retirement savings and financial insecurity in later life. Women are also more likely to work part-time or take career breaks to care for children or ageing parents, which can impact their pension entitlements. This can lead to poverty and social exclusion in later life [47,49]. Mental health issues, such as depression, anxiety, and cognitive impairment, can also impact ageing experiences. Depression is a common mental health issue among older adults and can lead to social isolation, physical illness, and suicide. Anxiety can affect quality of life and daily functioning. Cognitive impairment, including dementia, can result in memory loss, decision-making difficulties, and loss of independence and increased caregiving needs [50].

Studies have shown that gender and mental health can interact to influence ageing experiences [28,36,43,44,51,52]. Women with depression may be more prone to physical disability and cognitive decline in later life compared to men with depression. Similarly, men with higher levels of anxiety may be more likely to experience cognitive decline than women with anxiety [43]. Addressing gender and mental health in ageing policies and practices is crucial to ensure that older adults receive appropriate support and services. This includes promoting gender equity, addressing mental health stigma, and providing accessible and affordable mental health care for older adults [22].

3. Components of Mushrooms and Their Anti-Aging Effects

Mushrooms have long been recognized for their nutritional value and potential health benefits. Edible mushrooms are not only rich in protein, fiber, vitamins, and minerals but also have low levels of fat, making them highly nutritious [53,54]. They contain all the essential amino acids and have a higher protein content compared to most vegetables, making them particularly beneficial for vegetarians. In addition to their nutritional value, edible mushrooms, as fungi, have the ability to produce a wide range of chemical compounds known as mycochemicals. These mycochemicals can act as bioactive substances with various advantages for human health [55]. Mushrooms have been found to contain significant levels of mycochemicals that serve as bioactive compounds, offering a range of health benefits against aging and age-related diseases [53,54].

3.1. Bioactive Compounds in Mushrooms

Bioactive compounds extracted from mushrooms have been extensively studied for their ability to enhance cellular functions and provide health benefits. The following text summarizes four representative categories of bioactive compounds found in mushrooms: carbohydrates, proteins, lipids, and phenolic compounds.

3.1.1. Carbohydrates

Carbohydrates derived from mushrooms have been extensively studied for their anti-tumor, anti-inflammatory, and immunomodulatory activities [56,57]. Numerous monosaccharides found in mushrooms, including arabinose, fructose, fucose, galactose, glucose, mannose, mannitol, rhamnose, trehalose, and xylose, have been identified as exhibiting these activities. They primarily achieve this through the activation of cytokines, such as interferons and interleukins, and involve cellular pathways that include dendritic cells, natural killer cells, neutrophils, and cytotoxic macrophages [57,58]. β-Glucans, the main type of carbohydrates found in mushrooms, have been shown to possess antioxidative, anti-cancer, immunomodulatory, and neuroprotective properties. They are considered potent agents for stimulating the immune system and protecting against carcinogens, pathogens, and toxins [59–64]. The biological activity and health benefits of β-glucans isolated from mushrooms, particularly in relation to immune health, are crucial for healthy aging. Supplementation with mushroom carbohydrates, which contain β-glucans, could be an effective strategy for anti-aging. Table 1 provides a list of various mushrooms that contain bioactive carbohydrates.

Table 1. Bioactive carbohydrates in selected mushrooms.

Mushrooms	Common Names	Bioactive Compounds	Source and Yield	Bioactivities	References
Agaricus bisporus	Button mushroom	Heteropolysaccharide Abnp1001, Abnp1002, Abap1001, Abap1002	Concentrated industrial wastewater of *A. bisporus*; 0.989 mg/g, 1.849 mg/g, 0.128 mg/g, and 0.68 mg/g (Abnp1001, Abnp1002, Abap1001, Abap1002)	Hepatoprotective	[65]
		Heteropolysaccharide AcAPS, AcAPS-1, AcAPS-2, AcAPS-3, with rhamnose and glucose as major monosaccharide	Dried fruiting body; yield n.s.	Hepatoprotective, nephroprotective, antioxidative	[66]
		Polysaccharide extracts, main components n.s.	Whole mushroom; yield n.s.	Anti-tumor, immunostimulatory	[67]
		Heteropolysaccharide/Mannogalacoglucan mannose, galactose, glucose	Freeze-dried fresh fruiting body; 41.4% yield (*w/w* dry weight)	Anti-tumor	[68]
		β-glucan	Dried fresh fruiting body; yield n.s.	Immunostimulatory	[69]
		Fructose, mannitol, trehalose	Fresh fruiting body; 5.79% (white mushroom) & 4.27% (brown mushroom) (*w/w* fresh weight)	n.s.	[70]
Calocybe indica	Milky mushroom	Polysaccharide extracts, main components n.s.	Fresh fruiting body; 3.27% (*w/w* dry weight)	Anti-oxidant, neuroprotective	[71]
Flammulina velutipes	Enoki/Golden needle mushroom	Polysaccharide extracts, main components n.s.	Base of stipe; yield n.s.	Anti-tumor	[72]
		Polysaccharide extracts, main components n.s.	Fresh whole-mushroom; yield n.s.	Neuroprotective	[73]
		Fructose, mannitol, sucrose, trehalose	Fresh fruiting body; 8.29% (*w/w* fresh weight)	n.s.	[70]
Ganoderma lucidum	Ling Zhi	Polysaccharide extracts, main components n.s.	Mycelia; 71.99% (*w/w* dry weight)	Anti-inflammation, ameliorating insulin resistance, suppressing lipid accumulation, regulation of gut microbiota	[74]
		Polysaccharide extracts, main components n.s.	Commercialized spray dried mycelia; 91.48% (*w/w* dry weight)	Improving intestinal barrier functions	[75]
		Arabinose, galactose, glucose, xylose	Whole mushroom; yield n.s.	Anti-tumor	[76]
		Polysaccharide extracts, main components n.s.	Dried conidial powder; 2% (*w/w* dry weight, crude extracts)	Promote cognitive function and neural progenitor proliferation	[77]

Table 1. *Cont.*

Mushrooms	Common Names	Bioactive Compounds	Source and Yield	Bioactivities	References
Lentinula edodes	Shiitake mushroom	Glucose, galactose, mannose, arabinose	Fruiting body; 1.3% (*w/w* dry weight, purified polysaccharide cLEP1)	Therapeutic to cervical carcinoma	[78]
		Rhamnose	Residue/byproduct; yield n.s.	Anti-inflammatory, anti-oxidant	[79]
		Pyranose, β-D-glucans (β-(1→3)-D-glucose as backbone & β-(1→6)-D-glucose as side chains)	Dried fruiting body; 0.76% (*w/w* dry weight)	Anti-tumor	[80]
		Mannogalactoglucan-type polysaccharides WPLE-N-2, WPLE-A0.5-2	Fruiting body; yield n.s.	Anti-cancer, immunomodulatory	[81]
		Lentinan (β-(1,3)-glucan with β-(1,6) branches)	Dried fruiting body (commercial product); 2.6% (*w/w* dry weight)	Anti-tumor	[82]
		Mannitol, trehalose, arabinose	Dried powder; 23.3% (mannitol), 13.2% (trehalose), 1.79% (arabinose) (*w/w* dry weight)	n.s.	[83]
Pleurotus eryngii	King oyster mushroom	Mannose, glucose, galactose	Fresh whole-mushroom; 5.4% (*w/w* dry weight)	Anti-tumor	[84]
		Heteropolysaccharides, novel fractions PEPE-1, PEPE-2, PEPE-3 (mannose, glucose, galactose, xylose)	Fresh mushroom residue; yield n.s.	Anti-tumor	[85]
		Mannose, glucose, galactose	Fresh whole-mushroom; 28.3% (*w/w* dry weight)	Immunomodulatory	[86]
Pleurotus ostreatus	Oyster mushroom	Crude polysaccharide extracts	Fresh whole-mushroom; 61% (*w/w*)	Alleviation of cognitive impairment	[87]
		Crude polysaccharide extracts	Fresh whole-mushroom; 63.98% (*w/w*)	Regulation of dislipidemia	[88]
		Homogeneous polysaccharides, fractions POMP1, POMP2, POMP3	Mycelia; yield n.s.	Anti-tumor	[89]

n.s., not specified; Abnp, *Agaricus bisporus* polysaccharides between 5 kDa and 100 kDa; Abap, *Agaricus bisporus* polysaccharides under 5 kDa; AcAPS, purified fractions of acidic-extractable polysaccharides; WPLE, mannogalactoglucan-type polysaccharides from *Lentinus edodes*; POMP, *Pleurotus ostreatus* mycelium polysaccharide.

3.1.2. Proteins

Compared to other food sources, mushrooms contain higher levels of bioactive proteins such as lectins, ribosome inactivating proteins, fungal immunomodulatory proteins, and laccases, which possess various biological activities (Table 2) including antioxidative, immunomodulatory, anti-inflammatory, and anti-cancer properties [90]. Lectins are non-immune proteins or glycoproteins that bind to specific carbohydrates on cell surfaces, acting as nutraceuticals with immunomodulatory, anti-tumor, and anti-proliferative properties [90]. Other mushroom proteins, such as laccase, fungal immunomodulatory protein, and ribosome inactivating proteins, have distinct bioactive activities. Laccases are considered multicopper oxidases implicated in processes such as pathogenesis, morphogenesis, and immunogenesis of an organism [90]. Fungal immunomodulatory proteins purified from mushrooms, such as *Ganoderma lucidum*, *Ganoderma tsugae*, *Poria cocos, and Trametes versicolor*, have been suggested as potential adjuvants for tumor therapy due to their structural similarity to human antibodies and their ability to suppress tumor metastasis and invasion [91–95].

The ribosome inactivating protein family acts as rRNA N-glycosylase, inactivating 60S ribosomal subunits through an N-glycosidic cleavage that eliminates one or more adenosine residues from rRNA to inhibit protein synthesis [112]. Members of the ribosome inactivating protein family, such as trichosanthin, luffin, ricin, and abrin, have been of considerable interest due to their potent activity against viral infections and their potential use as immunotoxins for cancer treatment by conjugating with monoclonal antibodies [113–115]. However, it is noteworthy that some mushroom ribosome inactivating proteins may be hazardous and pose adverse effects on health. For instance, hypsin from *Hypsizigus mamoreus* has been reported to increase in vitro cell death [116]. Therefore, it is important to elucidate the structure-functional properties of mushroom proteins as they may be toxic to humans when consumed. Table 2 lists various bioactive proteins derived from mushrooms.

Table 2. Bioactive proteins in mushrooms.

Mushrooms	Common Names	Bioactive Compounds/Substances *	Bioactivities	References
Agaricus bisporus	Button mushroom	Lectin	Immunomodulatory	[96]
Cerrena unicolor	Mossy maze polypore	Laccase	Anti-tumor	[97]
Coprinus comatus	Shaggy mane/chicken drumstick mushroom	Laccase	Anti-viral	[98]
Flammulina velutipes	Enoki/Golden needle mushroom	FIP	Anti-inflammatory	[99]
		RIP	Anti-viral	[100]
Ganoderma applanatum	Artist's conk	Lectin	Anti-tumor	[101]
Ganoderma lucidum	Lingzhi	Laccase	Anti-viral	[102]
Ganoderma tsugae	Hemlock reishi	FIP	Immunomodulatory	[103]
Hypsizygus marmoreus	Jade mushroom	RIPs (hypsin, marmorin)	Anti-fungal, anti-tumor	[104,105]
Inonotus baumii	Sanghuang	Laccase	Anti-tumor	[106]
Macrolepiota procera	Parasol mushroom	Lectin	Anti-tumor	[107]
Pleurotus cornucopiae	Golden oyster	Laccase	Anti-viral, anti-tumor	[108]
Pleurotus eryngii	King oyster mushroom	Laccase	Anti-viral	[109]
Pleurotus ostreatus	Oyster mushroom	Lectin	Immunomodulatory	[110]
Sparassis latifolia	Cauliflower mushroom	Lectin	Anti-fungal, anti-bacteria	[111]

* Include various categories and sub-categories of proteins. FIP, fungal immunomodulatory protein. RIP, ribosome inactivating protein.

3.1.3. Lipids

Although mushrooms have a low fat content ranging from 0.1% to 16.3%, they are a good source of high-quality essential fatty acids such as oleic acid (1–60.3% of total fatty acids in 100 g), linoleic acid (0–81.1% of total fatty acids in 100 g), and linolenic acid (0–28.8% of total fatty acids in 100 g) [117]. Table 3 summarizes the lipid profiles of various mushrooms in terms of the content of saturated fatty acids (SFA), monounsaturated fatty acids (MUFA), and polyunsaturated fatty acids (PUFA). Mushrooms are good sources of unsaturated fatty acids, as observed in a study by Günç Ergönül et al. [118] who investigated the fatty acid compositions of six wild edible mushroom species and found that unsaturated fatty acids predominated over saturated ones. In most nutritional characterization studies, mushroom fatty acids are commonly determined using gas-liquid chromatography coupled with a flame ionization detector. However, the sample extraction method used prior to measurement may impact the final outcome of lipid profiles. For instance, a study by Sinanoglou et al. [119] investigated the lipid profiles of *Laetiporus sulphureus* using different combinations of extraction methods and two individual solvents and found variations among the four combinations [119]. Ergosterol, the major sterol found in mushrooms, accounts for the major lipid component of fungal extracellular vesicles as well [120]. Ergosterol extracted from medicinal mushroom *Ganoderma lucidum* has been shown to exert anti-oxidant effects and reduce the risk of cardiovascular diseases while extending lifespan [55,121,122]. Compared to lipids from animal sources, edible mushrooms are advantageous due to their high levels of polyunsaturated fatty acids, which may regulate various physiological functions in age-related diseases, such as decreasing blood pressure and triglyceride levels, and reducing the risks of age-related cardiovascular diseases, arthritis, and neurodegenerative diseases [64,123]. Therefore, mushrooms may play a significant role in human nutrition and anti-aging regimens based on their fatty acid profiles.

Table 3. Lipid profiles in mushrooms.

Mushrooms	Common Name	Total SFA (% of Total FA)	Total MUFA (% of Total FA)	Total PUFA (% of Total FA)	Measurement Techniques	References
Agaricus blazei	Almond mushroom	24.4	2.0	73.6	GC-FID	[83]
Agaricus bisporus	White button mushroom	20.3	1.4	78.3	Capillary GLC-FID	[70]
	Brown button mushroom	18.4	1.8	79.8		
Agrocybe cylindracea	Poplar mushroom	28.1	2.83	69.1	Capillary GLC-FID	[124]
Boletus reticulatus	Summer cep	21.1	40.3	38.4	GLC-FID	[118]
Coprinus comatus	Shaggy mane/Lawyer's cap	23.8	11.4	64.8	Capillary GLC-FID	[124]
Flammulina velutipes	Enoki/Golden needle mushroom	18.5	7.2	74.3	Capillary GLC-FID	[70]
		20.7	18.6	60.7	GLC-FID	[118]
Lactarius deliciocus	Saffron milkcap	20.8	42.0	37.3	Capillary GLC-FID	[124]
Lactarius salmonicolor	Salmon milkcap	19.0	19.6	61.6	GLC-FID	[118]
Lentinus edodes	Shiitake mushroom	16.7	3.5	79.8	GC-FID	[83]
		15.1	2.9	82.0	Capillary GLC-FID	[70]
Pleurotus eryngii	King oyster mushroom	17.4	13.1	69.4	Capillary GLC-FID	[70]
Pleurotus ostreatus	Oyster mushroom	17.0	13.6	69.4	Capillary GLC-FID	[70]
		21.8	11.4	66.5	GLC-FID	[118]
Polyporus squamosus	Dryad's saddle	25.2	34.3	40.6	GLC-FID	[118]
Russula anthracina	-	23.7	53.3	22.9	GLC-FID	[118]
Laetiporus sulphureus	Sulphur polypore	21.6	17.6	60.8	GC-FID, TLC-FID	[119]
Suillus collinitus	-	17.5	34.4	47.4	Capillary GLC-FID	[124]
Tricholoma myomyces	Grey knight mushroom	15.8	46.3	37.8	Capillary GLC-FID	[124]

SFA, saturated fatty acid. MUFA, monosaturated fatty acid. PUFA, polysaturated fatty acid. GC-FID, gas chromatography coupled with flame ionization detector. GLC-FID, gas-liquid chromatography coupled with flame ionization detection. TLC-FID, thin layer chromatography–flame ionization detection.

3.1.4. Phenolic Compounds

Phenolic compounds found in mushrooms are typically considered secondary metabolites. The most prominent phenolic compounds in mushrooms include heteroglycans, lectins, phenolic acids (such as ferulic, gallic, and cinnamic acids), flavonoids (including hesperetin, quercetin, kaempferol, and naringenin), steroids, alkaloids, tannins, chitinous substances, terpenoids, and tocopherols. These compounds exhibit various biological activities, including anti-oxidant, anti-tumor, anti-inflammatory, anti-hyperglycemic, antiosteoporotic, anti-tyrosinase, and anti-microbial effects, primarily due to their strong antioxidative properties [125–128]. Some of the preferred mushroom species for extracting phenolic compounds include *Agaricus brasiliensis* (almond mushroom), *Cantharellus cibarius* (chanterelle), *Lactarius indigo* (indigo milk cap), *Inonotus obliquus* (chaga mushroom), and *Melanoleuca cognate* [126,129–131]. Table 4 provides a summary of representative phenolic compounds extracted from various mushroom species.

Table 4. Extractable phenolic compounds in mushrooms.

Phenolic Compound Categories	Phenolic Compounds	Mushroom Sources	References
Phenolic acids	Ferulic acid	*Agaricus brasiliensis, Agrocybe aegerita, Calocybe indica, Cantharellus cibarius*	[126,127,132–134]
	Gallic acid	*Agaricus brasiliensis, Agrocybe aegerita, Calocybe indica, Cantharellus cibarius, Ganoderma lucidum, Pleurotus citrinopileatus, Pleurotus pulmonarius, Russula aurora*	[126,130,132–138]
	Cinnamic acid	*Amanita crocea, Ganoderma lucidum, Pleurotus ostreatus, Suilus belinii*	[135,139–141]
	Caffeic acid	*Calocybe indica, Cantharellus cibarius, Hyphodontia paradoxa, Inonotus obliquus, Pleurotus citrinopileatus, Pleurotus pulmonarius,*	[127,130,133,134,142,143]
	p-Coumaric acid	*Agaricus brasiliensis, Agaricus subrufescens, Amanita crocea, Hyphodontia paradoxa, Laccaria amethystea, Melanoleuca cognate, Pleurotus ostreatus*	[56,126,129,139,140,142,144]
	p-Hydroxybenzoic acid	*Agaricus brasilensis, Amanita crocea, Cantharellus cibarius, Lactarius indigo, Lentinus edodes, Melanoleuca cognate, Suillus belinii*	[126,129,134,138,139,141]
	Fumaric acid	*Agaricus brasiliensis*	[126]
	Vanillic acid	*Morchella esculenta* (L.) *Pers., Russula emetic*	[136,137]
	Syringic acid	*Hyphodontia paradoxa, Morchella esculenta* (L.) *Pers.*	[129,130,136,142]
	Protocatechuic acid	*Agrocybe aegerita, Calocybe indica, Cantharellus cibarius, Hyphodontia paradoxa, Inonotus obliquus, Melanoleuca, Morchella esculenta* (L.) *Pers., Suillus belinii, Russula emetic*	[129,130,132–134,136,137,141,142]
	Rosmarinic acid	*Hyphodontia paradoxa, Russula aurora, Russula emetic*	[137,142,145]
Flavonoids	Quercetin	*Ganoderma lucidum, Laccaria amethystea, Pleurotus citrinopileatus,*	[135,143]
	Kaempferol	*Ganoderma lucidum, Lactarius indigo*	[135,146]
	Hesperetin	*Calocybe indica, Ganoderma lucidum*	[133,135]
	Naringenin	*Calocybe indica, Ganoderma lucidum*	[133,135]
	Catechin	*Laccaria amethystea, Russula emetic*	[137,144]
	Myricetin	*Cantharellus cibarius, Lactarius indigo*	[134,146]
	Procyanidin	*Lactarius indigo*	[146]
	Rutin	*Pleurotus citrinopileatus, Russula emetic*	[137,143]
Tannins	Tannic acid	*Agaricus silvaticus, Hydnum rufescens, Meripilus giganteus, Pleurotus citrinopileatus, Pleurotus ostreatus, Pleurotus tuber-regium(fries)*	[147–149]
Tocopherols	α-Tocopherol	*Agaricus bisporus, Boletus badius, Lepista inversa, Pleurotus ostreatus, Russula delica*	[150,151]
	β-Tocopherol	*Laccaria laccata*	[150]
	γ-Tocopherol	*Clitocybe alexandri*	[150]
	δ-Tocopherol	*Lepista inversa*	[150]

4. Effects of Mushrooms and Their Anti-Aging Properties

Indeed, numerous studies have investigated the composition of mushrooms and their potential anti-aging effects. Various components extracted from mushrooms, including polysaccharides, phenolics, terpenes, lipids, vitamins, and minerals, have been found to possess anti-oxidant, anti-wrinkle, and anti-aging properties [152,153]. However, it is important to note that the anti-aging effects of mushrooms are primarily focused on skin aging and age-related diseases. The following provides an overview of these two aspects. The disruption of the collagen and elastin network in the skin due to excessive oxidative stress or free radicals is a characteristic of aging. As a result, anti-aging cosmetics are developed to repair and maintain the skin barrier. Many studies have highlighted the potential of bioactive compounds derived from mushrooms to serve as anti-aging ingredients in serums, topical creams, and other cosmetics, primarily due to their anti-oxidant and anti-wrinkle properties [10,58,154–162]. These compounds can help protect the skin from oxidative damage, reduce the appearance of wrinkles, and improve overall skin health.

4.1. Anti-Oxidant Activity

Oxidative stress is a condition that occurs when there is an imbalance between the production of ROS and the body's ability to neutralize them with anti-oxidants. ROS can damage cellular components, including DNA, proteins, and lipids, leading to cellular dysfunction and aging. Mushrooms have been investigated for their potential anti-oxidative properties and their ability to mitigate oxidative stress [163,164]. The anti-oxidant activity of mycochemicals derived from mushrooms plays a significant role in the defense and repair systems against oxidative damage and free radicals, which accelerate the aging process. Extracts from shiitake mushrooms (*Lentinula edodes*) have been found to act as inducers of anti-oxidant enzymes, such as glutathione peroxidase and superoxide dismutase. These enzymes stimulate the conversion of myofibroblasts to fibroblasts, reversing fibrosis and protecting the skin from oxidative damage [154]. Furthermore, L-ergothioneine, isolated from shiitake mushrooms, has been shown to scavenge free radicals, particularly those affecting the mitochondrial membrane, thus reducing oxidative stress on the skin [155]. L-ergothioneine, a thiourea derivative of histidine, is found in high concentrations in various mushrooms, including *Pleurotus ostreatus* (oyster mushroom), *Pleurotus eryngii* (King oyster mushroom), brown *Agaricus bisporus* (brown button mushroom), and *Grifola frondose* [155]. Additionally, mushroom glucan, extracted from *Phellinus ribis* and the somatic hybrid mushroom of *Pleurotus florida* and *Calocybe indica* var. APK2, has been found to activate immune cells and act as an anti-aging and anti-oxidant agent for the skin [156,157]. The anti-oxidant properties of mushrooms have also been demonstrated by *Ganoderma lucidum* (lingzhi) and *Phellinus linteus* (black hoof mushroom) in both in vitro assays and in vivo when consumed as food [58,158].

Mushrooms are rich in various anti-oxidants, including phenolic compounds, polysaccharides, and ergothioneine, that can scavenge free radicals and reduce oxidative damage. For example, polysaccharides from mushrooms such as *Grifola frondosa* (maitake), *Agaricus blazei* (almond mushroom), and *Pleurotus ostreatus* (oyster mushroom) have been shown to possess potent anti-oxidant activity [155,165,166]. A study retrospectively examined 37 participants who underwent a dietary intervention featuring daily consumption of 100 g of *A. bisorus* for 16 weeks [165]. Significant improvements in serum markers associated with inflammation and oxidative stress were observed after 16 weeks, including increases in ergothioneine levels and oxygen radical absorption capacity and reductions in oxidative stress-inducing factors carboxymethyllysine and methylglyoxal, suggesting potential anti-inflammatory and anti-oxidant benefits of *A. bisporus* consumption [165]. Ethanolic extract of oyster mushroom demonstrated potent radical-scavenging activity. At a maximum concentration of 10 mg/mL, the extract showed the highest level of radical-scavenging activity, with scavenging rates of 56.20% and 60.02% observed for hydroxyl and superoxide radicals, respectively [151]. The results show great potential of oyster mushroom as a readily available source of natural anti-oxidants for dietary supplementation or pharmaceutical use. Ergothioneine, also found in various mushrooms like *Pleurotus eryngii* (king trumpet mushroom) and *Lactarius deliciosus* (saffron milk cap), has been shown to have powerful anti-oxidant properties and reduce oxidative stress [84,87,151,167].

Mushrooms have also been found to protect against oxidative stress-induced damage to mitochondria, the organelles responsible for energy production within cells. For instance, polysaccharides from *Cordyceps sinensis* (caterpillar fungus) have been shown to enhance mitochondrial function and reduce oxidative damage in aging mice [39,168,169]. This suggests that mushroom bioactive compounds may help preserve mitochondrial function and mitigate age-related decline. Mushroom bioactive compounds can modulate signaling pathways involved in oxidative stress. For example, polysaccharides from mushrooms such as *Ganoderma lucidum* (lingzhi) and *Lentinula edodes* (shiitake) have been found to inhibit the production of ROS and increase the activity of anti-oxidant enzymes, such as SOD and CAT [77,158,170,171]. By regulating ROS production and anti-oxidant enzyme activity, mushrooms may help reduce oxidative stress and associated tissue damage.

4.2. Anti-Wrinkle Effects

One of the primary signs of skin aging is the formation of wrinkles, which is primarily caused by the loss of structural proteins in the dermis and elastase-induced degradation of elastin, leading to the expression of matrix metalloproteinases [159]. Lee, Lee, Kim, Yoo and Yang [10] discovered that Clitocybin A, an isoindolinone derived from the Korean mushroom *Clitocybe aurantiaca*, exhibited scavenging activity against ROS and inhibitory effects on elastase in human primary dermal fibroblast-neonatal cells. This suggests the potential of clitocybin A as an effective ingredient in anti-wrinkle cosmetic products. Similarly, the extract of the mycelium of the pine mushroom (*Tricholoma matsutake*) was found to inhibit elastase activity and the expression of matrix metalloproteinases in human fibroblasts [160]. In addition to the factors mentioned above, targeting the pro-inflammatory enzyme cyclooxygenase-2 (COX-2) may also be a strategy for anti-wrinkle treatments. COX-2 is associated with the production of ROS and inflammation in normal skin tissue. Therefore, COX-2 inhibitors are applied in anti-wrinkle cosmetics [161].

Notably, several bioactive compounds extracted from mushrooms have been found to effectively inhibit COX-2 activity. Stanikunaite, Khan, Trappe and Ross [161] reported that the ethanol extract of fruiting bodies of the truffle-like fungus *Elaphomyces granulatus* exhibited a 68% inhibition of COX-2 activity at a concentration of 50 mg/mL in mouse macrophages (RWA 264.7). Further investigation led to the identification of two bioactive compounds in *E. granulatus*, namely syringic acid and syringaldehyde acid, which were suggested to be responsible for the COX-2 inhibitory property [161]. Moreover, an extract of *Ganoderma lucidum* containing spores and fruiting bodies in a ratio of 30:8 was found to attenuate UV-induced epidermis thickening and inhibit the expression of COX-2 in non-tumor skin tissues of mice. This highlights the potential of the extract as a key component in cosmetic products for skin maintenance [162].

4.3. Immunomodulatory Effects

Immunosenescence refers to the gradual deterioration of various components in the immune system due to natural age advancement, which can lead to irregular immune responses against viruses or pathogens and increased vulnerability to illnesses such as chronic inflammation, autoimmune diseases, and cancer [172]. Reinforcing the immune system is essential for longevity. Extracts of the medicinal mushroom *Agaricus blazei* Murill have been found to enhance the functions of phagocytic cells [8], contributing to anti-tumor effects by strengthening innate immunity. When exposed to *A. blazei* Murill extracts, the phagocytic cells interact and remove invasive pathogens, further triggering innate and adaptive immune responses through the release of chemokines and cytokines. Short-term oral supplementation of the extracts at doses of 0.5–5% has been shown to exert an immunostimulatory effect characterized by increased secretion of cytokines in whole blood [173]. 1,3-β-Glucans found in medicinal mushrooms are effective in stimulating the immune system by modulating T cells, macrophages, and natural killer cells, along with the production of cytokines [174].

Several edible mushrooms, including *Agaricus bisporus*, *Flammulina velutipes*, *Lentinus edodes*, *Pleurotus florida*, and *Trametes pubescens*, have been found to possess anti-inflammatory properties, as assessed by levels of lipopolysaccharide and interferon that activate macrophages, indicating their immunomodulatory ability [9,175,176]. Additionally, the secondary metabolite lectin purified from *Latiporus sulphureus* could promote immune cell proliferation and phagocytosis and activate cytokines, suggesting its potential immunopotentiation in pharmacology and functional foods [177]. The use of cultured Sanghuang mushroom (*Inonotus sanghuang*) extracts at doses of 8 mg/kg or 16 mg/kg in immunodeficient mice has exhibited immunoregenerative functions, suggesting the potential of these extracts as an alternative for nutraceutical medicine concerning cancer chemotherapy [178]. From a more recent point of view, the bioactivity of mushrooms is closely related to its interaction with the gut microbiota, where gut microbial metabolites play a key role in bridging the gap between immunomodulatory effect of mushrooms and the host after consumption. Vlassopoulou et al. [179] selected *Pleurotus eryngii* as a substrate for in vitro fermentation using gut microbiota sampled from healthy elderly volunteers. The fermentation supernatants, which comprised a group of gut microbial metabolites, were subjected to cellular assays in U937-derived human macrophages. Interestingly, improved immune response was observed in treatment of gut microbial fermentation supernatants from each individual, characterized by altered gene expression and levels of pro- and anti-inflammatory cytokines in the macrophages, and further verified using peripheral blood mononuclear cells of the volunteers [179]. Boulaka et al. [180] also assessed the immunomodulatory property of *P. eryngii* through in vitro fermentation using fecal sample collected from both male and female elderly subjects. While not observed in pre-fermentation supernatant treatment, post-fermentation supernatant exhibited protective effects against mitomycin C-induced DNA damage for human lymphocytes in a dose-dependent manner, suggesting its significant role in maintaining genome integrity via metabolites-gut microbiome-host interaction during aging, which attributes to immunomodulatory and anti-oxidant activities [180].

4.4. Cardioprotective Effects

The circulatory system is essential for the transportation of oxygenated blood and nutrients to tissues and organs. The aging process can significantly impact the cardiovascular system, leading to the development of cardiovascular diseases such as hypertension, cardiac hypertrophy, atherosclerosis, myocardial infarction, and stroke [181]. One of the factors responsible for high blood pressure and cardiac hypertrophy is the vasopressor octapeptide angiotensin II (Figure 2b), which is converted from angiotensin I in the presence of angiotensin I converting enzyme [167]. D-glucopyranose mannitol extracted from the mushroom *Pleurotus cornucopiae* (Tamogi-take mushroom) has been found to alleviate hypertension in spontaneously hypertensive rat models by inhibiting angiotensin I converting enzyme and lowering blood pressure [167,182]. Similarly, bioactive peptides extracted from the fruiting body of *Tricholoma matsutakei* also disrupt the function of angiotensin responsible for hypertension [183]. Atherosclerosis, a disease commonly associated with hypercholesterolemia, high levels of low-density lipoprotein (LDL) (Figure 2a), and low levels of high-density lipoprotein (HDL), is prevalent among older populations and poses risks of stroke. Regularly consuming mushrooms has been shown in various animal studies to have significant benefits in reducing hypertension, atherosclerosis, dyslipidemia, inflammation, and obesity [184,185].

Several mushrooms with medicinal properties, including *Hypsizygus marmoreus* (bunashimeji), *Grifola frondosa* (maitake), and *Pleurotus eryngii* (eringi), show potential in treating atherosclerosis [186,187]. In an atherosclerosis mouse model, the application of mushroom extracts decreased the incidence of atherosclerosis lesions, suggesting their potential use in treatment [187]. In a rat model fed a high-cholesterol diet, oral administration of *Pleurotus florida* powder extracts increased fecal lipid excretion while effectively decreasing serum triglycerides, total cholesterol, LDL, and very low-density lipoprotein levels when compared to control mice [186]. Additionally, the ethanol extract of lion's mane mushroom and hot water extract from the mycelia of *Cordyceps sinensis* (caterpillar fungus) can enhance lipid metabolism by suppressing platelet aggregation, lowering LDL levels, and increasing HDL levels, acting as therapeutic agents for atherosclerosis and potentially decreasing the risk of myocardial infarction [61,188].

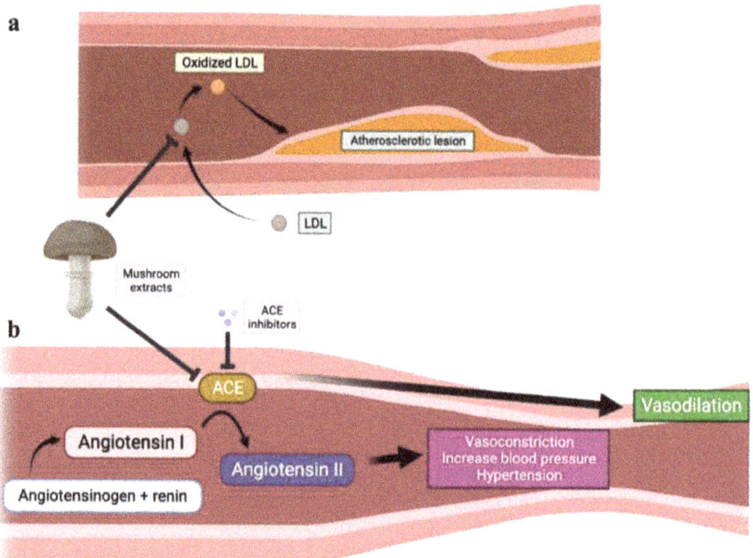

Figure 2. Schematic illustration of effects of mushroom extracts against cardiovascular diseases regarding lowering blood LDL and blood pressure. (**a**) Application of mushroom extracts decrease LDL levels and inhibit platelet aggregation. (**b**) Application of mushroom extracts inhibit ACE activity similarly to ACE inhibitors, thus leading to vasodilation and lower blood pressure. ACE, angiotensin converting enzyme. LDL, low-density lipoprotein.

4.5. Neuroprotective Effects

Brain aging is a significant risk factor for neurodegenerative diseases and cognitive decline, including dementia, Alzheimer's disease (AD), and Parkinson's disease (PD). Excessive oxidative stress is a major contributor to brain aging. Research has been conducted on the effects of mushroom extracts on the oxidative state of the brain during aging. For example, an aqueous extract of *Agaricus blazei* was found to maintain the ROS levels in the brain of rats at a level that did not accelerate brain aging when administered daily at a dose of 50 mg/kg [189]. However, long-term and continuous treatment with the extract showed a tendency to be less effective in rats aged above 12 months, suggesting that intermittent treatment with short-term doses may be more beneficial [189]. In experiments using the roundworm *Caenorhabditis elegans*, an ethanolic extract of cloud ear fungus (*Auricularia polytricha*) attenuated glutamate-induced cytotoxicity and increased the expression of anti-oxidant enzyme genes, promoting longevity and health in the worms [190]. This suggests that cloud ear fungus could serve as a natural source of neuroprotective and anti-brain-aging agents.

Figure 3 illustrates the neuroprotective properties of lion's mane mushroom in four preclinical study models. Ethanol extracts of *H. erinaceus* demonstrated neuroprotective effects in mouse hippocampal neurons and microglia, protecting against oxidative damage and inflammation [191]. In the context of PD, *H. erinaceus* and *Grifola frondosa* (maitake mushroom) extracts have shown anti-aging effects in yeast by reducing α-synuclein toxicity and levels of ROS, as well as lowering α-synuclein membrane localization [192]. *H. erinaceus* has also shown beneficial effects in improving cognitive function and behavioral deficits in animal models of AD, as well as enhancing recognition memory and inducing neurogenesis in frail aging mice [193,194]. While studies have indicated the neuroprotective effects of edible and medicinal mushrooms, it is important to carefully verify their efficacy and potential adverse effects in human trials as the effects may vary.

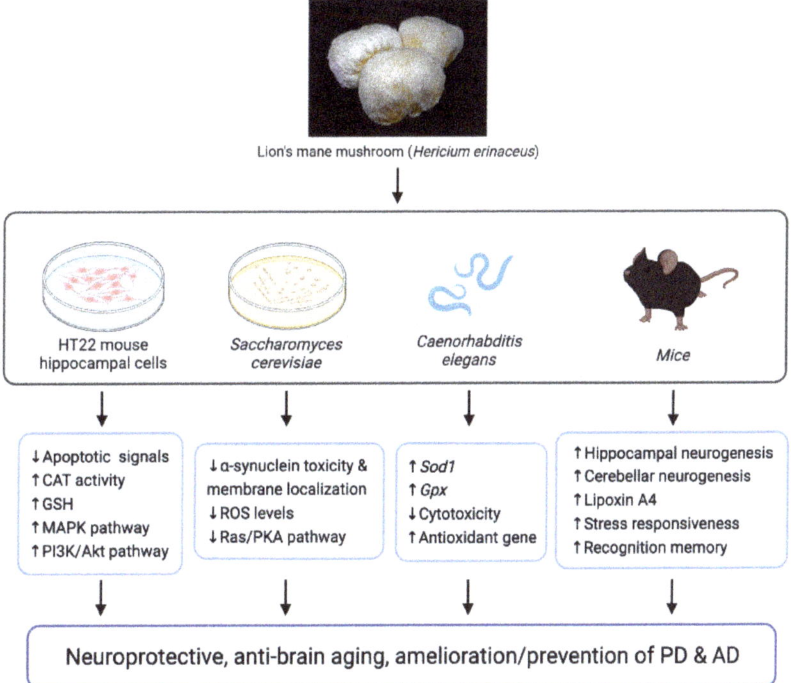

Figure 3. Anti-brain aging properties of Lion's mane mushroom (*Hericium erinaceus*) observed in four experimental models. CAT, catalase; GSH, glutathione; MAPK, mitogen-activated protein kinase; PI3K, phosphatidylinositol 3-kinase; Akt, protein kinase B; ROS, reactive oxygen species; Ras/PKA, Ras protein/protein kinase A; *Sod1*, Cu/Zn superoxide dismutase; *Gpx*, glutathione peroxidase. Symbol ↑ denotes increase; symbol ↓ denotes decrease.

4.6. Anti-Diabetic Effects

According to recent research, age-related type 2 diabetes is primarily caused by pathological changes in pancreatic beta cells. These changes include decreased proliferation and regeneration potential, disrupted transcriptome and proteostasis, increased accumulation of senescent cells, and the impact of systemic environmental stress. These factors result in the loss of functional cell mass and impaired insulin secretion and action [195]. Mushrooms, specifically polysaccharides like β-glucans, have been found to play a role in restoring pancreatic function. They boost insulin secretion by pancreatic beta cells, lower blood glucose levels, and improve the insulin response in peripheral tissues [196]. Exopolysaccharides isolated from cultured mycelium of *Phellinus baumii* and *Tremella fuciformis* (snow fungus) have shown blood glucose-lowering effects in mice with obesity-induced diabetes [197].

Various mushroom-derived extracts and bioactive compounds, including glycoproteins and β-glucans from *Agaricus blazei* (almond mushroom), polysaccharides from *Phellinus linteus* (black hoof mushroom), lectins from *Agaricus bisporus* (white button mushroom), and extracts from *Pleurotus osteratus* (oyster mushroom) and *Ganoderma lucidum* (lingzhi), have demonstrated blood glucose reduction abilities in diabetic animal model [198–201]. Notably, lectins from white button mushrooms have been found to promote the proliferation of islet beta cells in mice with partial pancreatic removal, suggesting their potential use in the treatment of type 2 diabetes [201]. Furthermore, a retrospective study suggested that the consumption of white button mushrooms may be correlated with anti-inflammatory and anti-oxidant health benefits in individuals predisposed to type 2 diabetes [165]. Figure 4 provides an overview of the mechanisms underlying the antidiabetic activities of mushrooms.

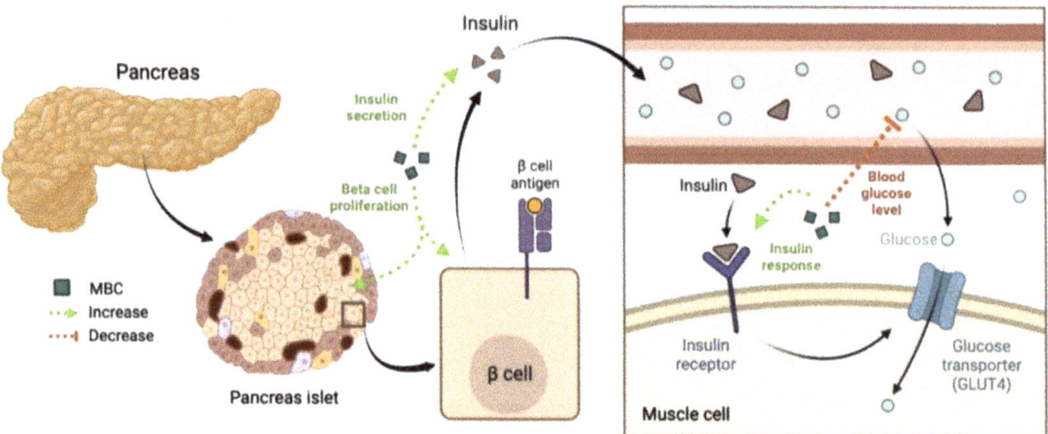

Figure 4. Schematic illustration of anti-diabetic properties of mushroom. Application of mushroom extracts or mushroom-derived bioactive compounds may improve insulin secretion and response by promoting pancreatic beta-cell proliferation, which increases performance of glucose take up by cells and lowering blood glucose. MBC, mushroom-derived bioactive compounds. Red dashed line indicates inhibitory effects while green dashed lines indicate promoting/strengthening effects.

4.7. Beneficial for Age-Related Diseases

Mushrooms are not only valued for their nutritional content but also considered functional foods that can enhance biological function and promote overall health [13,202,203]. Additionally, mushrooms have been found to possess pharmacological and medicinal properties that can be beneficial in age-related diseases. These properties include immunomodulatory, anti-inflammatory, anti-cancer, anti-diabetic, and neuroprotective effects, among others [13,202,204]. The following summary will highlight several representative age-related diseases or conditions that can be influenced by mushroom extracts and bioactive compounds derived from the mycelium or fruiting body of mushrooms. For a comprehensive overview of the medicinal properties of mushrooms, including the responsible compounds and proposed mechanisms, please refer to Table 5.

Table 5. Medicinal properties of mushrooms.

Properties	Mushroom Species	Bioactive Compounds	Study Type/Model/Effective Dosage	Mechanisms of Action	References
Immunomodulatory	*Agaricus blazei*	β-glucans (from pure AbM extracts or commercial mushroom extracts mixture AndoSan™ containing 85% of AbM)	Ex vivo/human whole blood/0.1–15% for 6 h; In vivo/human/20 mL thrice per day orally for 12 days	Anti-oxidant activities, enhance immune cells function and innate immune responses, trigger release of cytokines, chemokines, and leukocyte growth factors	[173]
	Pleurotus cornucopiae	β-glucans	Clinical trial/human/24 mg per meal for 8 weeks	Th1 phenotype potentiation via macrophage-IL-12-IFN-γ pathway, up-regulation of NK cell activity	[176]
	Latiporus sulphureus	Lectin (LSL4)	In vitro/RAW264.7 cells/0–650 μg·mL^{-1} (IC$_{50}$ = 1004.6 μg·mL^{-1})	Cell phagocytosis via TLR4 signaling pathway, triggers release of NO, iNOS, TNF-α, IL-1β, IL-6, and IL-10	[177]
	Inonotus sanghuang	Extract containing polysaccharides and amino acids	In vivo/mice/4 and 8 mg·kg^{-1} once a day orally for 12 days	Stimulation of T lymphocytes, natural killer cells, and B cells; inhibition of cytochrome P450 isozymes	[178]
	Ganoderma lucidum	Polysaccharides extract (Ganoderan, heteroglycan, mannoglucan, glycopeptide)	In vivo/mice/2.5 mg·kg^{-1} intraperitoneal injection once per day for 7 days	Stimulation of TNF-α, IL-1, IFN-γ production, activate NF-κB	[205]
	Ganoderma Microsporum	FIP	In vitro/human alveolar epithelial A549 cells/4 and 16 μg·mL^{-1}	Down-regulation of TNF-α via NF-κB pathway	[206]
Anti-cardiovascular diseases	*Tricholoma matsutake*	Functional peptides	In vivo/rats/50 mg·kg^{-1} acute oral dose	Alleviated hypertension via inhibition of angiotensin I converting enzyme	[183]
	Pleurotus florida	Aqueous extract containing 80% soluble fiber, 44% protein, 1.4% soluble sugars, 0.2% polyphenols (*w/w* dry weight)	In vivo/rats/5 and 7.5% of 100 g basal diet for 4 weeks	Suppression of hepatic biosynthesis of cholesterol by inhibiting activity of liver enzyme HMG-CoA	[186]
	Cordyceps sinensis	Aqueous extract containing 83.9% carbohydrates (glucose, mannose, galactose, arabinose), 11.8% protein, *w/w* dry weight	In vivo/mice/150 and 300 mg·kg^{-1} per day orally for 7 days	Suppression of hepatic biosynthesis of cholesterol by inhibiting activity of liver enzyme HMG-CoA	[188]
Neuroprotective	*Agaricus blazei*	Extract, composition not specified	In vivo/rats/50 mg·kg^{-1} per day intragastrically at the age of 7–23 months	Free-radical scavenging ability, cytoprotective action, antioxidation reaction	[189]
	Hericium erinaceus	Aqueous and ehthanol extracts, composition not specified	In vitro/HT22 mouse hippocampal neurons/ethanol extracts at 400 μg·mL^{-1}	Inhibition of mitochondria-dependent apoptotic cellular signals activation; elevated CAT activity and GSH content; up-regulation of MAPK and PI3K/Akt pathway	[191]
		Extract containing erinacine A, hericenones C and D	in vivo/mice/1 mg (solubilized in water) per day for 2 months	Promoting hippocampal neurogenesis; up-regulation of lipoxin A4 and modulation of stress responsive proteins	[194]
	Auricularia polytricha	Ethanolic extract containing flavonoids, phenols, linoleic acid	In vitro/HT22 mouse hippocampal cells/5, 10, 20, and 40 μg·mL^{-1}; In vivo/*Caenorhabditis elegans*/20, 40 μg·mL^{-1}	Anti-oxidant activity via Nrf2 signaling pathway; up-regulation of *Sod1* and *Gpx* gene expressions	[190]
	Grifola frondose	Aqueous extract containing, β-glucan, chitin, amino acids, unsaturated fatty acids, monosaccharides	In vivo/*Saccharomyces cerevisiae*/0.2 and 0.5% in culture medium; In vivo/*Drosophila melanogaster*/0.2% in culture medium	Increase of heat shock proteins expression by inhibition of Ras/PKA pathway; reduce levels of ROS	[192]

Table 5. *Cont.*

Properties	Mushroom Species	Bioactive Compounds	Study Type/Model/Effective Dosage	Mechanisms of Action	References
Antidiabetic	*Tremella fuciformis, Phellinus baumii*	Exopolysaccharides, composition not specified	In vivo/mice/200 mg·kg^{-1} per day orally for 52 days	Improve insulin sensitivity via regulating PPAR-γ-mediated lipid metabolism	[197]
	Agaricus bisporus	Not specified	In vivo/rats/200 mg·kg^{-1} per day orally for 3 weeks	Stimulate secretion of insulin from pancreatic beta cells	[198]
		Lectins	In vivo/mice/10 mg·kg^{-1} for 2 weeks	Induce beta-cell proliferation	[201]
	Phellinus linteus	Aqueous extract containing 13.2% peptide, 82.5% carbohydrates (*w/w* dry weight)	In vivo/mice/30 mg·kg^{-1} intraperitoneally daily from 8 to 24 weeks of age	Inhibit expression of inflammatory cytokines (IFN-γ, IL-2, and TNF-α); up-regulation of IL-4 expression	[199]
	Agaricus blazei Murill	Isoflavovoids (genistein, genistin, daidzein, daidzin)	In vivo/rats/400 mg·kg^{-1} per day orally for 2 weeks	Improve beta-cell function; increase lipid peroxidation via enhanced fatty acyl CoA activity	[200]

AbM, *Agaricus blazei* Murill; IFN, interferon. NK cell, natural killer cell; LSL4, one of the lectins yields from *Latiporus sulphureus*, a glycoprotein containing 6.32% sugar; TNF, tumor necrosis factor; IL, inter-leukin; FIP, fungal immunomodulatory protein; HMG-CoA, 3-hydroxy-3-methylglutaryl coenzyme A; CAT, catalase; GSH, glutathione; MAPK, mitogen-activated protein kinase; PI3K, phosphatidylinositol 3-kinase; Akt, protein kinase B; Nrf2, nuclear factor erythroid 2-related factor 2; *Sod1*, Cu/Zn superoxide dismutase; *Gpx*, glutathione peroxidase; Ras/PKA, Ras protein/protein kinase A; PPAR-γ, peroxisome proliferator-activated receptor gamma.

4.8. Structure–Activity Relationship

The structure–activity relationship of mushroom bioactive compounds refers to the relationship between the molecular structure of these compounds and their biological activities in addressing the mechanisms of aging. Mushroom bioactive compounds, such as polysaccharides, phenolic compounds, triterpenoids, and ergothioneine, exhibit diverse chemical structures that contribute to their bioactivity [185]. The specific structural features, such as the presence of specific functional groups or the arrangement of atoms, can influence their anti-oxidant, anti-inflammatory, and immunomodulatory properties [207,208]. Exploring the structure–activity relationship of mushroom bioactive compounds provides insights into their potential mechanisms of action and aids in the design of novel compounds with enhanced anti-aging properties.

Understanding the mechanisms and structure–activity relationship of aging is vital for developing effective interventions to mitigate the aging process and its associated diseases. Mushroom bioactive compounds have shown significant potential in addressing the mechanisms of aging through their anti-oxidant, anti-inflammatory, and immunomodulatory properties. Harnessing the power of mushroom bioactive compounds may pave the way for innovative strategies to promote healthy aging and improve the quality of life in the aging population.

5. Molecular and Cellular Mechanisms Underlying Aging Processes

Aging is influenced by a multitude of interconnected molecular and cellular mechanisms. These mechanisms include DNA damage and repair, telomere shortening, epigenetic changes, cellular senescence, mitochondrial dysfunction, oxidative stress, and chronic inflammation. Each of these mechanisms contributes to the aging process and the development of age-related diseases. Understanding the intricate interactions among these mechanisms is crucial for developing effective anti-aging strategies. This includes understanding how mushroom compounds interact with these mechanisms and influence their progression. The overall mechanisms of the anti-aging properties of mushrooms are depicted in Figure 5.

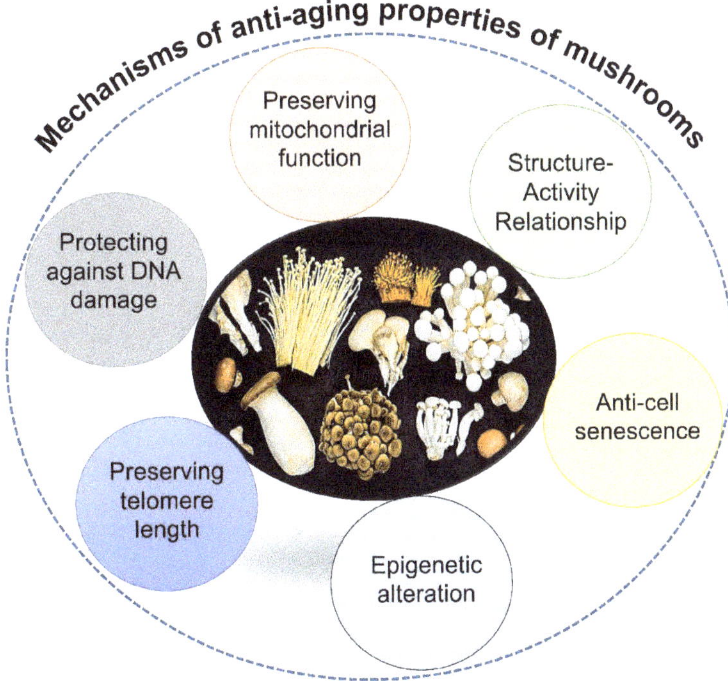

Figure 5. The overall mechanisms of the anti-aging properties of mushrooms.

5.1. Cell Senescence

Cell senescence is a key hallmark of aging and is defined as the irreversible loss of cell division potential and the acquisition of a senescence-associated secretory phenotype. Senescent cells accumulate with age and contribute to tissue dysfunction and inflammation, which are characteristic of aging [16]. The accumulation of senescent cells has been linked to a variety of age-related diseases, including cancer, cardiovascular disease, and neurodegenerative disorders [29]. Various interventions, including the use of mushroom bioactive compounds, have been explored as potential strategies to delay or mitigate the accumulation of senescent cells and promote healthy aging [101].

Mushroom bioactive compounds have been investigated for their potential to delay or mitigate the accumulation of senescent cells and promote healthy aging. Several studies have reported the anti-senescence effects of mushroom bioactive compounds, including polysaccharides, peptides, and phenolic compounds [170,171,209]. For example, polysaccharides from *Ganoderma lucidum* (lingzhi) have been shown to reduce senescence-associated β-galactosidase activity and decrease the expression of senescence-associated markers in aging human dermal fibroblasts [171,210]. Polysaccharides from *Hericium erinaceus* (lion's mane mushroom) have also been found to reduce senescence-associated β-galactosidase activity and increase the expression of anti-senescence markers in senescent human dermal fibroblasts [169,211].

Mushroom bioactive compounds have also been found to have anti-inflammatory and anti-oxidant properties, which may contribute to their anti-senescence effects [212]. For example, polysaccharides from *Lentinus edodes* (shiitake mushroom) have been shown to reduce oxidative stress and inflammation in aging mice, which may help delay the accumulation of senescent cells [78,213,214]. Overall, mushroom bioactive compounds have shown promise as potential interventions to delay or mitigate the accumulation of senescent cells and promote healthy aging through various mechanisms. In addition to their anti-senescence effects, mushroom bioactive compounds have been investigated for their potential to promote healthy aging through other mechanisms, such as preserving telomere length, protecting against DNA damage, and preserving mitochondrial function.

5.2. Telomere Maintenance

Telomeres are specific structures found at the ends of linear chromosomes. They are composed of repeated sequences of TTAGGG, known as hexanucleotides, and a protein complex called shelterin. These components work together to create a protective loop structure that prevents chromosome fusion and degradation [215]. When telomeres become shortened or damaged, and the protective loop is opened, it triggers an uncapped state that activates a DNA damage response. This response can lead to cellular senescence or programmed cell death. Traditionally, average telomere length, often measured in human blood lymphocytes, has been considered a biomarker for aging, survival, and mortality [216]. This shortening is a natural part of the aging process and is primarily caused by the inability of DNA replication machinery to fully replicate the ends of linear chromosomes. Telomerase is an enzyme that plays a critical role in maintaining telomere length, which protects the ends of chromosomes from degradation and fusion. Telomere shortening, caused by telomerase deficiency, is a hallmark of aging [217]. Shortened telomeres have been linked to cellular dysfunction, inflammation, age-related diseases and the overall decline in tissue and organ function [218].

Several factors can influence telomere maintenance and the rate of telomere shortening. These include genetic factors, lifestyle choices (such as diet, exercise, and stress levels), and environmental exposures [16,215,216,218]. Certain lifestyle modifications, such as regular physical activity, a healthy diet, and stress reduction techniques, have been associated with better telomere maintenance and potentially slower aging.

Mushroom bioactive compounds have been investigated for their potential to preserve telomere length and delay or mitigate age-related decline. While specific studies focusing on the effects of mushrooms on telomerase deficiency are limited, some studies have explored the broader anti-aging mechanisms of mushrooms that may indirectly contribute to telomere maintenance. For instance, polysaccharides from *Agaricus blazei* (almond mushroom) have been found to enhance telomerase activity and preserve telomere length in aging mice [219]. Another study found that *Ganoderma lucidum* (lingzhi) polysaccharides increased telomerase activity and extended the lifespan of fruit flies [217]. These findings suggest that mushroom bioactive compounds may have the potential to counteract telomerase deficiency and promote healthy aging.

5.3. Mitochondrial Dysfunction

Mitochondria are organelles responsible for producing energy in cells. Mitochondrial dysfunction, characterized by impaired energy production and increased production of ROS, is a key aspect of aging. Mitochondrial dysfunction has been linked to a variety of age-related diseases, including neurodegenerative disorders, cardiovascular disease, metabolic disorders, and impaired immune function [220–222]. Mitochondrial dysfunction is closely linked to the process of aging. Several factors contribute to mitochondrial dysfunction during aging. One major factor is the accumulation of mitochondrial DNA (mtDNA) mutations, which can impair the production of energy and increase the generation of harmful ROS [223]. ROS can cause oxidative damage to cellular components, including mtDNA itself, leading to a vicious cycle of further mitochondrial dysfunction [224]. It can affect various tissues and organs, including the brain, heart, muscles, and immune system.

Researchers are actively investigating strategies to mitigate mitochondrial dysfunction and its impact on aging. Approaches include improving mitochondrial quality control mechanisms, enhancing cellular anti-oxidant defenses, and exploring interventions that can promote mitochondrial biogenesis and function [225]. Understanding the complex relationship between mitochondrial dysfunction and aging is crucial for developing interventions to maintain mitochondrial health and potentially delay age-related diseases.

By targeting mitochondrial function, it may be possible to enhance overall health span and improve the quality of life in older individuals.

Mushroom bioactive compounds have been investigated for their potential to preserve mitochondrial function and mitigate age-related decline. Several studies have reported the anti-mitochondrial dysfunction effects of mushroom bioactive compounds, including polysaccharides, peptides, and phenolic compounds. For example, polysaccharides from *Grifola frondosa* (maitake mushroom) have been found to preserve mitochondrial function and increase anti-oxidant enzyme activity in aging mice [226]. Polysaccharides from *Agaricus blazei* (almond mushroom) have also been shown to improve mitochondrial function and increase ATP production in aging mice [227]. The extract of *A. blazei* was found to effectively restore lipid peroxidation levels (measured by TBARS) in old rats to levels comparable to those observed in young rats [228]. This effect is likely due to the ability of various constituents in *A. blazei*, such as phenolics, to scavenge free radicals. Among the phenolics identified in *A. blazei*, gallic acid, syringic acid, and pyrogallol have demonstrated significant anti-oxidant activities [229]. Considering their hydrophilic nature, it is probable that these phenolics are present in the aqueous extract used in the study.

In addition, the treatment with *A. blazei* was effective in elevating the activity levels of various mitochondrial enzymes in old rats, including succinate dehydrogenase, α-ketoglutarate dehydrogenase, NADH dehydrogenase, and cytochrome c oxidase. Notably, the cytochrome c oxidase activity was nearly doubled by the *A. blazei* treatment. These findings are consistent with a previous study in which old rats were treated with *Ganoderma lucidum* extracts using a similar experimental protocol [230]. In addition, the *A. blazei* treatment resulted in improved membrane energization of the mitochondrial membrane, both in the presence of succinate and ATP [231]. Succinate-driven respiration in the presence of exogenous ADP was significantly increased, approaching the respiration rates observed in the brain mitochondria of young rats. This effect is likely due to the stimulation of succinate dehydrogenase by the *A. blazei* treatment, which represents a benefit in terms of rat brain energetics. The aqueous extract of *A. blazei* has shown potential in improving the oxidative state of brain tissue and reversing certain detrimental effects of aging on mitochondrial oxidative enzymes [231]. Overall, mushroom bioactive compounds have shown promise as potential interventions to preserve mitochondrial function and mitigate age-related decline through various mechanisms.

5.4. DNA Damage

DNA damage is a natural consequence of aging. Over time, the genetic material in our cells can accumulate various types of damage, such as DNA strand breaks, oxidative damage, and the formation of DNA adducts. This damage can result from both endogenous factors, such as metabolic processes and ROS, as well as exogenous factors, such as exposure to environmental toxins and radiation [221,232,233]. The accumulation of DNA damage is believed to contribute to the aging process and age-related diseases. When DNA damage is not properly repaired, it can lead to mutations and genomic instability, which can affect cellular function and increase the risk of diseases such as cancer. Various mechanisms are in place to repair DNA damage, such as base excision repair, nucleotide excision repair, and homologous recombination. However, as we age, the efficiency of these repair mechanisms can decline, leading to a higher accumulation of unrepaired DNA damage [234–236]. Additionally, chronic inflammation and oxidative stress, which are associated with aging, can further contribute to DNA damage. These processes can generate ROS that can directly damage DNA and interfere with DNA repair mechanisms. DNA damage accumulates with age and contributes to cellular dysfunction. DNA damage can be caused by a variety of factors, including oxidative stress, radiation, and environmental toxins. DNA damage has been linked to a variety of age-related diseases, including cancer, cardiovascular disease, and neurodegenerative disorders [237–239].

Mushroom bioactive compounds have been investigated as potential interventions to protect against DNA damage and promote healthy aging. Several studies have reported the anti-DNA damage effects of mushroom bioactive compounds, including polysaccharides, peptides, and phenolic compounds [238]. Polysaccharides from *Phellinus linteus* (black hoof mushroom) have also been shown to protect against DNA damage in aging mice [169]. Furthermore, polysaccharides from *Grifola frondosa* (maitake mushroom) have been found to protect against DNA damage and increase anti-oxidant enzyme activity in aging mice [192]. Polysaccharides from *Ganoderma lucidum* (lingzhi) have also been shown to protect against DNA damage in human liver cells exposed to oxidative stress [240]. Mushroom bioactive compounds have also been found to have anti-oxidant and anti-inflammatory properties, which may contribute to their anti-DNA damage effects [241]. For example, polysaccharides from *Pleurotus ostreatus* (oyster mushroom) have been shown to reduce oxidative stress and inflammation in aging mice, which may help protect against DNA damage [242]. As mentioned earlier in Section 4.3, gut microbial fermentation product of *P. eryngii* (king oyster mushroom), which carries higher bioactivity than pre-fermentated original substrate, also exerts genoprotective effect via the metabolites-gut microbiome-host pathway, illustrated by its ability to protect cyclophosphamide-induced DNA damage in bone marrow and whole blood cells in young and elderly female and male mice [180]. Furthermore, mushrooms contain bioactive compounds such as ergothioneine, which has been shown to have potent anti-oxidant properties that protect DNA from oxidative damage [243]. Ergothioneine can be found in various mushrooms, including oyster mushrooms, shiitake mushrooms, and king trumpet mushrooms. Overall, mushroom bioactive compounds have shown promise as potential interventions to protect against DNA damage and promote healthy aging through various mechanisms.

5.5. Epigenetic Changes

Epigenetic changes refer to modifications in gene expression that do not involve alterations to the underlying DNA sequence. These changes can have a significant impact on aging and age-related diseases. One of the key epigenetic changes associated with aging is DNA methylation. DNA methylation involves the addition of a methyl group to the DNA molecule, typically at specific sites called CpG sites [244]. Methylation patterns can change over time, and certain regions of the genome can become more methylated or less methylated with age. Global DNA hypomethylation, which is a decrease in overall DNA methylation levels, is commonly observed in aging tissues. This hypomethylation can lead to genomic instability and the activation of normally silenced genes. On the other hand, specific genomic regions, such as gene promoters, can become hypermethylated with age, resulting in the repression of gene expression [245].

Another important epigenetic modification associated with aging is histone modification. Histones are proteins that help package DNA into a compact structure called chromatin. Different modifications, such as acetylation, methylation, and phosphorylation, can occur on histones and influence gene expression [246]. Age-related changes in histone modifications can impact gene expression patterns and cellular function. For example, decreased histone acetylation levels have been observed in aging tissues, leading to a more compact chromatin structure and reduced gene expression. These epigenetic changes can be influenced by various factors, including environmental factors, lifestyle choices, and genetic predisposition [247]. They can have wide-ranging effects on cellular processes, such as DNA repair, cellular senescence, and inflammation, which are all associated with aging and age-related diseases [248].

While specific studies on the effects of mushrooms on epigenetic changes and aging are limited, some research suggests that mushroom bioactive compounds may have potential anti-aging effects through epigenetic mechanisms. For example, a study demonstrated that polysaccharides from *Ganoderma lucidum* (lingzhi) can inhibit DNA methyltransferase activity, leading to DNA hypomethylation and reactivation of tumor suppressor genes in cancer cells [249]. This suggests that mushroom polysaccharides may influence epigenetic processes that regulate gene expression. Additionally, certain mushroom bioactive compounds have been found to modulate histone modifications. For instance, extracts from

Trametes versicolor (Turkey tail mushroom) have been shown to increase the acetylation of histone proteins, which can result in changes in gene expression [250]. These changes in histone modifications may have implications for aging and age-related diseases. Furthermore, some mushrooms contain microRNAs, which are small non-coding RNA molecules that can regulate gene expression. For instance, *Pleurotus ostreatus* (oyster mushroom) has been found to contain microRNAs that have anti-inflammatory effects by targeting specific genes involved in inflammation [251]. As chronic inflammation is associated with aging, the anti-inflammatory effects of mushroom microRNAs may have potential anti-aging benefits. It is important to note that the field of epigenetics and the effects of mushrooms on epigenetic changes and aging are still emerging areas of research.

5.6. Chronic Low-Grade Inflammation

Inflammation and aging are interconnected processes that have been the subject of extensive research in recent years. Chronic low-grade inflammation, often referred to as "inflammaging", is now recognized as a hallmark of aging. As we age, our immune system undergoes changes, leading to a state of chronic inflammation. This persistent low-level inflammation can contribute to the development of various age-related diseases, including cardiovascular disease, neurodegenerative disorders, and certain types of cancer [252].

Several factors contribute to the age-related increase in inflammation. One of the key factors is the accumulation of senescent cells in tissues throughout the body. Senescent cells are damaged or dysfunctional cells that no longer divide and can produce pro-inflammatory molecules. Their accumulation over time contributes to chronic inflammation [253]. Another factor is the dysregulation of the immune system with age. This dysregulation, often referred to as immunosenescence, leads to a state of chronic immune activation and increased production of pro-inflammatory cytokines [254].

Additionally, changes in the gut microbiota, the collection of microorganisms residing in our intestines, have been linked to age-related inflammation. Alterations in the composition of the gut microbiota can lead to increased gut permeability and the release of bacterial components into the bloodstream, triggering an immune response and inflammation [255]. The consequences of chronic inflammation in aging are far-reaching. In addition to contributing to the development of age-related diseases, inflammation can also accelerate the aging process itself. It can lead to tissue damage and impair the function of organs, such as the brain, heart, and joints [256].

Efforts to mitigate age-related inflammation are being actively explored. Lifestyle factors, such as regular exercise, a healthy diet, and stress management, have been shown to reduce inflammation and promote healthy aging [257–259]. Certain medications and dietary supplements, such as anti-inflammatory drugs and anti-oxidants, are also being studied for their potential to modulate age-related inflammation [260]. Understanding the complex relationship between inflammation and aging is crucial for developing interventions that can promote healthy aging and reduce the burden of age-related diseases. Ongoing research in this field holds promise for improving the quality of life in older adults.

Mushrooms, particularly certain species, have been investigated for their potential anti-inflammatory properties and their ability to mitigate inflammaging. Here is some comprehensive information on inflammation and the anti-aging mechanisms of mushrooms: Mushrooms contain bioactive compounds, including polysaccharides, phenolic compounds, and triterpenoids, that have demonstrated anti-inflammatory effects [261]. For example, polysaccharides from various mushroom species, such as *Ganoderma lucidum* (lingzhi), *Lentinula edodes* (shiitake), and *Pleurotus ostreatus* (oyster mushroom), have been shown to inhibit pro-inflammatory cytokines, such as tumor necrosis factor-α (TNF-α) and interleukin-6 (IL-6) [261–263]. These compounds can help reduce inflammation and associated tissue damage.

Furthermore, mushrooms have been found to modulate the immune response, which is closely linked to inflammation. For instance, mushroom polysaccharides have been shown to enhance the activity of natural killer cells, macrophages, and other immune cells, thus promoting a balanced immune response, and reducing chronic inflammation [264]. Oxidative stress plays a significant role in inflammation and aging. Mushrooms contain various anti-oxidants, including phenolic compounds and ergothioneine, which can scavenge free radicals and reduce oxidative damage. Ergothioneine, specifically found in mushrooms like *Pleurotus eryngii* (king trumpet mushroom) and *Lactarius deliciosus* (saffron milk cap), has been shown to possess potent anti-oxidant and anti-inflammatory properties [265,266].

Mushroom bioactive compounds can modulate signaling pathways involved in inflammation. For instance, polysaccharides from mushrooms like *Grifola frondosa* (maitake) and *Agaricus bisporus* (white button mushroom) have been found to inhibit the nuclear factor-kappa B (NF-κB) pathway, which is a key regulator of inflammation [267,268]. By suppressing NF-κB activation, mushrooms may help alleviate chronic inflammation. It is worth noting that while mushrooms have shown promising anti-inflammatory effects, more research is needed to fully understand their mechanisms and establish their efficacy and safety in the context of aging and age-related diseases.

6. Concluding Remarks and Future Perspective

As the population ages, there is an increasing demand for strategies to promote healthy aging. Dietary interventions and nutrient supplementation have been identified as effective ways to extend both health span and lifespan among the elderly. Among various food sources, mushrooms have demonstrated promising anti-aging potential due to the presence of bioactive compounds such as polysaccharides, proteins and peptides, lipids, and phenolic compounds, which have been shown to have anti-inflammatory, anti-oxidant, immunomodulatory, neuroprotective, anti-diabetic, and cardiovascular disease-ameliorating properties.

Mushrooms can be used as functional foods and may serve as valuable source materials for drug and functional food development. While the majority of studies have used mushroom extracts in aging models and demonstrated their effectiveness in expanding lifespan, a minority of studies have identified individual compounds responsible for their anti-aging properties. It is important to identify the chemical structure of these compounds to gain insight into how they interact with cells and develop more effective anti-aging strategies. However, most studies have been performed in vivo or in vitro, with limited clinical trials, and results from different studies are not always consistent or supportive. Furthermore, individuals who consume mushrooms may also consume a variety of different self-selected meals or prepare mushrooms in different ways, which may counteract the proposed health benefits of mushroom bioactive compounds and limit their effectiveness. Therefore, meal plans for healthy aging should be designed with this factor in mind. Additionally, safety, dosage, and effectiveness of the bioactive compounds should be verified. If mushroom extracts are to be applied in the treatment of diseases among the elderly, particularly vulnerable populations, further research, particularly clinical or nutritional trials, will be highly required.

Author Contributions: Methodology, software, formal analysis, investigation, validation, data curation, writing—original draft preparation, writing—review and editing, visualization, J.L. and K.G.; Conceptualization, resources, supervision, project administration, funding acquisition, writing—review and editing, B.X. All authors have read and agreed to the published version of the manuscript.

Funding: This research was funded by BNU-HKBU United International College, grant number UICR0200007-23.

Institutional Review Board Statement: Not applicable.

Informed Consent Statement: Not applicable.

Data Availability Statement: Not applicable.

Conflicts of Interest: The authors declare no conflicts of interest.

References

1. Murray, C.; Barber, R.M.; Foreman, K.J.; Ozgoren, A.A.; Abdallah, F.; Abera, S.F.; Aboyans, V.; Abraham, J.P.; Abubakar, I.; Aburaddad, L.J. Global, regional, and national disability-adjusted life years (DALYs) for 306 diseases and injuries and healthy life expectancy (HALE) for 188 countries, 1990–2013: Quantifying the epidemiological transition. *Lancet* **2015**, *386*, 2145–2191. [CrossRef]
2. Foreman, K.J.; Marquez, N.; Dolgert, A.; Fukutaki, K.; Fullman, N.; McGaughey, M.; Pletcher, M.A.; Smith, A.E.; Tang, K.; Yuan, C.W.; et al. Forecasting life expectancy, years of life lost, and all-cause and cause-specific mortality for 250 causes of death: Reference and alternative scenarios for 2016-40 for 195 countries and territories. *Lancet* **2018**, *392*, 2052–2090. [CrossRef]
3. World Health Organization. Preventing chronic diseases: A vital investment. *Prev. Chronic Dis. A Vital Invest.* **2008**, *126*, 95.
4. de Cabo, R.; Mattson, M.P. Effects of intermittent fasting on health, aging, and disease. *N. Engl. J. Med.* **2019**, *381*, 2541–2551. [CrossRef]
5. Dominguez, L.J.; Veronese, N.; Baiamonte, E.; Guarrera, M.; Parisi, A.; Ruffolo, C.; Tagliaferri, F.; Barbagallo, M. Healthy aging and dietary patterns. *Nutrients* **2022**, *14*, 889. [CrossRef]
6. Gao, Y.; Wei, Y.; Wang, Y.; Gao, F.; Chen, Z. *Lycium barbarum*: A traditional Chinese herb and a promising anti-aging agent. *Aging Dis.* **2017**, *8*, 778–791. [CrossRef] [PubMed]
7. Kunugi, H.; Mohammed Ali, A. Royal jelly and its components promote healthy aging and longevity: From animal models to humans. *Int. J. Mol. Sci.* **2019**, *20*, 4662. [CrossRef] [PubMed]
8. Hetland, G.; Johnson, E.; Lyberg, T.; Bernardshaw, S.; Tryggestad, A.M.A.; Grinde, B. Effects of the medicinal mushroom *Agaricus blazei* Murill on immunity, infection and cancer. *Scand. J. Immunol.* **2008**, *68*, 157015. [CrossRef] [PubMed]
9. Im, K.H.; Nguyen, T.K.; Choi, J.; Lee, T.S. In vitro antioxidant, anti-diabetes, anti-dementia, and inflammation inhibitory effect of *Trametes pubescens* fruiting body extracts. *Molecules* **2016**, *21*, 639. [CrossRef]
10. Lee, J.E.; Lee, I.S.; Kim, K.C.; Yoo, I.D.; Yang, H.M. ROS scavenging and anti-wrinkle effects of clitocybin A isolated from the mycelium of the mushroom *Clitocybe aurantiaca*. *J. Microbiol. Biotechnol.* **2017**, *27*, 933–938. [CrossRef] [PubMed]
11. Yuan, F.; Gao, Z.; Liu, W.; Li, H.; Zhang, Y.; Feng, Y.; Song, X.; Wang, W.; Zhang, J.; Huang, C.; et al. Characterization, antioxidant, anti-aging and organ protective effects of sulfated polysaccharides from *Flammulina velutipes*. *Molecules* **2019**, *24*, 3517. [CrossRef]
12. Jo Feeney, M.; Miller, A.M.; Roupas, P. Mushrooms-biologically distinct and nutritionally unique: Exploring a "Third Food Kingdom". *Nutr. Today* **2014**, *49*, 301–307. [CrossRef]
13. Elkhateeb, W.A. What medicinal mushroom can do? *J. Chem. Res.* **2020**, *5*, 106–118.
14. Franceschi, C.; Garagnani, P.; Morsiani, C.; Conte, M.; Santoro, A.; Grignolio, A.; Monti, D.; Capri, M.; Salvioli, S. The continuum of aging and age-related diseases: Common mechanisms but different rates. *Front. Med.* **2018**, *5*, 61. [CrossRef]
15. Harman, D. The aging process. *Proc. Natl. Acad. Sci. USA* **1981**, *78*, 7124–7128. [CrossRef]
16. Lopez-Otin, C.; Blasco, M.A.; Partridge, L.; Serrano, M.; Kroemer, G. The hallmarks of aging. *Cell* **2013**, *153*, 1194–1217. [CrossRef] [PubMed]
17. Harman, D. Aging: A theory based on free radical and radiation chemistry. *J. Gerontol.* **1956**, *11*, 298–300. [CrossRef] [PubMed]
18. Harman, D. Free radical theory of aging: Dietary implications. *Am. J. Clin. Nutr.* **1972**, *25*, 839–843. [CrossRef]
19. Van Remmen, H.; Ikeno, Y.; Hamilton, M.; Pahlavani, M.; Wolf, N.; Thorpe, S.R.; Alderson, N.L.; Baynes, J.W.; Epstein, C.J.; Huang, T.T.; et al. Life-long reduction in MnSOD activity results in increased DNA damage and higher incidence of cancer but does not accelerate aging. *Physiol. Genom.* **2003**, *16*, 29–37. [CrossRef]
20. Chen, Q.; Xu, B.J.; Huang, W.S.; Amrouche, A.T.; Maurizio, B.; Simal-Gandara, J.; Tundis, R.; Xiao, J.B.; Zou, L.; Lu, B.Y. Edible flowers as functional raw materials: A review on anti-aging properties. *Trends Food Sci. Technol.* **2020**, *106*, 30–47. [CrossRef]
21. Liguori, I.; Russo, G.; Curcio, F.; Bulli, G.; Aran, L.; Della-Morte, D.; Gargiulo, G.; Testa, G.; Cacciatore, F.; Bonaduce, D.; et al. Oxidative stress, aging, and diseases. *Clin. Interv. Aging* **2018**, *13*, 757–772. [CrossRef]
22. Belikov, A.V. Age-related diseases as vicious cycles. *Ageing Res. Rev.* **2019**, *49*, 11–26. [CrossRef] [PubMed]
23. Lopez-Otin, C.; Galluzzi, L.; Freije, J.M.P.; Madeo, F.; Kroemer, G. Metabolic control of longevity. *Cell* **2016**, *166*, 802–821. [CrossRef] [PubMed]
24. Fontana, L.; Partridge, L.; Longo, V.D. Extending healthy life span—From yeast to humans. *Science* **2010**, *328*, 321–326. [CrossRef]
25. Lee, S.H.; Min, K.J. Caloric restriction and its mimetics. *BMB Rep.* **2013**, *46*, 181–187. [CrossRef]
26. McCay, C.M.; Crowell, M.F.; Maynard, L.A. The effect of retarded growth upon the length of life span and upon the ultimate body size. 1935. *Nutrition* **1989**, *5*, 155–171, discussion 172.
27. Martins, I.; Galluzzi, L.; Kroemer, G. Hormesis, cell death and aging. *Aging* **2011**, *3*, 821–828. [CrossRef]
28. Kenyon, C.J. The genetics of ageing. *Nature* **2010**, *464*, 504–512. [CrossRef]
29. Miller, R.A. Cell stress and aging: New emphasis on multiplex resistance mechanisms. *J. Gerontol. A Biol. Sci. Med. Sci.* **2009**, *64*, 179–182. [CrossRef]
30. Cava, E.; Fontana, L. Will calorie restriction work in humans? *Aging* **2013**, *5*, 507–514. [CrossRef]
31. de Magalhaes, J.P. The scientific quest for lasting youth: Prospects for curing aging. *Rejuvenation Res.* **2014**, *17*, 458–467. [CrossRef]
32. Dirks, A.J.; Leeuwenburgh, C. Caloric restriction in humans: Potential pitfalls and health concerns. *Mech. Ageing Dev.* **2006**, *127*, 1–7. [CrossRef] [PubMed]
33. Di Francesco, A.; Di Germanio, C.; Bernier, M.; de Cabo, R. A time to fast. *Science* **2018**, *362*, 770–775. [CrossRef] [PubMed]

34. Mattson, M.P.; Moehl, K.; Ghena, N.; Schmaedick, M.; Cheng, A. Intermittent metabolic switching, neuroplasticity and brain health. *Nat. Rev. Neurosci.* **2018**, *19*, 63–80. [CrossRef] [PubMed]

35. Hu, D.; Xie, F.; Xiao, Y.; Lu, C.; Zhong, J.; Huang, D.; Chen, J.; Wei, J.; Jiang, Y.; Zhong, T. Metformin: A potential candidate for targeting aging mechanisms. *Aging Dis.* **2021**, *12*, 480–493. [CrossRef]

36. Zhang, Y.; Zhang, J.; Wang, S. The role of rapamycin in healthspan extension via the delay of organ aging. *Ageing Res. Rev.* **2021**, *70*, 101376. [CrossRef]

37. Li, J.; Kim, S.G.; Blenis, J. Rapamycin: One drug, many effects. *Cell Metab.* **2014**, *19*, 373–379. [CrossRef]

38. Soukas, A.A.; Hao, H.; Wu, L. Metformin as anti-aging therapy: Is it for everyone? *Trends Endocrinol. Metab.* **2019**, *30*, 745–755. [CrossRef]

39. Martel, J.; Ko, Y.F.; Liau, J.C.; Lee, C.S.; Ojcius, D.M.; Lai, H.C.; Young, J.D. Myths and realities surrounding the mysterious caterpillar fungus. *Trends Biotechnol.* **2017**, *35*, 1017–1021. [CrossRef]

40. Martel, J.; Ko, Y.F.; Ojcius, D.M.; Lu, C.C.; Chang, C.J.; Lin, C.S.; Lai, H.C.; Young, J.D. Immunomodulatory properties of plants and mushrooms. *Trends Pharmacol. Sci.* **2017**, *38*, 967–981. [CrossRef]

41. Martel, J.; Ojcius, D.M.; Chang, C.J.; Lin, C.S.; Lu, C.C.; Ko, Y.F.; Tseng, S.F.; Lai, H.C.; Young, J.D. Anti-obesogenic and antidiabetic effects of plants and mushrooms. *Nat. Rev. Endocrinol.* **2017**, *13*, 149–160. [CrossRef]

42. Petrovska, B.B. Historical review of medicinal plants' usage. *Pharmacogn. Rev.* **2012**, *6*, 1–5. [CrossRef]

43. Kiely, K.M.; Brady, B.; Byles, J. Gender, mental health and ageing. *Maturitas* **2019**, *129*, 76–84. [CrossRef] [PubMed]

44. Cedrone, F.; Catalini, A.; Stacchini, L.; Berselli, N.; Caminiti, M.; Mazza, C.; Cosma, C.; Minutolo, G.; Di Martino, G. The role of gender in the association between mental health and potentially preventable hospitalizations: A single-center retrospective observational study. *Int. J. Environ. Res. Public Health* **2022**, *19*, 14691. [CrossRef] [PubMed]

45. Rinsky-Halivni, L.; Brammli-Greenberg, S.; Christiani, D.C. Ageing workers' mental health during COVID-19: A multilevel observational study on the association with the work environment, perceived workplace safety and individual factors. *BMJ Open* **2022**, *12*, e064590. [CrossRef] [PubMed]

46. Bockting, W.; Coleman, E.; Deutsch, M.B.; Guillamon, A.; Meyer, I.; Meyer, W., 3rd; Reisner, S.; Sevelius, J.; Ettner, R. Adult development and quality of life of transgender and gender nonconforming people. *Curr. Opin. Endocrinol. Diabetes Obes.* **2016**, *23*, 188–197. [CrossRef]

47. Thomas Tobin, C.S.; Erving, C.L.; Hargrove, T.W.; Satcher, L.A. Is the Black-White mental health paradox consistent across age, gender, and psychiatric disorders? *Aging Ment. Health* **2022**, *26*, 196–204. [CrossRef]

48. Zhang, X.; Yan, Y.; Ye, Z.; Xie, J. Descriptive analysis of depression among adolescents in Huangshi, China. *BMC Psychiatry* **2023**, *23*, 176. [CrossRef]

49. Ulep, V.G.T.; Uy, J.; Casas, L.D. What explains the large disparity in child stunting in the Philippines? A decomposition analysis. *Public Health Nutr.* **2022**, *25*, 2995–3007. [CrossRef]

50. Xu, C.; Ganesan, K.; Liu, X.; Ye, Q.; Cheung, Y.; Liu, D.; Zhong, S.; Chen, J. Prognostic value of negative emotions on the incidence of breast cancer: A systematic review and meta-analysis of 129,621 patients with breast cancer. *Cancers* **2022**, *14*, 475. [CrossRef]

51. Lorenzo, E.C.; Kuchel, G.A.; Kuo, C.L.; Moffitt, T.E.; Diniz, B.S. Major depression and the biological hallmarks of aging. *Ageing Res. Rev.* **2023**, *83*, 101805. [CrossRef]

52. Yeap, B.B. Hormonal changes and their impact on cognition and mental health of ageing men. *Maturitas* **2014**, *79*, 227–235. [CrossRef]

53. Barros, L.; Correia, D.M.; Ferreira, I.C.; Baptista, P.; Santos-Buelga, C. Optimization of the determination of tocopherols in *Agaricus sp.* edible mushrooms by a normal phase liquid chromatographic method. *Food Chem.* **2008**, *110*, 1046–1050. [CrossRef] [PubMed]

54. Mattila, P.; Konko, K.; Eurola, M.; Pihlava, J.M.; Astola, J.; Vahteristo, L.; Hietaniemi, V.; Kumpulainen, J.; Valtonen, M.; Piironen, V. Contents of vitamins, mineral elements, and some phenolic compounds in cultivated mushrooms. *J. Agric. Food Chem.* **2001**, *49*, 2343–2348. [CrossRef] [PubMed]

55. Kalac, P. A review of chemical composition and nutritional value of wild-growing and cultivated mushrooms. *J. Sci. Food Agric.* **2013**, *93*, 209–218. [CrossRef] [PubMed]

56. Ferreira, I.C.; Barros, L.; Abreu, R.M. Antioxidants in wild mushrooms. *Curr. Med. Chem.* **2009**, *16*, 1543–1560. [CrossRef] [PubMed]

57. Wasser, S.P. Current findings, future trends, and unsolved problems in studies of medicinal mushrooms. *Appl. Microbiol. Biotechnol.* **2011**, *89*, 1323–1332. [CrossRef] [PubMed]

58. Valverde, M.E.; Hernandez-Perez, T.; Paredes-Lopez, O. Edible mushrooms: Improving human health and promoting quality life. *Int. J. Microbiol.* **2015**, *2015*, 376387. [CrossRef] [PubMed]

59. Falch, B.H.; Espevik, T.; Ryan, L.; Stokke, B.T. The cytokine stimulating activity of (1→3)-beta-D-glucans is dependent on the triple helix conformation. *Carbohydr. Res.* **2000**, *329*, 587–596. [CrossRef]

60. Kataoka, K.; Muta, T.; Yamazaki, S.; Takeshige, K. Activation of macrophages by linear (1→3)-beta-D-glucans. Impliations for the recognition of fungi by innate immunity. *J. Biol. Chem.* **2002**, *277*, 36825–36831. [CrossRef]

61. Khan, M.A.; Tania, M.; Liu, R.; Rahman, M.M. *Hericium erinaceus*: An edible mushroom with medicinal values. *J. Complement. Integr. Med.* **2013**, *10*, 253–258. [CrossRef] [PubMed]

62. Vetvicka, V.; Yvin, J.C. Effects of marine beta-1,3 glucan on immune reactions. *Int. Immunopharmacol.* **2004**, *4*, 721–730. [CrossRef] [PubMed]

63. Zaidman, B.Z.; Yassin, M.; Mahajna, J.; Wasser, S.P. Medicinal mushroom modulators of molecular targets as cancer therapeutics. *Appl. Microbiol. Biotechnol.* **2005**, *67*, 453–468. [CrossRef] [PubMed]
64. Heleno, S.A.; Barros, L.; Martins, A.; Queiroz, M.J.; Santos-Buelga, C.; Ferreira, I.C. Phenolic, polysaccharidic, and lipidic fractions of mushrooms from northeastern Portugal: Chemical compounds with antioxidant properties. *J. Agric. Food Chem.* **2012**, *60*, 4634–4640. [CrossRef] [PubMed]
65. Huang, J.; Ou, Y.; Yew, T.W.; Liu, J.; Leng, B.; Lin, Z.; Su, Y.; Zhuang, Y.; Lin, J.; Li, X.; et al. Hepatoprotective effects of polysaccharide isolated from *Agaricus bisporus* industrial wastewater against CCl$_4$-induced hepatic injury in mice. *Int. J. Biol. Macromol.* **2016**, *82*, 678–686. [CrossRef]
66. Li, S.; Liu, H.; Wang, W.; Wang, X.; Zhang, C.; Zhang, J.; Jing, H.; Ren, Z.; Gao, Z.; Song, X.; et al. Antioxidant and anti-aging effects of acidic-extractable polysaccharides by *Agaricus bisporus*. *Int. J. Biol. Macromol.* **2018**, *106*, 1297–1306. [CrossRef]
67. Zhang, Y.; Ma, G.; Fang, L.; Wang, L.; Xie, J. The immunostimulatory and anti-tumor activities of polysaccharide from *Agaricus bisporus* (brown). *J. Food Nutr. Res.* **2014**, *2*, 122–126. [CrossRef]
68. Pires, A.; Ruthes, A.C.; Cadena, S.; Iacomini, M. Cytotoxic effect of a mannogalactoglucan extracted from *Agaricus bisporus* on HepG2 cells. *Carbohydr. Polym.* **2017**, *170*, 33–42. [CrossRef]
69. Smiderle, F.R.; Alquini, G.; Tadra-Sfeir, M.Z.; Iacomini, M.; Wichers, H.J.; Van Griensven, L.J. *Agaricus bisporus* and *Agaricus brasiliensis* (1→6)-β-D-glucans show immunostimulatory activity on human THP-1 derived macrophages. *Carbohydr. Polym.* **2013**, *94*, 91–99. [CrossRef]
70. Reis, F.S.; Barros, L.; Martins, A.; Ferreira, I.C. Chemical composition and nutritional value of the most widely appreciated cultivated mushrooms: An inter-species comparative study. *Food Chem. Toxicol.* **2012**, *50*, 191–197. [CrossRef]
71. Govindan, S.; Johnson, E.E.; Christopher, J.; Shanmugam, J.; Thirumalairaj, V.; Gopalan, J. Antioxidant and anti-aging activities of polysaccharides from *Calocybe indica* var. APK2. *Exp. Toxicol. Pathol.* **2016**, *68*, 329–334. [CrossRef] [PubMed]
72. Chen, G.T.; Fu, Y.X.; Yang, W.J.; Hu, Q.H.; Zhao, L.Y. Effects of polysaccharides from the base of *Flammulina velutipes* stipe on growth of murine RAW264.7, B16F10 and L929 cells. *Int. J. Biol. Macromol.* **2018**, *107*, 2150–2156. [CrossRef] [PubMed]
73. Yang, W.; Yu, J.; Zhao, L.; Ma, N.; Fang, Y.; Pei, F.; Mariga, A.M.; Hu, Q. Polysaccharides from *Flammulina velutipes* improve scopolamine-induced impairment of learning and memory of rats. *J. Funct. Foods* **2015**, *18*, 411–422. [CrossRef]
74. Xu, S.; Dou, Y.; Ye, B.; Wu, Q.; Wang, Y.; Hu, M.; Ma, F.; Rong, X.; Guo, J. *Ganoderma lucidum* polysaccharides improve insulin sensitivity by regulating inflammatory cytokines and gut microbiota composition in mice. *J. Funct. Foods* **2017**, *38*, 545–552. [CrossRef]
75. Jin, M.; Zhu, Y.; Shao, D.; Zhao, K.; Xu, C.; Li, Q.; Yang, H.; Huang, Q.; Shi, J. Effects of polysaccharide from mycelia of *Ganoderma lucidum* on intestinal barrier functions of rats. *Int. J. Biol. Macromol.* **2017**, *94*, 1–9. [CrossRef]
76. Yang, G.; Yang, L.; Zhuang, Y.; Qian, X.; Shen, Y. *Ganoderma lucidum* polysaccharide exerts anti-tumor activity via MAPK pathways in HL-60 acute leukemia cells. *J. Recept. Signal Transduct. Res.* **2016**, *36*, 6–13. [CrossRef]
77. Huang, S.; Mao, J.; Ding, K.; Zhou, Y.; Zeng, X.; Yang, W.; Wang, P.; Zhao, C.; Yao, J.; Xia, P.; et al. Polysaccharides from *Ganoderma lucidum* promote cognitive function and neural progenitor proliferation in mouse model of Alzheimer's disease. *Stem Cell Rep.* **2017**, *8*, 84–94. [CrossRef]
78. Ya, G. A *Lentinus edodes* polysaccharide induces mitochondrial-mediated apoptosis in human cervical carcinoma HeLa cells. *Int. J. Biol. Macromol.* **2017**, *103*, 676–682. [CrossRef]
79. Ren, Z.; Li, J.; Song, X.; Zhang, J.; Wang, W.; Wang, X.; Gao, Z.; Jing, H.; Li, S.; Jia, L. The regulation of inflammation and oxidative status against lung injury of residue polysaccharides by *Lentinula edodes*. *Int. J. Biol. Macromol.* **2018**, *106*, 185–192. [CrossRef]
80. Wang, J.; Li, W.; Huang, X.; Liu, Y.; Li, Q.; Zheng, Z.; Wang, K. A polysaccharide from *Lentinus edodes* inhibits human colon cancer cell proliferation and suppresses tumor growth in athymic nude mice. *Oncotarget* **2017**, *8*, 610–623. [CrossRef] [PubMed]
81. Jeff, I.B.; Fan, E.; Tian, M.; Song, C.; Yan, J.; Zhou, Y. In vivo anticancer and immunomodulating activities of mannogalactoglucan-type polysaccharides from *Lentinus edodes* (Berkeley) Singer. *Cent. Eur. J. Immunol.* **2016**, *41*, 47–53. [CrossRef]
82. Xu, H.; Zou, S.; Xu, X.; Zhang, L. Anti-tumor effect of β-glucan from *Lentinus edodes* and the underlying mechanism. *Sci. Rep.* **2016**, *6*, 288–302. [CrossRef]
83. Carneiro, A.A.; Ferreira, I.C.; Dueñas, M.; Barros, L.; da Silva, R.; Gomes, E.; Santos-Buelga, C. Chemical composition and antioxidant activity of dried powder formulations of *Agaricus blazei* and *Lentinus edodes*. *Food Chem.* **2013**, *138*, 2168–2173. [CrossRef]
84. Ren, D.; Wang, N.; Guo, J.; Yuan, L.; Yang, X. Chemical characterization of *Pleurotus eryngii* polysaccharide and its tumor-inhibitory effects against human hepatoblastoma HepG-2 cells. *Carbohydr. Polym.* **2016**, *138*, 123–133. [CrossRef]
85. Ma, G.; Yang, W.; Mariga, A.M.; Fang, Y.; Ma, N.; Pei, F.; Hu, Q. Purification, characterization and antitumor activity of polysaccharides from *Pleurotus eryngii* residue. *Carbohydr. Polym.* **2014**, *114*, 297–305. [CrossRef]
86. Xu, D.; Wang, H.; Zheng, W.; Gao, Y.; Wang, M.; Zhang, Y.; Gao, Q. Charaterization and immunomodulatory activities of polysaccharide isolated from *Pleurotus eryngii*. *Int. J. Biol. Macromol.* **2016**, *92*, 30–36. [CrossRef]
87. Zhang, Y.; Yang, X.; Jin, G.; Yang, X.; Zhang, Y. Polysaccharides from *Pleurotus ostreatus* alleviate cognitive impairment in a rat model of Alzheimer's disease. *Int. J. Biol. Macromol.* **2016**, *92*, 935–941. [CrossRef] [PubMed]
88. Zhang, Y.; Wang, Z.; Jin, G.; Yang, X.; Zhou, H. Regulating dyslipidemia effect of polysaccharides from *Pleurotus ostreatus* on fat-emulsion-induced hyperlipidemia rats. *Int. J. Biol. Macromol.* **2017**, *101*, 107–116. [CrossRef] [PubMed]

89. Cao, X.Y.; Liu, J.L.; Yang, W.; Hou, X.; Li, Q.J. Antitumor activity of polysaccharide extracted from *Pleurotus ostreatus* mycelia against gastric cancer in vitro and in vivo. *Mol. Med. Rep.* **2015**, *12*, 2383–2389. [CrossRef] [PubMed]
90. Xu, X.; Yan, H.; Chen, J.; Zhang, X. Bioactive proteins from mushrooms. *Biotechnol. Adv.* **2011**, *29*, 667–674. [CrossRef]
91. Chang, H.H.; Sheu, F. Anti-tumor mechanisms of orally administered a fungal immunomodulatory protein from *Flammulina velutipes* in mice. *FASEB J.* **2006**, *20*, 297–306. [CrossRef]
92. Lin, C.H.; Sheu, G.T.; Lin, Y.W.; Yeh, C.S.; Huang, Y.H.; Lai, Y.C.; Chang, J.G.; Ko, J.L. A new immunomodulatory protein from *Ganoderma microsporum* inhibits epidermal growth factor mediated migration and invasion in A549 lung cancer cells. *Process Biochem.* **2010**, *45*, 1537–1542. [CrossRef]
93. Peek, H.W.; Halkes, S.B.A.; Tomassen, M.M.M.; Mes, J.J.; Landman, W.J.M. In vivo screening of five phytochemicals/extracts and a fungal immunomodulatory protein against colibacillosis in broilers. *Avian Pathol.* **2013**, *42*, 235–247. [CrossRef]
94. Lin, W.H.; Hung, C.H.; Hsu, C.-N.; Lin, J.Y. Dimerization of the N-terminal amphipathic helix domain of the fungal immunomodulatory protein from *Ganoderma tsugae* defined by a yeast two-hybrid system and site-directed mutagenesis. *J. Biol. Chem.* **1997**, *272*, 20044–20048. [CrossRef] [PubMed]
95. Chang, H.H.; Yeh, C.H.; Sheu, F. A novel immunomodulatory protein from *Poria cocos* induces toll-like receptor 4-dependent activation within mouse peritoneal macrophages. *J. Agric. Food Chem.* **2009**, *57*, 6129–6139. [CrossRef]
96. Ditamo, Y.; Rupil, L.L.; Sendra, V.G.; Nores, G.A.; Roth, G.A.; Irazoqui, F.J. In vivo immunomodulatory effect of the lectin from edible mushroom *Agaricus bisporus*. *Food Funct.* **2016**, *7*, 262–279. [CrossRef]
97. Matuszewska, A.; Karp, M.; Jaszek, M.; Janusz, G.; Osińska-Jaroszuk, M.; Sulej, J.; Stefaniuk, D.; Tomczak, W.; Giannopoulos, K. Laccase purified from *Cerrena unicolor* exerts antitumor activity against leukemic cells. *Oncol. Lett.* **2016**, *11*, 2009–2018. [CrossRef] [PubMed]
98. Zhao, S.; Rong, C.B.; Kong, C.; Liu, Y.; Xu, F.; Miao, Q.J.; Wang, S.X.; Wang, H.X.; Zhang, G.Q. A novel laccase with potent antiproliferative and HIV-1 reverse transcriptase inhibitory activities from mycelia of mushroom *Coprinus comatus*. *Biomed. Res. Int.* **2014**, *2014*, 417461. [CrossRef]
99. Chu, P.Y.; Sun, H.L.; Ko, J.L.; Ku, M.S.; Lin, L.J.; Lee, Y.T.; Liao, P.F.; Pan, H.H.; Lu, H.L.; Lue, K.H. Oral fungal immunomodulatory protein-*Flammulina velutipes* has influence on pulmonary inflammatory process and potential treatment for allergic airway disease: A mouse model. *J. Microbiol. Immunol. Infect.* **2017**, *50*, 297–306. [CrossRef]
100. Wang, H.; Ng, T.B. Isolation and characterization of velutin, a novel low-molecular-weight ribosome-inactivating protein from winter mushroom (*Flammulina velutipes*) fruiting bodies. *Life Sci.* **2001**, *68*, 2151–2168. [CrossRef]
101. Kumaran, S.; Pandurangan, A.K.; Shenbhagaraman, R.; Esa, N.M. Isolation and characterization of lectin from the Artist's Conk medicinal mushroom, *Ganoderma applanatum* (Agaricomycetes), and evaluation of its antiproliferative activity in HT-29 colon cancer cells. *Int. J. Med. Mushrooms* **2017**, *19*, 675–684. [CrossRef]
102. Wang, H.X.; Ng, T.B. A laccase from the medicinal mushroom *Ganoderma lucidum*. *Appl. Microbiol. Biotechnol.* **2006**, *72*, 508–513. [CrossRef]
103. Hsin, I.L.; Wang, S.C.; Li, J.R.; Ciou, T.C.; Wu, C.H.; Wu, H.M.; Ko, J.L. Immunomodulatory proteins FIP-gts and chloroquine induce caspase-independent cell death via autophagy for resensitizing cisplatin-resistant urothelial cancer cells. *Phytomedicine* **2016**, *23*, 1566–1573. [CrossRef]
104. Lam, S.K.; Ng, T.B. Hypsin, a novel thermostable ribosome-inactivating protein with antifungal and antiproliferative activities from fruiting bodies of the edible mushroom *Hypsizigus marmoreus*. *Biochem. Biophys. Res. Commun.* **2001**, *285*, 1071–1085. [CrossRef]
105. Wong, J.H.; Wang, H.X.; Ng, T.B. Marmorin, a new ribosome inactivating protein with antiproliferative and HIV-1 reverse transcriptase inhibitory activities from the mushroom *Hypsizigus marmoreus*. *Appl. Microbiol. Biotechnol.* **2008**, *81*, 669–674. [CrossRef] [PubMed]
106. Sun, J.; Chen, Q.J.; Zhu, M.-U.; Wang, H.X.; Zhang, G.Q. An extracellular laccase with antiproliferative activity from the sanghuang mushroom *Inonotus baumii*. *J. Mol. Catal. B Enzym.* **2014**, *99*, 20–25. [CrossRef]
107. Žurga, S.; Nanut, M.P.; Kos, J.; Sabotič, J. Fungal lectin MpL enables entry of protein drugs into cancer cells and their subcellular targeting. *Oncotarget* **2017**, *8*, 26896–26910. [CrossRef]
108. Wu, X.; Huang, C.; Chen, Q.; Wang, H.; Zhang, J. A novel laccase with inhibitory activity towards HIV-I reverse transcriptase and antiproliferative effects on tumor cells from the fermentation broth of mushroom *Pleurotus cornucopiae*. *Biomed. Chromatogr.* **2014**, *28*, 548–553. [CrossRef] [PubMed]
109. Wang, H.X.; Ng, T.B. Purification of a laccase from fruiting bodies of the mushroom *Pleurotus eryngii*. *Appl. Microbiol. Biotechnol.* **2006**, *69*, 521–535. [CrossRef] [PubMed]
110. He, M.; Su, D.; Liu, Q.; Gao, W.; Kang, Y. Mushroom lectin overcomes hepatitis B virus tolerance via TLR6 signaling. *Sci. Rep.* **2017**, *7*, 5814. [CrossRef] [PubMed]
111. Chandrasekaran, G.; Lee, Y.C.; Park, H.; Wu, Y.; Shin, H.J. Antibacterial and antifungal activities of lectin extracted from fruiting bodies of the Korean cauliflower medicinal mushroom, *Sparassis latifolia* (Agaricomycetes). *Int. J. Med. Mushrooms* **2016**, *18*, 291–309. [CrossRef]
112. May, M.J.; Hartley, M.R.; Roberts, L.M.; Krieg, P.A.; Osborn, R.W.; Lord, J.M. Ribosome inactivation by ricin A chain: A sensitive method to assess the activity of wild-type and mutant polypeptides. *EMBO J.* **1989**, *8*, 301–308. [CrossRef]
113. Zhou, K.; Fu, Z.; Chen, M.; Lin, Y.; Pan, K. Structure of trichosanthin at 1.88 Å resolution. *Proteins Struct. Funct. Bioinform.* **1994**, *19*, 4–13. [CrossRef]

114. Zhu, F.; Zhou, Y.K.; Ji, Z.L.; Chen, X.R. The plant ribosome-inactivating proteins play important roles in defense against pathogens and insect pest attacks. *Front. Plant Sci.* **2018**, *9*, 146. [CrossRef]

115. Domashevskiy, A.V.; Goss, D.J. Pokeweed antiviral protein, a ribosome inactivating protein: Activity, inhibition and prospects. *Toxins* **2015**, *7*, 274–298. [CrossRef] [PubMed]

116. Ng, T.B.; Lam, J.S.; Wong, J.H.; Lam, S.K.; Ngai, P.H.; Wang, H.X.; Chu, K.T.; Chan, W.Y. Differential abilities of the mushroom ribosome-inactivating proteins hypsin and velutin to perturb normal development of cultured mouse embryos. *Toxicol. In Vitro* **2010**, *24*, 1250–1257. [CrossRef] [PubMed]

117. Sande, D.; Oliveira, G.P.d.; Moura, M.A.F.E.; Martins, B.d.A.; Lima, M.T.N.S.; Takahashi, J.A. Edible mushrooms as a ubiquitous source of essential fatty acids. *Food Res. Int.* **2019**, *125*, 108524. [CrossRef]

118. Günç Ergönül, P.; Akata, I.; Kalyoncu, F.; Ergönül, B. Fatty acid compositions of six wild edible mushroom species. *Sci. World J.* **2013**, *2013*, 163964. [CrossRef] [PubMed]

119. Sinanoglou, V.J.; Zoumpoulakis, P.; Heropoulos, G.; Proestos, C.; Ćirić, A.; Petrovic, J.; Glamoclija, J.; Sokovic, M. Lipid and fatty acid profile of the edible fungus *Laetiporus sulphurous*. Antifungal and antibacterial properties. *J. Food Sci. Technol.* **2015**, *52*, 3264–3272. [CrossRef] [PubMed]

120. Rodrigues, M.L.; Nimrichter, L.; Oliveira, D.L.; Frases, S.; Miranda, K.; Zaragoza, O.; Alvarez, M.; Nakouzi, A.; Feldmesser, M.; Casadevall, A. Vesicular polysaccharide export in *Cryptococcus neoformans* is a eukaryotic solution to the problem of fungal trans-cell wall transport. *Eukaryot. Cell* **2007**, *6*, 48–59. [CrossRef] [PubMed]

121. Guillamón, E.; García-Lafuente, A.; Lozano, M.; D'Arrigo, M.; Martínez, J. Edible mushrooms: Role in the prevention of cardiovascular diseases. *Fitoterapia* **2010**, *81*, 715–723. [CrossRef]

122. Weng, Y.; Xiang, L.; Matsuura, A.; Zhang, Y.; Huang, Q.; Qi, J. Ganodermasides A and B, two novel anti-aging ergosterols from spores of a medicinal mushroom *Ganoderma lucidum* on yeast via UTH1 gene. *Bioorg. Med. Chem.* **2010**, *18*, 999–1002. [CrossRef] [PubMed]

123. Saiki, P.; Kawano, Y.; Griensven, L.; Miyazaki, K. The anti-inflammatory effect of *Agaricus brasiliensis* is partly due to its linoleic acid content. *Food Funct.* **2017**, *8*, 4150–4158. [CrossRef] [PubMed]

124. Ergon, P.; Ergonul, B.; Kalyoncu, F.; Akata, I. Fatty acid compositions of five wild edible mushroom species collected from Turkey. *Int. J. Pharmacol.* **2012**, *8*, 463–466. [CrossRef]

125. Abdelshafy, A.M.; Belwal, T.; Liang, Z.; Wang, L.; Li, D.; Luo, Z.; Li, L. A comprehensive review on phenolic compounds from edible mushrooms: Occurrence, biological activity, application and future prospective. *Critic. Rev. Food Sci. Nutr.* **2022**, *62*, 6204–6224. [CrossRef] [PubMed]

126. Bach, F.; Zielinski, A.A.F.; Helm, C.V.; Maciel, G.M.; Pedro, A.C.; Stafussa, A.P.; Ávila, S.; Haminiuk, C.W.I. Bio compounds of edible mushrooms: In vitro antioxidant and antimicrobial activities. *LWT* **2019**, *107*, 214–220. [CrossRef]

127. Contato, A.G.; Inácio, F.D.; de Araújo, C.A.V.; Brugnari, T.; Maciel, G.M.; Haminiuk, C.W.I.; Bracht, A.; Peralta, R.M.; de Souza, C.G.M. Comparison between the aqueous extracts of mycelium and basidioma of the edible mushroom *Pleurotus pulmonarius*: Chemical composition and antioxidant analysis. *J. Food Meas. Charact.* **2020**, *14*, 830–837. [CrossRef]

128. Mutukwa, I.B.; Hall, C.A., III; Cihacek, L.; Lee, C.W. Evaluation of drying method and pretreatment effects on the nutritional and antioxidant properties of oyster mushroom (*Pleurotus ostreatus*). *J. Food Process. Preserv.* **2019**, *43*, e13910. [CrossRef]

129. Bahadori, M.B.; Sarikurkcu, C.; Yalcin, O.U.; Cengiz, M.; Gungor, H. Metal concentration, phenolics profiling, and antioxidant activity of two wild edible *Melanoleuca mushrooms* (*M. cognata* and *M. stridula*). *Microchem. J.* **2019**, *150*, 104172. [CrossRef]

130. Hwang, A.Y.; Yang, S.C.; Kim, J.; Lim, T.; Cho, H.; Hwang, K.T. Effects of non-traditional extraction methods on extracting bioactive compounds from chaga mushroom (*Inonotus obliquus*) compared with hot water extraction. *LWT* **2019**, *110*, 80–84. [CrossRef]

131. Stojkovic, D.; Smiljkovic, M.; Ciric, A.; Glamoclija, J.; Griensven, L.V.; Ferreira, I.; Sokovic, M. An insight into antidiabetic properties of six medicinal and edible mushrooms: Inhibition of α-amylase and α-glucosidase linked to type-2 diabetes. *S. Afr. J. Bot.* **2019**, *120*, 100–103. [CrossRef]

132. Lin, S.; Ching, L.T.; Lam, K.; Cheung, P.C.K. Anti-angiogenic effect of water extract from the fruiting body of *Agrocybe aegerita*. *LWT* **2017**, *75*, 155–163. [CrossRef]

133. Alam, N.; Sikder, M.M.; Karim, M.A.; Amin, S. Antioxidant and antityrosinase activities of milky white mushroom. *Bangladesh J. Bot.* **2019**, *48*, 1065–1073. [CrossRef]

134. Palacios, I.; Lozano, M.; Moro, C.; D'Arrigo, M.; Rostagno, M.A.; Martínez, J.A.; García-Lafuente, A.; Guillamón, E.; Villares, A. Antioxidant properties of phenolic compounds occurring in edible mushrooms. *Food Chem.* **2011**, *128*, 674–678. [CrossRef]

135. Veljović, S.; Veljović, M.; Nikićević, N.; Despotović, S.; Radulović, S.; Nikšić, M.; Filipović, L. Chemical composition, antiproliferative and antioxidant activity of differently processed *Ganoderma lucidum* ethanol extracts. *J. Food Sci. Technol.* **2017**, *54*, 1312–1320. [CrossRef] [PubMed]

136. Gąsecka, M.; Siwulski, M.; Mleczek, M. Evaluation of bioactive compounds content and antioxidant properties of soil-growing and wood-growing edible mushrooms. *J. Food Process. Preserv.* **2018**, *42*, e13386. [CrossRef]

137. Kaewnarin, K.; Suwannarach, N.; Kumla, J.; Lumyong, S. Phenolic profile of various wild edible mushroom extracts from Thailand and their antioxidant properties, anti-tyrosinase and hyperglycaemic inhibitory activities. *J. Funct. Foods* **2016**, *27*, 352–364. [CrossRef]

138. Islam, T.; Yu, X.; Xu, B. Phenolic profiles, antioxidant capacities and metal chelating ability of edible mushrooms commonly consumed in China. *LWT* **2016**, *72*, 423–431. [CrossRef]

139. Alkan, S.; Uysal, A.; Kasik, G.; Vlaisavljevic, S.; Berežni, S.; Zengin, G. Chemical characterization, antioxidant, enzyme inhibition and antimutagenic properties of eight mushroom species: A comparative study. *J. Fungi* **2020**, *6*, 166. [CrossRef]
140. de Souza Campos Junior, F.A.; Petrarca, M.H.; Meinhart, A.D.; de Jesus Filho, M.; Godoy, H.T. Multivariate optimization of extraction and validation of phenolic acids in edible mushrooms by capillary electrophoresis. *Food Res. Int.* **2019**, *126*, 108685. [CrossRef]
141. Souilem, F.; Fernandes, Â.; Calhelha, R.C.; Barreira, J.C.M.; Barros, L.; Skhiri, F.; Martins, A.; Ferreira, I.C.F.R. Wild mushrooms and their mycelia as sources of bioactive compounds: Antioxidant, anti-inflammatory and cytotoxic properties. *Food Chem.* **2017**, *230*, 40–48. [CrossRef] [PubMed]
142. Nowacka-Jechalke, N.; Olech, M.; Nowak, R. Chapter 11—Mushroom polyphenols as chemopreventive agents. In *Polyphenols: Prevention and Treatment of Human Disease*, 2nd ed.; Watson, R.R., Preedy, V.R., Zibadi, S., Eds.; Academic Press: Cambridge, MA, USA, 2018; pp. 137–150.
143. Gogoi, P.; Chutia, P.; Singh, P.; Mahanta, C.L. Effect of optimized ultrasound-assisted aqueous and ethanolic extraction of *Pleurotus citrinopileatus* mushroom on total phenol, flavonoids and antioxidant properties. *J. Food Proc. Engin.* **2019**, *42*, e13172. [CrossRef]
144. Liu, Y.T.; Sun, J.; Luo, Z.Y.; Rao, S.Q.; Su, Y.J.; Xu, R.R.; Yang, Y.J. Chemical composition of five wild edible mushrooms collected from Southwest China and their antihyperglycemic and antioxidant activity. *Food Chem. Toxicol.* **2012**, *50*, 1238–1244. [CrossRef]
145. Çayan, F.; Deveci, E.; Tel-Çayan, G.; Duru, M.E. Identification and quantification of phenolic acid compounds of twenty-six mushrooms by HPLC–DAD. *J. Food Measu. Charact.* **2020**, *14*, 1690–1698. [CrossRef]
146. Yahia, E.M.; Gutiérrez-Orozco, F.; Moreno-Pérez, M.A. Identification of phenolic compounds by liquid chromatography-mass spectrometry in seventeen species of wild mushrooms in Central Mexico and determination of their antioxidant activity and bioactive compounds. *Food Chem.* **2017**, *226*, 14–22. [CrossRef]
147. Akindahunsi, A.A.; Oyetayo, F.L. Nutrient and antinutrient distribution of edible mushroom, *Pleurotus tuber*-regium (fries) singer. *LWT* **2006**, *39*, 548–553. [CrossRef]
148. Garrab, M.; Edziri, H.; El Mokni, R.; Mastouri, M.; Mabrouk, H.; Douki, W. Phenolic composition, antioxidant and anti-cholinesterase properties of the three mushrooms *Agaricus silvaticus* Schaeff., *Hydnum rufescens* Pers. and *Meripilus giganteus* (Pers.) Karst. in Tunisia. *S. Afr. J. Bot.* **2019**, *124*, 359–363. [CrossRef]
149. Pavithra, M.; Sridhar, K.R.; Greeshma, A.A.; Tomita-Yokotani, K. Bioactive potential of the wild mushroom *Astraeus hygrometricus* in South-west India. *Mycology* **2016**, *7*, 191–202. [CrossRef]
150. Heleno, S.A.; Barros, L.; Sousa, M.J.; Martins, A.; Ferreira, I.C.F.R. Tocopherols composition of Portuguese wild mushrooms with antioxidant capacity. *Food Chem.* **2010**, *119*, 1443–1450. [CrossRef]
151. Jayakumar, T.; Thomas, P.A.; Geraldine, P. In-vitro antioxidant activities of an ethanolic extract of the oyster mushroom, *Pleurotus ostreatus*. *Innov. Food Sci. Emerg. Technol.* **2009**, *10*, 228–234. [CrossRef]
152. Wu, Y.; Moon-Hee, C.; Li, J.; Yang, H.; Hyun-Jae, S. Mushroom cosmetics: The present and future. *Cosmetics* **2016**, *3*, 22. [CrossRef]
153. Hyde, K.D.; Bahkali, A.H.; Moslem, M.A. Fungi—An unusual source for cosmetics. *Fungal Divers.* **2010**, *43*, 1–9. [CrossRef]
154. Cheung, L.M.; Cheung, P.; Ooi, V. Antioxidant activity and total phenolics of edible mushroom extracts. *Food Chem.* **2003**, *81*, 249–255. [CrossRef]
155. Dubost, N.J.; Beelman, R.B.; Peterson, D.; Royse, D.J. Identification and quantification of ergothioneine in cultivated mushrooms by liquid chromatography-mass spectroscopy. *Int. J. Med. Mush.* **2006**, *8*, 215–222. [CrossRef]
156. Maity, K.; Kar Mandal, E.; Maity, S.; Gantait, S.K.; Das, D.; Maiti, S.; Maiti, T.K.; Sikdar, S.R.; Islam, S.S. Structural characterization and study of immunoenhancing and antioxidant property of a novel polysaccharide isolated from the aqueous extract of a somatic hybrid mushroom of *Pleurotus florida* and *Calocybe indica* variety APK2. *Int. J. Biol. Macromol.* **2011**, *48*, 304–310. [CrossRef]
157. Yuan, C.; Huang, X.; Cheng, L.; Bu, Y.; Liu, G.; Yi, F.; Yang, Z.; Song, F. Evaluation of antioxidant and immune activity of *Phellinus ribis* glucan in mice. *Food Chem.* **2009**, *115*, 581–584. [CrossRef]
158. Sun, J.; He, H.; Xie, B.J. Novel antioxidant peptides from fermented mushroom *Ganoderma lucidum*. *J. Agric. Food Chem.* **2004**, *52*, 6646–6652. [CrossRef]
159. Lupo, M.P.; Cole, A.L. Cosmeceutical peptides. *Dermatol. Ther.* **2007**, *20*, 343–359. [CrossRef]
160. Kim, S.Y.; Go, K.C.; Song, Y.S.; Jeong, Y.S.; Kim, E.J.; Kim, B.J. Extract of the mycelium of *T. matsutake* inhibits elastase activity and TPA-induced MMP-1 expression in human fibroblasts. *Int. J. Mol. Med.* **2014**, *34*, 1613–1621. [CrossRef]
161. Stanikunaite, R.; Khan, S.I.; Trappe, J.M.; Ross, S.A. Cyclooxygenase-2 inhibitory and antioxidant compounds from the truffle *Elaphomyces granulatus*. *Phytother. Res.* **2009**, *23*, 575–588. [CrossRef]
162. Shahid, A.; Huang, M.; Liu, M.; Shamim, M.A.; Parsa, C.; Orlando, R.; Huang, Y. The medicinal mushroom *Ganoderma lucidum* attenuates UV-induced skin carcinogenesis and immunosuppression. *PLoS ONE* **2022**, *17*, e0265615. [CrossRef]
163. Thangthaeng, N.; Miller, M.G.; Gomes, S.M.; Shukitt-Hale, B. Daily supplementation with mushroom (*Agaricus bisporus*) improves balance and working memory in aged rats. *Nutr. Res.* **2015**, *35*, 1079–1084. [CrossRef]
164. Yang, X.; Wang, X.; Lin, J.; Lim, S.; Cao, Y.; Chen, S.; Xu, P.; Xu, C.; Zheng, H.; Fu, K.C.; et al. Structure and anti-inflammatory activity relationship of ergostanes and lanostanes in *Antrodia cinnamomea*. *Foods* **2022**, *11*, 1831. [CrossRef]
165. Calvo, M.S.; Mehrotra, A.; Beelman, R.B.; Nadkarni, G.; Wang, L.; Cai, W.; Goh, B.C.; Kalaras, M.D.; Uribarri, J. A retrospective study in adults with metabolic syndrome: Diabetic risk factor response to daily consumption of *Agaricus bisporus* (white button mushrooms). *Plant Foods Hum. Nutr.* **2016**, *71*, 245–251. [CrossRef]

166. Song, T.Y.; Yang, N.C.; Chen, C.L.; Thi, T.L.V. Protective effects and possible mechanisms of ergothioneine and hispidin against methylglyoxal-induced injuries in rat pheochromocytoma cells. *Oxid. Med. Cell. Longev.* **2017**, *2017*, 4824371. [CrossRef]
167. Hagiwara, S.-Y.; Takahashi, M.; Shen, Y.; Kaihou, S.; Tomiyama, T.; Yazawa, M.; Tamai, Y.; Sin, Y.; Kazusaka, A.; Terazawa, M. A phytochemical in the edible tamogitake mushroom (*Pleurotus cornucopiae*), D-mannitol, inhibits ACE activity and lowers the blood pressure of spontaneously hypertensive rats. *Biosci. Biotechnol. Biochem.* **2005**, *69*, 1603–1605. [CrossRef]
168. Patel, S.; Goyal, A. Recent developments in mushrooms as anti-cancer therapeutics: A review. *3 Biotech* **2012**, *2*, 1–15. [CrossRef] [PubMed]
169. Łysakowska, P.; Sobota, A.; Wirkijowska, A. Medicinal mushrooms: Their bioactive components, nutritional value and application in functional food production—A review. *Molecules* **2023**, *28*, 5393. [CrossRef] [PubMed]
170. Lu, J.; He, R.; Sun, P.; Zhang, F.; Linhardt, R.J.; Zhang, A. Molecular mechanisms of bioactive polysaccharides from *Ganoderma lucidum* (Lingzhi), a review. *Int. J. Biol. Macromol.* **2020**, *150*, 765–774. [CrossRef] [PubMed]
171. Seweryn, E.; Ziała, A.; Gamian, A. Health-promoting of polysaccharides extracted from *Ganoderma lucidum*. *Nutrients* **2021**, *13*, 2725. [CrossRef] [PubMed]
172. Ikeda, H.; Togashi, Y. Aging, cancer, and antitumor immunity. *Int. J. Clin. Oncol.* **2022**, *27*, 316–322. [CrossRef]
173. Johnson, E.; Førland, D.T.; Sætre, L.; Bernardshaw, S.V.; Lyberg, T.; Hetland, G. Effect of an extract based on the medicinal mushroom *Agaricus blazei* Murill on release of cytokines, chemokines and leukocyte growth factors in human blood ex vivo and in vivo. *Scand. J. Immunol.* **2009**, *69*, 242–250. [CrossRef] [PubMed]
174. Roupas, P.; Keogh, J.; Noakes, M.; Margetts, C.; Taylor, P. The role of edible mushrooms in health: Evaluation of the evidence. *J. Funct. Foods* **2012**, *4*, 687–709. [CrossRef]
175. Pandimeena, M.; Prabu, M.; Sumathy, R.; Kumuthakalavalli, R. Evaluation of phytochemicals and in vitro anti-inflammatiry, anti-diabetic activity of the white oyster mushroom, *Pleurotus florida*. *Int. Res. J. Pharm. Appl. Sci.* **2015**, *5*, 16–21.
176. Tanaka, A.; Nishimura, M.; Sato, Y.; Sato, H.; Nishihira, J. Enhancement of the Th1-phenotype immune system by the intake of Oyster mushroom (Tamogitake) extract in a double-blind, placebo-controlled study. *J. Tradit. Complement. Med.* **2016**, *6*, 424–430. [CrossRef] [PubMed]
177. Wang, Y.; Zhang, Y.; Shao, J.; Wu, B.; Li, B. Potential immunomodulatory activities of a lectin from the mushroom *Latiporus sulphureus*. *Int. J. Biol. Macromol.* **2019**, *130*, 399–406. [CrossRef] [PubMed]
178. Wen, Y.; Wan, Y.Z.; Qiao, C.X.; Xu, X.F.; Shen, Y. Immunoregenerative effects of the bionically cultured Sanghuang mushrooms (*Inonotus sanghuagn*) on the immunodeficient mice. *J. Ethnopharmacol.* **2019**, *245*, 112047. [CrossRef] [PubMed]
179. Vlassopoulou, M.; Paschalidis, N.; Savvides, A.L.; Saxami, G.; Mitsou, E.K.; Kerezoudi, E.N.; Koutrotsios, G.; Zervakis, G.I.; Georgiadis, P.; Kyriacou, A.; et al. Immunomodulating activity of *Pleurotus eryngii* mushrooms following their in vitro fermentation by human fecal microbiota. *J. Fungi* **2022**, *8*, 329. [CrossRef]
180. Boulaka, A.; Mantellou, P.; Stanc, G.-M.; Souka, E.; Valavanis, C.; Saxami, G.; Mitsou, E.; Koutrotsios, G.; Zervakis, G.I.; Kyriacou, A.; et al. Genoprotective activity of the *Pleurotus eryngii* mushrooms following their in vitro and in vivo fermentation by fecal microbiota. *Front. Nutr.* **2022**, *9*, 988517. [CrossRef]
181. Lakatta, E.G.; Levy, D. Arterial and cardiac aging: Major shareholders in cardiovascular disease enterprises: Part I: Aging arteries: A "set up" for vascular disease. *Circulation* **2003**, *107*, 139–146. [CrossRef]
182. Adachi, K.; Nanba, H.; Otsuka, M.; Kuroda, H. Blood pressure-lowering activity present in the fruit body of *Grifola frondosa* (maitake). I. *Chem. Pharm. Bull.* **1988**, *36*, 1000–1006. [CrossRef] [PubMed]
183. Geng, X.; Tian, G.; Zhang, W.; Zhao, Y.; Zhao, L.; Wang, H.; Ng, T.B. A *Tricholoma matsutake* peptide with angiotensin converting enzyme inhibitory and antioxidative activities and antihypertensive effects in spontaneously hypertensive rats. *Sci. Rep.* **2016**, *6*, 24130. [CrossRef]
184. Ganesan, K.; Xu, B. Polyphenol-rich lentils and their health promoting effects. *Int. J. Mol. Sci.* **2017**, *18*, 2390. [CrossRef]
185. Ganesan, K.; Xu, B. Anti-obesity effects of medicinal and edible mushrooms. *Molecules* **2018**, *23*, 2880. [CrossRef] [PubMed]
186. Fombang, E.; Ee, L. *Pleurotus florida* aqueous extracts and powder influence lipid profile and suppress weight gain in rats fed high cholesterol diet. *J. Nutr. Food Sci.* **2016**, *6*, 2. [CrossRef]
187. Mori, K.; Kobayashi, C.; Tomita, T.; Inatomi, S.; Ikeda, M. Antiatherosclerotic effect of the edible mushrooms *Pleurotus eryngii* (Eringi), *Grifola frondosa* (Maitake), and *Hypsizygus marmoreus* (Bunashimeji) in apolipoprotein E-deficient mice. *Nutr. Res.* **2008**, *28*, 335–342. [CrossRef]
188. Koh, J.H.; Kim, J.M.; Chang, U.J.; Suh, H.J. Hypocholesterolemic effect of hot-water extract from mycelia of *Cordyceps sinensis*. *Biol. Pharm. Bull.* **2003**, *26*, 84–87. [CrossRef]
189. de Sá-Nakanishi, A.B.; Soares, A.A.; Natali, M.R.; Comar, J.F.; Peralta, R.M.; Bracht, A. Effects of the continuous administration of an *Agaricus blazei* extract to rats on oxidative parameters of the brain and liver during aging. *Molecules* **2014**, *19*, 18590–18603. [CrossRef]
190. Sillapachaiyaporn, C.; Rangsinth, P.; Nilkhet, S.; Ung, A.T.; Chuchawankul, S.; Tencomnao, T. Neuroprotective effects against glutamate-induced HT-22 hippocampal cell damage and *Caenorhabditis elegans* lifespan/healthspan enhancing activity of *Auricularia polytricha* mushroom extracts. *Pharmaceuticals* **2021**, *14*, 1001. [CrossRef]
191. Kushairi, N.; Phan, C.W.; Sabaratnam, V.; David, P.; Naidu, M. Lion's mane mushroom, *Hericium erinaceus* (Bull.: Fr.) Pers. suppresses H_2O_2-induced oxidative damage and LPS-induced inflammation in HT22 hippocampal neurons and BV2 microglia. *Antioxidants* **2019**, *8*, 261. [CrossRef]

192. Tripodi, F.; Falletta, E.; Leri, M.; Angeloni, C.; Beghelli, D.; Giusti, L.; Milanesi, R.; Sampaio-Marques, B.; Ludovico, P.; Goppa, L.; et al. Anti-aging and neuroprotective properties of *Grifola frondosa* and *Hericium erinaceus* extracts. *Nutrients* **2022**, *14*, 4368. [CrossRef] [PubMed]

193. Yanshree; Yu, W.S.; Fung, M.L.; Lee, C.W.; Lim, L.W.; Wong, K.H. The monkey head mushroom and memory enhancement in Alzheimer's disease. *Cells* **2022**, *11*, 2284. [CrossRef]

194. Ratto, D.; Corana, F.; Mannucci, B.; Priori, E.C.; Cobelli, F.; Roda, E.; Ferrari, B.; Occhinegro, A.; Di Iorio, C.; De Luca, F.; et al. *Hericium erinaceus* improves recognition memory and induces hippocampal and cerebellar neurogenesis in frail mice during aging. *Nutrients* **2019**, *11*, 715. [CrossRef] [PubMed]

195. Zhu, M.; Liu, X.; Liu, W.; Lu, Y.; Cheng, J.; Chen, Y. Beta cell aging and age-related diabetes. *Aging* **2021**, *13*, 7691–7706. [CrossRef] [PubMed]

196. Kumar, K.; Mehra, R.; Guiné, R.P.F.; Lima, M.J.; Kumar, N.; Kaushik, R.; Ahmed, N.; Yadav, A.N.; Kumar, H. Edible mushrooms: A comprehensive review on bioactive compounds with health benefits and processing aspects. *Foods* **2021**, *10*, 2996. [CrossRef] [PubMed]

197. Cho, E.J.; Hwang, H.J.; Kim, S.W.; Oh, J.Y.; Baek, Y.M.; Choi, J.W.; Bae, S.H.; Yun, J.W. Hypoglycemic effects of exopolysaccharides produced by mycelial cultures of two different mushrooms *Tremella fuciformis* and *Phellinus baumii* in ob/ob mice. *Appl. Microbiol. Biotechnol.* **2007**, *75*, 1257–1265. [CrossRef] [PubMed]

198. Jeong, S.C.; Jeong, Y.T.; Yang, B.K.; Islam, R.; Koyyalamudi, S.R.; Pang, G.; Cho, K.Y.; Song, C.H. White button mushroom (*Agaricus bisporus*) lowers blood glucose and cholesterol levels in diabetic and hypercholesterolemic rats. *Nutr. Res.* **2010**, *30*, 49–56. [CrossRef]

199. Kim, H.M.; Kang, J.S.; Kim, J.Y.; Park, S.K.; Kim, H.S.; Lee, Y.J.; Yun, J.; Hong, J.T.; Kim, Y.; Han, S.B. Evaluation of antidiabetic activity of polysaccharide isolated from *Phellinus linteus* in non-obese diabetic mouse. *Int. Immunopharmacol.* **2010**, *10*, 72–78. [CrossRef]

200. Oh, T.W.; Kim, Y.A.; Jang, W.J.; Byeon, J.I.; Ryu, C.H.; Kim, J.O.; Ha, Y.L. Semipurified fractions from the submerged-culture broth of *Agaricus blazei* Murill reduce blood glucose levels in streptozotocin-induced diabetic rats. *J. Agric. Food Chem.* **2010**, *58*, 4113–4129. [CrossRef]

201. Wang, Y.; Liu, H.; Li, C.; Qi, P.; Bao, J. *Agaricus bisporus* lectins mediates islet β-cell proliferation through regulation of cell cycle proteins. *Exp. Biol. Med.* **2012**, *237*, 287–296. [CrossRef]

202. Guggenheim, A.G.; Wright, K.M.; Zwickey, H.L. Immune modulation from five major mushrooms: Application to integrative oncology. *Integr. Med.* **2014**, *13*, 32–44.

203. Wasser, S.P. Medicinal mushroom science: Current perspectives, advances, evidences, and challenges. *Biomed. J.* **2014**, *37*, 345–356. [CrossRef]

204. Jeitler, M.; Michalsen, A.; Frings, D.; Hübner, M.; Kessler, C.S. Significance of medicinal mushrooms in integrative oncology: A narrative review. *Front. Pharmacol.* **2020**, *11*, 580656. [CrossRef] [PubMed]

205. Zhu, X.L.; Chen, A.F.; Lin, Z.B. *Ganoderma lucidum* polysaccharides enhance the function of immunological effector cells in immunosuppressed mice. *J. Ethnopharmacol.* **2007**, *111*, 219–226. [CrossRef] [PubMed]

206. Lin, C.H.; Hsiao, Y.M.; Ou, C.C.; Lin, Y.W.; Chiu, Y.L.; Lue, K.H.; Chang, J.G.; Ko, J.L. GMI, a Ganoderma immunomodulatory protein, down-regulates tumor necrosis factor α-induced expression of matrix metalloproteinase 9 via NF-κB pathway in human alveolar epithelial A549 cells. *J. Agric. Food Chem.* **2010**, *58*, 12014–12021. [CrossRef] [PubMed]

207. Ganesan, K.; Jayachandran, M.; Xu, B. A critical review on hepatoprotective effects of bioactive food components. *Crit. Rev. Food Sci. Nutr.* **2018**, *58*, 1165–1229. [CrossRef] [PubMed]

208. Zhang, T.; Jayachandran, M.; Ganesan, K.; Xu, B. The black truffle, *Tuber melanosporum* (Ascomycetes), ameliorates hyperglycemia and regulates insulin signaling pathway in STZ-induced diabetic rats. *Int. J. Med. Mushrooms* **2020**, *22*, 1057–1066. [CrossRef] [PubMed]

209. Chang, C.-J.; Lin, C.-S.; Lu, C.-C.; Martel, J.; Ko, Y.-F.; Ojcius, D.M.; Tseng, S.-F.; Wu, T.-R.; Chen, Y.-Y.M.; Young, J.D.; et al. *Ganoderma lucidum* reduces obesity in mice by modulating the composition of the gut microbiota. *Nat. Commun.* **2015**, *6*, 7489. [CrossRef] [PubMed]

210. Ahmad, R.; Riaz, M.; Khan, A.; Aljamea, A.; Algheryafi, M.; Sewaket, D.; Alqathama, A. *Ganoderma lucidum* (Reishi) an edible mushroom; a comprehensive and critical review of its nutritional, cosmeceutical, mycochemical, pharmacological, clinical, and toxicological properties. *Phytother. Res.* **2021**, *35*, 6030–6062. [CrossRef]

211. He, X.; Wang, X.; Fang, J.; Chang, Y.; Ning, N.; Guo, H.; Huang, L.; Huang, X.; Zhao, Z. Structures, biological activities, and industrial applications of the polysaccharides from *Hericium erinaceus* (Lion's Mane) mushroom: A review. *Int. J. Biol. Macromol.* **2017**, *97*, 228–237. [CrossRef]

212. Jin, X.; Ruiz Beguerie, J.; Sze, D.M.; Chan, G.C. *Ganoderma lucidum* (Reishi mushroom) for cancer treatment. *Cochrane Database Syst. Rev.* **2016**, *4*, Cd007731. [CrossRef]

213. Shah, S.K.; Walker, P.A.; Moore-Olufemi, S.D.; Sundaresan, A.; Kulkarni, A.D.; Andrassy, R.J. An evidence-based review of a *Lentinula edodes* mushroom extract as complementary therapy in the surgical oncology patient. *JPEN J. Parenter. Enteral Nutr.* **2011**, *35*, 449–458. [CrossRef]

214. Gong, P.; Wang, X.; Liu, M.; Wang, M.; Wang, S.; Guo, Y.; Chang, X.; Yang, W.; Chen, X.; Chen, F. Hypoglycemic effect of a novel polysaccharide from *Lentinus edodes* on STZ-induced diabetic mice via metabolomics study and Nrf2/HO-1 pathway. *Food Funct.* **2022**, *13*, 3036–3049. [CrossRef]
215. Saretzki, G. Telomeres, telomerase and ageing. *Subcell. Biochem.* **2018**, *90*, 221–308. [CrossRef] [PubMed]
216. Bernardes de Jesus, B.; Blasco, M.A. Telomerase at the intersection of cancer and aging. *Trends Genet.* **2013**, *29*, 513–520. [CrossRef]
217. Yuen, J.W.; Gohel, M.D.; Au, D.W. Telomerase-associated apoptotic events by mushroom *Ganoderma lucidum* on premalignant human urothelial cells. *Nutr. Cancer* **2008**, *60*, 109–119. [CrossRef]
218. Ganesan, K.; Xu, B. Telomerase inhibitors from natural products and their anticancer potential. *Int. J. Mol. Sci.* **2017**, *19*, 13. [CrossRef]
219. Xu, B.; Li, C.; Sung, C. Telomerase inhibitory effects of medicinal mushrooms and lichens, and their anticancer activity. *Int. J. Med. Mushrooms* **2014**, *16*, 17–28. [CrossRef]
220. Sukalingam, K.; Ganesan, K.; Xu, B. *Trianthema portulacastrum* L. (giant pigweed): Phytochemistry and pharmacological properties. *Phytochem. Rev.* **2017**, *16*, 461–478. [CrossRef]
221. Jayasuriya, R.; Dhamodharan, U.; Ali, D.; Ganesan, K.; Xu, B.; Ramkumar, K.M. Targeting Nrf2/Keap1 signaling pathway by bioactive natural agents: Possible therapeutic strategy to combat liver disease. *Phytomedicine* **2021**, *92*, 153755. [CrossRef]
222. Sakshi, S.; Jayasuriya, R.; Ganesan, K.; Xu, B.; Ramkumar, K.M. Role of circRNA-miRNA-mRNA interaction network in diabetes and its associated complications. *Mol. Ther. Nucleic Acids* **2021**, *26*, 1291–1302. [CrossRef]
223. Lin, M.T.; Beal, M.F. Mitochondrial dysfunction and oxidative stress in neurodegenerative diseases. *Nature* **2006**, *443*, 787–795. [CrossRef]
224. Miwa, S.; Kashyap, S.; Chini, E.; von Zglinicki, T. Mitochondrial dysfunction in cell senescence and aging. *J. Clin. Investig.* **2022**, *132*, e158447. [CrossRef] [PubMed]
225. Amorim, J.A.; Coppotelli, G.; Rolo, A.P.; Palmeira, C.M.; Ross, J.M.; Sinclair, D.A. Mitochondrial and metabolic dysfunction in ageing and age-related diseases. *Nat. Rev. Endocrinol.* **2022**, *18*, 243–258. [CrossRef]
226. Kudryavtseva, A.V.; Krasnov, G.S.; Dmitriev, A.A.; Alekseev, B.Y.; Kardymon, O.L.; Sadritdinova, A.F.; Fedorova, M.S.; Pokrovsky, A.V.; Melnikova, N.V.; Kaprin, A.D.; et al. Mitochondrial dysfunction and oxidative stress in aging and cancer. *Oncotarget* **2016**, *7*, 44879–44905. [CrossRef]
227. Guo, Y.; Guan, T.; Shafiq, K.; Yu, Q.; Jiao, X.; Na, D.; Li, M.; Zhang, G.; Kong, J. Mitochondrial dysfunction in aging. *Ageing Res. Rev.* **2023**, *88*, 101955. [CrossRef] [PubMed]
228. Shahidi, F.; Wanasundara, P.K. Phenolic antioxidants. *Crit. Rev. Food Sci. Nutr.* **1992**, *32*, 67–103. [CrossRef] [PubMed]
229. Carvajal, A.E.S.S.; Koehnlein, E.A.; Soares, A.A.; Eler, G.J.; Nakashima, A.T.A.; Bracht, A.; Peralta, R.M. Bioactives of fruiting bodies and submerged culture mycelia of *Agaricus brasiliensis* (*A. blazei*) and their antioxidant properties. *LWT* **2012**, *46*, 493–499. [CrossRef]
230. Ajith, T.A.; Sudheesh, N.P.; Roshny, D.; Abishek, G.; Janardhanan, K.K. Effect of *Ganoderma lucidum* on the activities of mitochondrial dehydrogenases and complex I and II of electron transport chain in the brain of aged rats. *Exp. Gerontol.* **2009**, *44*, 219–223. [CrossRef]
231. de Sá-Nakanishi, A.B.; Soares, A.A.; de Oliveira, A.L.; Comar, J.F.; Peralta, R.M.; Bracht, A. Effects of treating old rats with an aqueous *Agaricus blazei* extract on oxidative and functional parameters of the brain tissue and brain mitochondria. *Oxid. Med. Cell. Longev.* **2014**, *2014*, 563179. [CrossRef]
232. Ganesan, K.; Wang, Y.; Gao, F.; Liu, Q.; Zhang, C.; Li, P.; Zhang, J.; Chen, J. Targeting engineered nanoparticles for breast cancer therapy. *Pharmaceutics* **2021**, *13*, 1829. [CrossRef] [PubMed]
233. Ganesan, K.; Nair, S.P.; Azalewor, H.G.; Letha, N.; Gani, S.B. Preliminary phytochemical screening and in vitro antioxidant activity of *Datura stramonium* L. collected from Jimma, South West Ethiopia. *Int. J. Pharma Bio Sci.* **2016**, *7*, P261–P266.
234. Ganesh, G.V.; Ganesan, K.; Xu, B.; Ramkumar, K.M. Nrf2 driven macrophage responses in diverse pathophysiological contexts: Disparate pieces from a shared molecular puzzle. *BioFactors* **2022**, *48*, 795–812. [CrossRef]
235. Nagarajan, S.; Mohandas, S.; Ganesan, K.; Xu, B.; Ramkumar, K.M. New insights into dietary pterostilbene: Sources, metabolism, and health promotion effects. *Molecules* **2022**, *27*, 6316. [CrossRef] [PubMed]
236. Islam, T.; Ganesan, K.; Xu, B. New insight into mycochemical profiles and antioxidant potential of edible and medicinal mushrooms: A review. *Int. J. Med. Mushrooms* **2019**, *21*, 237–251. [CrossRef]
237. Jayachandran, M.; Zhang, T.; Ganesan, K.; Xu, B.; Chung, S.S.M. Isoquercetin ameliorates hyperglycemia and regulates key enzymes of glucose metabolism via insulin signaling pathway in streptozotocin-induced diabetic rats. *Euro. J. Pharmacol.* **2018**, *829*, 112–120. [CrossRef]
238. Islam, T.; Ganesan, K.; Xu, B. Insights into health-promoting effects of Jew's ear (*Auricularia auricula*-judae). *Trends Food Sci. Technol.* **2021**, *114*, 552–569. [CrossRef]
239. Sukalingam, K.; Ganesan, K.; Das, S.; Thent, Z.C. An insight into the harmful effects of soy protein: A review. *Clin. Ter.* **2015**, *166*, 131–139. [CrossRef]
240. Gurovic, M.S.V.; Viceconte, F.R.; Pereyra, M.T.; Bidegain, M.A.; Cubitto, M.A. DNA damaging potential of *Ganoderma lucidum* extracts. *J. Ethnopharmacol.* **2018**, *217*, 83–88. [CrossRef]
241. Lee, Y.H.; Kim, J.H.; Song, C.H.; Jang, K.J.; Kim, C.H.; Kang, J.S.; Choi, Y.H.; Yoon, H.M. Ethanol extract of *Ganoderma lucidum* augments cellular anti-oxidant defense through activation of Nrf2/HO-1. *J. Pharmacopunct.* **2016**, *19*, 59–69. [CrossRef]
242. Lin, J.; Lu, Y.Y.; Shi, H.Y.; Lin, P. Chaga medicinal mushroom, *Inonotus obliquus* (Agaricomycetes), polysaccharides alleviate photoaging by regulating Nrf2 pathway and autophagy. *Int. J. Med. Mushrooms* **2023**, *25*, 49–64. [CrossRef]

243. Sevindik, M.; Akgul, H.; Selamoglu, Z.; Braidy, N. Antioxidant and antigenotoxic potential of *Infundibulicybe geotropa* mushroom collected from northwestern Turkey. *Oxid. Med. Cell. Longev.* **2020**, *2020*, 5620484. [CrossRef]
244. Lillycrop, K.A.; Hoile, S.P.; Grenfell, L.; Burdge, G.C. DNA methylation, ageing and the influence of early life nutrition. *Proc. Nutr. Soc.* **2014**, *73*, 413–421. [CrossRef]
245. Lau, C.E.; Robinson, O. DNA methylation age as a biomarker for cancer. *Int. J. Cancer* **2021**, *148*, 2652–2663. [CrossRef] [PubMed]
246. Wan, Q.L.; Meng, X.; Wang, C.; Dai, W.; Luo, Z.; Yin, Z.; Ju, Z.; Fu, X.; Yang, J.; Ye, Q.; et al. Histone H3K4me3 modification is a transgenerational epigenetic signal for lipid metabolism in *Caenorhabditis elegans*. *Nat. Commun.* **2022**, *13*, 768. [CrossRef] [PubMed]
247. Ilango, S.; Paital, B.; Jayachandran, P.; Padma, P.R.; Nirmaladevi, R. Epigenetic alterations in cancer. *Front. Biosci.* **2020**, *25*, 1058–1109. [CrossRef]
248. Saul, D.; Kosinsky, R.L. Epigenetics of aging and aging-associated diseases. *Int. J. Mol. Sci.* **2021**, *22*, 401. [CrossRef] [PubMed]
249. Lai, G.; Guo, Y.; Chen, D.; Tang, X.; Shuai, O.; Yong, T.; Wang, D.; Xiao, C.; Zhou, G.; Xie, Y.; et al. Alcohol extracts from *Ganoderma lucidum* delay the progress of Alzheimer's disease by regulating DNA methylation in rodents. *Front. Pharmacol.* **2019**, *10*, 272. [CrossRef]
250. Yi, S.J.; Kim, K. New insights into the role of histone changes in aging. *Int. J. Mol. Sci.* **2020**, *21*, 8241. [CrossRef] [PubMed]
251. Bhukel, A.; Beuschel, C.B.; Maglione, M.; Lehmann, M.; Juhász, G.; Madeo, F.; Sigrist, S.J. Autophagy within the mushroom body protects from synapse aging in a non-cell autonomous manner. *Nat. Commun.* **2019**, *10*, 1318. [CrossRef] [PubMed]
252. López-Otín, C.; Blasco, M.A.; Partridge, L.; Serrano, M.; Kroemer, G. Hallmarks of aging: An expanding universe. *Cell* **2023**, *186*, 243–278. [CrossRef]
253. Gulen, M.F.; Samson, N.; Keller, A.; Schwabenland, M.; Liu, C.; Glück, S.; Thacker, V.V.; Favre, L.; Mangeat, B.; Kroese, L.J.; et al. cGAS-STING drives ageing-related inflammation and neurodegeneration. *Nature* **2023**, *620*, 374–380. [CrossRef] [PubMed]
254. Zhao, Y.; Simon, M.; Seluanov, A.; Gorbunova, V. DNA damage and repair in age-related inflammation. *Nat. Rev. Immunol.* **2023**, *23*, 75–89. [CrossRef] [PubMed]
255. Li, X.; Li, C.; Zhang, W.; Wang, Y.; Qian, P.; Huang, H. Inflammation and aging: Signaling pathways and intervention therapies. *Signal Transduct. Target Ther.* **2023**, *8*, 239. [CrossRef] [PubMed]
256. López-Otín, C.; Pietrocola, F.; Roiz-Valle, D.; Galluzzi, L.; Kroemer, G. Meta-hallmarks of aging and cancer. *Cell Metab.* **2023**, *35*, 12–35. [CrossRef] [PubMed]
257. Kumar, G.; Sharmila Banu, G.; Murugesan, A.G.; Rajasekara Pandian, M. Preliminary toxicity and phytochemical studies of aqueous bark extract of *Helicteres isora* L. *Int. J. Pharmacol.* **2007**, *3*, 96–100. [CrossRef]
258. Sharmila Banu, G.; Kumar, G.; Murugesan, A.G. Ethanolic leaves extract of *Trianthema portulacastrum* L. ameliorates aflatoxin B1 induced hepatic damage in rats. *Indian J. Clin. Biochem.* **2009**, *24*, 250–256. [CrossRef] [PubMed]
259. Kumar, G.; Sharmila Banu, G.; Maheswaran, R.; Rema, S.; Rajasekara Pandian, M.; Murugesan, A.G. Effect of *Plumbago zeylanica* L. on blood glucose and plasma antioxidant status in STZ diabetic rats. *J. Nat. Rem.* **2007**, *7*, 66–71.
260. Harrington, J.S.; Ryter, S.W.; Plataki, M.; Price, D.R.; Choi, A.M.K. Mitochondria in health, disease, and aging. *Physiol. Rev.* **2023**, *103*, 2349–2422. [CrossRef]
261. Muszyńska, B.; Grzywacz-Kisielewska, A.; Kała, K.; Gdula-Argasińska, J. Anti-inflammatory properties of edible mushrooms: A review. *Food Chem.* **2018**, *243*, 373–381. [CrossRef]
262. Li, X.; He, Y.; Zeng, P.; Liu, Y.; Zhang, M.; Hao, C.; Wang, H.; Lv, Z.; Zhang, L. Molecular basis for *Poria cocos* mushroom polysaccharide used as an antitumour drug in China. *J. Cell. Mol. Med.* **2019**, *23*, 4–20. [CrossRef] [PubMed]
263. Kim, J.H.; Sim, H.A.; Jung, D.Y.; Lim, E.Y.; Kim, Y.T.; Kim, B.J.; Jung, M.H. *Poria cocus* wolf extract ameliorates hepatic steatosis through regulation of lipid metabolism, inhibition of ER stress, and activation of autophagy via AMPK ctivation. *Int. J. Mol. Sci.* **2019**, *20*, 4801. [CrossRef] [PubMed]
264. Sheng, K.; Wang, C.; Chen, B.; Kang, M.; Wang, M.; Liu, K.; Wang, M. Recent advances in polysaccharides from *Lentinus edodes* (Berk.): Isolation, structures and bioactivities. *Food Chem.* **2021**, *358*, 129883. [CrossRef]
265. Yang, T.K.; Lee, Y.H.; Paudel, U.; Bhattarai, G.; Yun, B.S.; Hwang, P.H.; Yi, H.K. Davallialactone from mushroom reduced premature senescence and inflammation on glucose oxidative stress in human diploid fibroblast cells. *J. Agric. Food Chem.* **2013**, *61*, 7089–7095. [CrossRef]
266. Sillapachaiyaporn, C.; Chuchawankul, S.; Nilkhet, S.; Moungkote, N.; Sarachana, T.; Ung, A.T.; Baek, S.J.; Tencomnao, T. Ergosterol isolated from cloud ear mushroom (*Auricularia polytricha*) attenuates bisphenol A-induced BV2 microglial cell inflammation. *Food Res. Int.* **2022**, *157*, 111433. [CrossRef]
267. Li, X.; Ma, L.; Zhang, L. Molecular basis for *Poria cocos* mushroom polysaccharide used as an antitumor drug in China. *Prog. Mol. Biol. Transl. Sci.* **2019**, *163*, 263–296. [CrossRef] [PubMed]
268. Sousa, P.; Tavares-Valente, D.; Amorim, M.; Azevedo-Silva, J.; Pintado, M.; Fernandes, J. β-Glucan extracts as high-value multifunctional ingredients for skin health: A review. *Carbohydr. Polym.* **2023**, *322*, 121329. [CrossRef] [PubMed]

Journal of
Fungi

Article

Improving and Streamlining Gene Editing in *Yarrowia lipolytica* via Integration of Engineered Cas9 Protein

Baixi Zhang [1,2,*,†] and Jiacan Cao [1,†]

1   School of Food Science and Technology, Jiangnan University, Wuxi 214122, China;
     6210113158@stu.jiangnan.edu.cn
2   National Engineering Research Center for Functional Food, Jiangnan University, Wuxi 214122, China
*   Correspondence: zbx@jiangnan.edu.cn
†   These authors contributed equally to this work.

Abstract: The oleaginous yeast *Yarrowia lipolytica* is a prominent subject of biorefinery research due to its exceptional performance in oil production, exogenous protein secretion, and utilization of various inexpensive carbon sources. Many CRISPR/Cas9 genome-editing systems have been developed for *Y. lipolytica* to meet the high demand for metabolic engineering studies. However, these systems often necessitate an additional outgrowth step to achieve high gene editing efficiency. In this study, we introduced the eSpCas9 protein, derived from the *Streptococcus pyogenes* Cas9(SpCas9) protein, into the *Y. lipolytica* genome to enhance gene editing efficiency and fidelity, and subsequently explored the optimal expression level of *eSpCas9* gene by utilizing different promoters and selecting various growth periods for yeast transformation. The results demonstrated that the integrated *eSpCas9* gene editing system significantly enhanced gene editing efficiency, increasing from 16.61% to 86.09% on *TRP1* and from 33.61% to 95.19% on *LIP2*, all without the need for a time-consuming outgrowth step. Furthermore, growth curves and dilution assays indicated that the consistent expression of eSpCas9 protein slightly suppressed the growth of *Y. lipolytica*, revealing that strong inducible promoters may be a potential avenue for future research. This work simplifies the gene editing process in *Y. lipolytica*, thus advancing its potential as a natural product synthesis chassis and providing valuable insights for other comparable microorganisms.

Keywords: *Yarrowia lipolytica*; genome editing; CRISPR/Cas9

Citation: Zhang, B.; Cao, J. Improving and Streamlining Gene Editing in *Yarrowia lipolytica* via Integration of Engineered Cas9 Protein. *J. Fungi* **2024**, *10*, 63. https://doi.org/10.3390/jof10010063

Academic Editor: Baojun Xu

Received: 14 December 2023
Revised: 6 January 2024
Accepted: 10 January 2024
Published: 12 January 2024

1. Introduction

As a result of recent advances in gene editing technology, metabolic engineering, and synthetic biology, an increasing number of nonconventional microorganisms with unique characteristics have become hosts for synthesizing natural products [1–3]. Among these highly potential microorganisms, the oleaginous yeast *Yarrowia lipolytica* is a prominent subject of recent biorefinery research [4]. *Y. lipolytica* is an ideal host for industrial biomanufacturing, due to its excellent performance in oil production, exogenous protein secretion, and its ability to utilize various inexpensive carbon sources such as glycerol, alkanes, and acetic acid [5–7]. So far, *Y. lipolytica* has been used to produce various valuable products, including polyunsaturated fatty acids EPA [8] and CLA [9,10], terpenoid α-farnesene [11], and flavonoid naringenin [12], with several reaching commercialization. Despite numerous successful applications, there is still a lack of fundamental knowledge about *Y. lipolytica*. Only 44.5% of the 6448 coding genes are considered to be confidently annotated [13]. Thus, there is a need for more convenient and highly efficient gene editing tools to uncover the unknown secrets in *Y. lipolytica* and facilitate the construction of natural product synthesis chassis.

In recent years, the application of type II CRISPR/Cas9 systems from *Streptococcus pyogenes* has enhanced gene editing flexibility and efficiency in a wide range of organisms [14].

The CRISPR/Cas9 system comprises a Cas9 protein and a corresponding sgRNA. The sgRNA recognizes targeted sequences, while the Cas9 protein catalyzes a double-strand break (DSB) in the targeted genome loci. After the DSB forms, cells initiate the repair process, such as non-homologous end joining (NHEJ) and homologous recombination (HR) to repair the break loci, potentially introducing insertions and deletions (indels), or heterologous DNA fragments [15,16]. As a result, the CRISPR/Cas9 system has quickly gained widespread adoption due to its precision and high efficiency. Moreover, as metabolic engineering studies rapidly increase, various CRISPR/Cas9 systems have also been developed in *Y. lipolytica* as well [17].

Schwartz et al. [18] were the first to establish the CRISPR/Cas9 system in *Y. lipolytica*. Their focus was on identifying the best promoters for synthesis of sgRNA to enhance gene editing efficiency, resulting in high efficiency during single gene disruption by NHEJ. Subsequently, Gao et al. [19] developed a similar all-in-one CRISPR/Cas9 system, and successfully achieved double and triple gene disruption through NHEJ in *Y. lipolytica*. However, the CRISPR/Cas9 systems for *Y. lipolytica* mentioned above require an extended outgrowth step of 2-4 days following transformation to achieve high gene editing efficiency. While the outgrowth step has significantly increased gene editing efficiency by three to four times, it also presents some issues. Firstly, the gene editing process is extended by at least two days, making it more time-consuming. Besides, with the extension of cultivation time after transformation, the persistence of Cas9-sgRNA complex might also lead to increased frequencies of off-target mutations [20,21]. Moreover, the Cas9 nucleases of the aforementioned systems are delivered by plasmid vector with the size over 10kb, resulting in inconveniences when carrying out multiple genes' editing [22].

In this study, we initially introduced an engineered Cas9 protein, eSpCas9 [23], to the CRISPR/Cas9 system of *Y. lipolytica*, to improve the fidelity and efficiency of gene editing. Subsequently, we integrated the eSpCas9 gene into the genome of *Y. lipolytica*, and investigated the optimal expression level of eSpCas9 by employing different promoters and selecting various growth periods for yeast transformation. The objective of this work is to propose a strategy for streamlining the gene editing workflow in *Y. lipolytica*, with the potential to expand the synthetic toolbox of *Y. lipolytica* and provide insights for other similar microorganisms.

2. Materials and Methods

2.1. Strains, Media, and Culture Conditions

Escherichia coli DH5α was used for plasmid construction and propagation. All *E. coli* cultures were cultured in Luria–Bertani (LB) medium at 37 °C. Suitable antibiotics (100 mg/L ampicillin or 50 mg/L kanamycin) were added to the LB medium when necessary. *Y. lipolytica* Po1f (Catherine Madzak, Paris-Saclay University, INRA, AgroParisTech, UMR SayFood, Palaiseau, France), a leucine and uracil auxotrophic strain, was used as the base strain in this study. All engineered strains are listed in Table 1. The *Y. lipolytica* cells stored in glycerol stock were initialized by streaking onto a YPD plate and cultivated at 28 °C overnight. A single colony from the plate was cultivated overnight in 5 mL YPD medium (1% yeast extract, 2% tryptone and 2% glucose) to create a seed culture, which was then transferred to culture flasks with a 1% inoculation dose for transformation. Post-transformation, the yeast cells were cultivated in appropriate selective media. YNBD medium (0.67% yeast nitrogen base (without amino acids) and 2% glucose) and YNBD-triglyceride medium (0.67% yeast nitrogen base (without amino acids) and 0.5% triglyceride) were used for *LIP2* disruption transformants. YNBD-leu medium (YNBD medium supplemented with 0.01% Leucine) was used for transformants with pINA1312 plasmid. YNBD-trp medium (YNBD supplemented with 0.01% tryptophan) was used for *TRP1* disruption transformants. Then, 2% agar was added to the solid medium. All *Y. lipolytica* culturing was conducted at 28 °C, with liquid cultures shaken at 200 rpm.

Table 1. Strains and plasmids used in this work.

Strains or Plasmids	Characteristics	Source or References
Strains		
Po1f	*MatA, leu2-270, ura3-302, xpr2-322, axp-2, Leu−, Ura−, ΔAEP, ΔAXP, Suc+*	[24]
Po1g	*MatA, leu2-270, ura3-302: :URA3, xpr2-322, axp-2, Leu−, ΔAEP, ΔAXP, Suc+, pBR*	[24]
Po1fe-3	Po1f, pINA1312-TEFin-eCAS9 (*URA3, eSpCas9*)	This work
Po1fe-4	Po1f, pINA1312-UAS16TEFin-eCAS9 (*URA3, eSpCas9*)	This work
Po1f-ura+	Po1f, pINA1312 (*URA3*)	This work
Plasmids		
pCASyl-trp	*Y. lipolytica*-replicative plasmid, containing *TRP1* Guide RNA module and Cas9 expression cassette	[19]
pINA1312	*Y. lipolytica*-integrative plasmid, *Kan, hp4d, XPR2t, ura3d1*	[25]
pUC57-UAS1B8	Containing 8 tandem copies of UAS1B flanked with *Cla*I-*Hind*III-*Spe*I siteson the 5′-end and *Mlu*I-*Xba*I sites on the 3′-end.	[9]
pUC57-USA1B16	Derived from pUC57-UAS1B8, containing 16 tandem copies of UAS1B	This work
peCASyl	Derived from pCASyl-trp, replacing Cas9 gene to codon-optimized *eSpCas9* gene	This work
peCASyl-trp	Derived from peCASyl, containing *TRP1* Guide RNA module	This work
pINA1312-TEFin-eCAS9	pINA1312 containing TEFin-eSpCas9 cassette	This work
pINA1312-UAS16TEFin-eCAS9	pINA1312 containing UAS16TEFin-eSpCas9 cassette	This work
pCEN1-Leu-gRNA	Derived from pCASyl-trp, containing empty sgRNA cassette	This work
pCEN1-Leu-gTRP1	Derived from pCEN1-Leu-gRNA, containing *TRP1* Guide RNA module	This work
pCEN1-Leu-gLIP2	Derived from pCEN1-Leu-gRNA, containing *LIP2* Guide RNA module	This work

2.2. Plasmid Construction

Plasmids are listed in Table 1; primers are listed in Table 2. Representative plasmid maps can be found in Figure S1. The plasmid peCASyl, derived from pCASyl-trp, was synthesized by GenScript Biotech Co., Ltd. (Nanjing, China), which contained a codon-optimized *eSpCas9* cassette and an empty sgRNA cassette. The sequence of the codon-optimized *eSpCas9* gene is available in Table S1. The plasmid pUC57-USA1B16 was derived from pUC57-USA1B8 [9]. The UAS1B8 fragment was obtained from pUC57-UAS1B8 by digestion with *Hind*III and *Xba*I restriction enzymes, and then ligated with pUC-UAS1B8 cut with *Hind*III and *Spe*I using T4 DNA Ligase (Vayzme, Nanjing, China), resulting in pUC57-USA1B16. All remaining plasmids assemblies were completed using a ClonExpress Ultra One Step Cloning Kit V2 (Vayzme, Nanjing, China). Plasmids pINA1312-TEFin-eCAS9 and pINA1312-UAS16TEFin-eCAS9 were derived from pINA1312. The *eSpCas9* cassette was amplified with primer pairs eCAS9-F10/eCAS9-R10 from peCASyl. The vector fragment from pINA1312 was amplified with primer pairs pINA1312-F/pINA1312-R for pINA1312-TEFin-eCAS9, and primer pairs pINA1312-F/pINA1312-R-2 for pINA1312-UAS16TEFin-eCAS9. The pUC57-USA1B16 plasmid was digested by restrictive endonucleases *Spe* I and *Xba* I to obtain 16 tandem copies of UAS1B for construction of pINA1312-UAS16TEFin-eCAS9. Plasmid pCEN1-Leu-gRNA was derived from peCASyl, which abandoned the *eSpCas9* cassette. Primer pairs Mig1t-F2/TEF-R2 and AmpR-F/ORI001-R were used to amplified the required fragments for construction of pCEN1-Leu-gRNA. As for the insertion of 20 bp gRNA in peCASyl-trp, pCEN1-Leu-gTrp, and pCEN1-Leu-gLip, primer pairs gRNA-Trp-F/gRNA-Trp-R were used for gRNA of *TRP1* (5′-CGATGGCGTCCTGATCCAGT-3′), and primer pairs gRNA-LIP-F/gRNA-LIP-R were used for gRNA of *TRP1* (5′-CAGTTGAAGGGCTTGAAGAT-3′). All products from PCR or restrictive endonuclease were recovered by gel extraction and electrophoresis using The Fast Pure Gel DNA Extraction Mini Kit (Vayzme, Nanjing, China). The recombination products were transformed into *E. coli* DH5α, and cultured at 37 °C overnight.

Single colonies were then picked for final verification. Plasmid DNA was extracted using TIANprep Mini Plasmid Kit from Tiangen Biotech Co., Ltd. (Beijing, China).

Table 2. Primers used in this work.

Primers	Sequence (5′-3′)
eCAS9-F10	ACATGGAATTCGGACACGGGccgacgcgtctgtacaga
eCAS9-R10	AGTCTGCAGCCCAAGCTAGCgcccttttgggtttgtcgac
pINA1312-F	CCCGTGTCCGAATTCCATGTGTAAC
pINA1312-R	GCTAGCTTGGGCTGCAGACTAAATT
pINA1312-R-2	GACACCTCAGCATGCACTAGgctagcttgggctgcagactaaatt
Mig1t-F2	agacatagcggccgcttcgaaaacccaaaagggccgaagg
TEF-R2	ggggttccgcacacatttccagagaccgggttggcgg
AmpR-F	ggaaatgtgtgcggaacccc
ORI001-R	tcgaagcggccgctatgtct
NdeI-F	GTGCGGTATTTCACACCGCAtatggtgcactctcagtacaatctgc
NdeI-R	tgcggtgtgaaataccgcac
gRNA-Trp-F	CGATGGCGTCCTGATCCAGTgacgagcttactcgtttcg
gRNA-Trp-R	ACTGGATCAGGACGCCATCGgttttagagctagaaatagcaagt
gRNA-LIP-F	CAGTTGAAGGGCTTGAAGATgacgagcttactcgtttcg
gRNA-LIP-R	ATCTTCAAGCCCTTCAACTGgttttagagctagaaatagcaagt
eCas9-F	ATCGCCACCGAGCTGACT
eCas9-R	GACAAGAAGTACTCAATCGGCCTGG
TRP-F	ATGGACTTTCTCTACTCTTCGACAT
TRP-R	TTACCCCCTGGCGTTTTTGAC
LIP-F	CCCAAGTACCAGCTCCCCTC
LIP-R	CCACAGACACCCTCGGTGAC
Suc2-F	CTACGGTTCAGCATTAGGTATTG
Suc2-R	GACCAGGGACCAGCATTAC
ura3-F	GTGCTTCTCTGGATGTTACC
ura3-R	CAATATCTGCGAACTTTCTGTC
ACT2-F	TCCAGGCCGTCCTCTCCC
ACT2-R	GGCCAGCCATATCGAGTCGCA
qeCAS9-F2	TAAGCAACCGTAGGGGAATC
qeCAS9-R2	GTGTGGGACAAGGGCAGAGA

2.3. Yeast Transformation

The transformation of *Y. lipolytica* utilized the Frozen-EZ Yeast Transformation II Kit (Zymo Research, Orange, CA, USA). The number of cells was adjusted to a range of 1×10^7 to 1×10^8 before transforming. Subsequently, 1 mL of YPD medium was introduced into the transformation mixture, followed by cultivation at 28 °C and 200 rpm for 2 h to facilitate cell recovery. Post-recovery, the cells were harvested via centrifugation at $13,400 \times g$ for 1 min and then plated onto suitable medium.

2.4. Phenotype Verification

The phenotype verification involved transferring single transformant colonies from the original plate to two plates. One plate served for selecting transformants with distinct phenotypes, while the other functioned as a non-selective control. YNBD and YNBD-trp media were utilized to confirm the *TRP1* disruption transformants, while YNBD and YNBD-triglyceride media were used to validate the *LIP2* disruption transformants.

2.5. Colony PCR Confirmation and Sequencing

To obtain the genome template, the single colony was dispersed in 20 µL of a 0.25% SDS solution, followed by incubation at 90 °C for 3 min. The PCR was performed using the 2× Phanta Max Master Mix (Dye Plus) from Vazyme Biotech Co., Ltd. (Nanjing, China). And the resulting products were verified by DNA electrophoresis or sequenced as necessary by Sangon Biotech Co., Ltd. (Shanghai, China). *eSpCas9* was confirmed using primer pairs eCAS9-F/eCAS9-R; *TRP1* was confirmed using primer pairs TRP-F/TRP-R; and *LIP2* was confirmed using primer pairs LIP-F/LIP-R.

2.6. Real-Time Quantitative PCR

Total RNA from the yeast was extracted and purified using the RNAprep Pure Plant Kit from Tiangen Biotech Co., Ltd. (Beijing, China). Subsequently, the cDNA was synthesized using the HiScript III 1st cDNA synthesis kit from Vazyme Biotech Co., Ltd. (Nanjing, China). For the qPCR reaction system, SYBR GREEN $2\times$PCR Master Mix (ABI, Carlsbad, CA, USA) was employed. In the relative copy number measurement, strain Po1g was utilized as the control organism with a single copy of both the *URA3* and *SUC2* target sequences. The primer pairs targeting *URA3* are ura3-F/ura3-R, and the primer pairs targeting *SUC2* are Suc2-F/Suc2-R. As for the relative expression level measurement, *ACT1* (YALI0D08272g) was employed as the reference gene due to its relatively constant expression. The primer pairs targeting *eSpCas9* are qeCAS9-F2/qeCAS9-R2, and the primer pairs targeting *ACT1* are ACT-F/ACT-R. Gene expression changes were assessed using the $2^{-\Delta\Delta Ct}$ method, and all samples were prepared in triplicate to obtain the Ct value.

3. Results and Discussion

3.1. Gene Editing Efficiency of eSpCas9 in Y. lipolytica

Off-target effects arise when the nuclease activity of Cas9 is triggered at sites where the RNA guide sequence exhibits imperfect complementarity with off-target genomic sites, presenting a significant challenge for genome editing applications [26]. To improve specificity and efficiency in gene editing, several variants of SpCas9 were initially created through mutagenesis of protein regions involved in target DNA binding [27]. One such variant, eSpCas9, was engineered with K848A, K1003A, and R1060A mutations using alanine mutagenesis of positively charged residues in the non-target strand binding groove. This was to reduce affinity and promote rehybridization between the target and non-target DNA strands, resulting in highly efficient and more specific gene editing in mammalian cells compared to wild-type cells [23]. Building upon the CRISPR/Cas9 system established by Gao [19], we developed all-in-one CRISPR plasmids, termed peCASyl, containing a de novo codon-optimized version of the eSpCas9 gene, with the aim of achieving higher gene editing efficiency in *Y. lipolytica*.

To assess the efficacy of peCASyl in gene editing, we targeted the *TRP1* gene for editing, as its absence renders strains unable to grow in tryptophan-deficient media (Figure 1A). As the results demonstrate, a disruption efficiency of 35.5% was observed in strains transformed with peCASyl, and the disruption efficiency further increased to 71.23% and 77.21% after outgrowth periods of 2 and 4 days, respectively (Figure 2). It is worth noting that the introduction of eSpCas9 significantly enhances the efficiency of gene editing compared to the original CRISPR/Cas9 system, whether engaging in an additional outgrowth step or not (Figure 2). The efficient expression of the recombinant protein, facilitated by codon optimization and minimal off-target effects, likely contributes to this improvement in gene editing efficiency. Furthermore, there appears to be no substantial difference in gene editing efficiency for peCASyl between 2 and 4 days of outgrowth analyzed with an unpaired Student's *t*-test (Figure 2), suggesting the potential to shorten the gene editing workflow while maintaining high efficiency. These findings indicate that the introduction of *eSpCas9* into *Y. lipolytica* enables normal function and enhances gene editing efficiencies, akin to its performance in mammalian cells. The use of *eSpCas9* is particularly suitable for specific experimental requirements or for expanding the capabilities and applications of CRISPR toolboxes, such as single-nucleotide conversion, which may provide potential insights into the synthetic toolbox of *Y. lipolytica* [28].

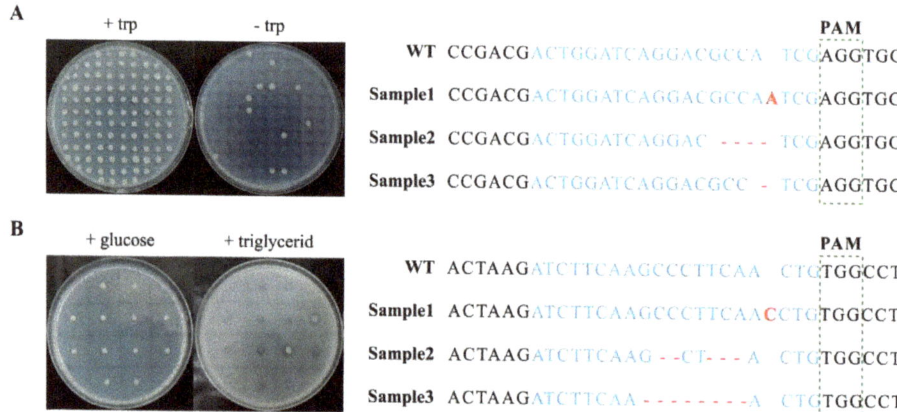

Figure 1. Phenotype and sequence confirmation of *TRP1* and *LIP2* disruption strains. (**A**) Verification of *TRP1* disruption strains. Transformants that failed to grow on YNBD were *trp*− mutants. A representative sample is shown for the alignments of the *TRP1* gene sequence from selected *trp*− mutants. (**B**) Verification of *LIP2* disruption strains. Transformants that exhibited a reduced halo of triglyceride hydrolysis on YNBD-triglyceride were *lip*− mutans. A representative sample is shown for the alignments of the *LIP2* gene sequence from selected *lip*− mutants. The blue represents gRNA sequence, the red represents indels, and the sequence in green rectangle represents PAM sequence.

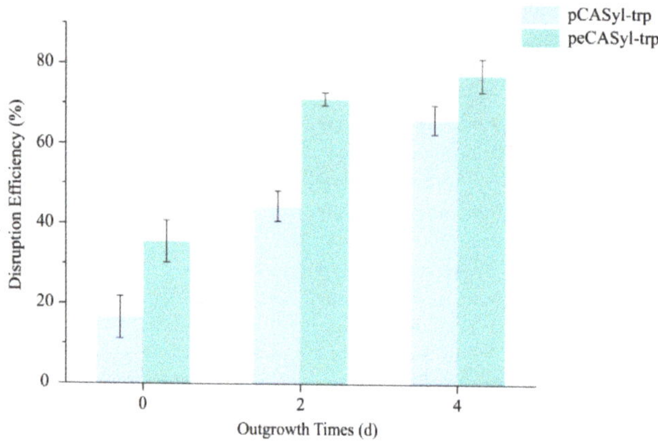

Figure 2. The disruption efficiency of pCASyl-trp and peCASyl-trp on gene *TRP1*. The values and error bars represent the average readings and standard deviations for three independent experiments.

3.2. Integration of eSpCas9 in Y. lipolytica

As previously mentioned, most CRISPR/Cas9 systems in *Y. lipolytica* use plasmids to deliver Cas9 into cells, often necessitating additional outgrowth to achieve high gene editing efficiency, and potentially resulting in oversized plasmids. Therefore, we opted to develop a recombinant yeast strain with integrated *eSpCas9* for more efficient and convenient gene editing [29]. Taking into account the subsequent marker rescue, we selected the single-copy vector pINA1312, which contains the non-defective *ura3d1* gene for mono-copy expression, to integrate the *eSpCas9* gene into *Y. lipolytica* [25]. Initially, we constructed the plasmid pINA1312-TEFin-eCas9, which inserted an *eSpCas9* cassette into vector pINA1312. Subsequently, we linearized the plasmid with restrictive endonucleases *Not*I to obtain a DNA fragment containing ura3d1 marker, *eSpCas9* cassette and Zeta sequence, which was subsequently introduced into yeast genome through transformation.

The agarose gel electrophoresis results confirmed the successful integration of the *eSpCas9* gene into the yeast strains, designated as po1fe-3 (Figure 3A). Following the integration of the expression cassette, we confirmed the copy numbers using real-time quantitative PCR. *Y. lipolytica* Po1g, which contains a single copy of the *URA3* and *SUC2* target sequences, was employed as a control organism. Given the co-existence of *URA3* and *eSpCas9* within the expression cassette, their relative copy numbers were assumed to be equivalent. As depicted in Figure 3B, all transformants exhibited less than two copies.

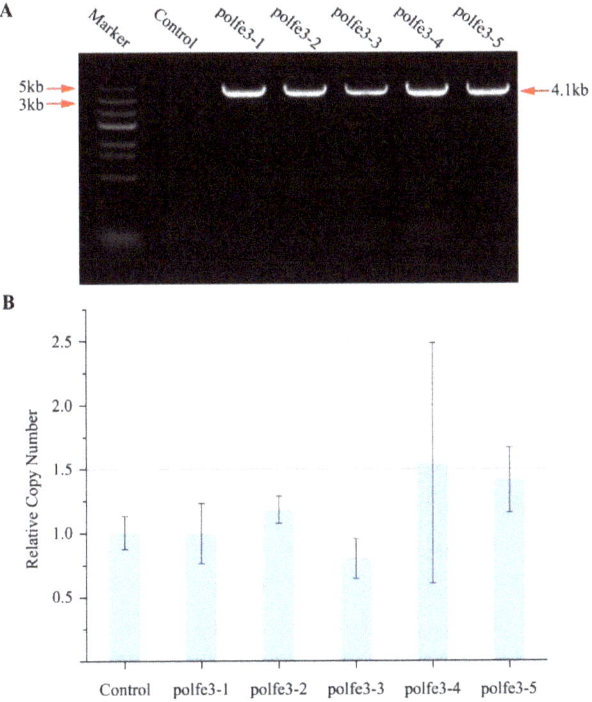

Figure 3. Confirmation of *eSpCas9* integration. (**A**) Agarose gel electrophoresis of an *eSpCAS9* fragment amplified with primer pairs eCas9-F/eCas9-R. (**B**) Relative copy number of *eSpCAS9* in transformants via real time PCR; *Y. lipolytica* Po1g containing a single copy of both the *URA3* and *SUC2* was set as a control. The values and error bars represent the average readings and standard deviations for three independent experiments.

To assess the functionality of integrated *eSpCas9*, we introduced a replicative plasmid containing the 20bp gRNA of the *TRP1* gene, named pCEN1-Leu-gTrp, which abandoned the *eSpCas9* cassette on peCASyl. As depicted in Table 3, the disruption efficiency achieved with the system po1fe3-1/pCEN1-Leu-gTrp reached 86.09% ± 3.28 on *TRP1* without the need for outgrowth, compared to a disruption efficiency of 35.50% ± 5.22 for po1f/peCASyl-trp and 16.61% ± 5.31 for po1f/pCASyl-trp without outgrowth as well. As anticipated, the heterologous expression of *eSpCas9* in *Y. lipolytica* significantly enhances gene editing efficiency. This is due to the pre-existing abundance of Cas9 protein in the cell before the delivery and transcription of the sgRNA cassette. This abundance immediately facilitates the formation of Cas9-sgRNA complexes within the cell, ultimately leading to a higher rate of double-strand break formation at the target site. Consequently, we achieved a five-fold increase in gene editing efficiency compared to the original CRISPR/Cas9 system developed by Gao [19], without the need for outgrowth. This implies that we can forego the time-consuming outgrowth step and reduce the risk of off-target effects caused by the continuous presence of Cas9-sgRNA complexes through use of the high-fidelity eSpCas9 protein [23,27].

Table 3. The disruption efficiency and transformation efficiency of different gene editing systems targeting *TRP1* and *LIP2* without outgrowth.

Targeted Gene	Strain/Palsmid	Disruption Efficiency (%)	Transformation Efficiency (CFU/ng)
TRP1	po1f/pCASyl-trp	16.61 ± 5.31	689.67 ± 62.85
	po1f/peCASyl-trp	35.50 ± 5.22	536.67 ± 38.52
	po1fe3-1/pCEN1-Leu-gTrp	86.09 ± 3.28	626.00 ± 57.20
LIP2	po1f/pCASyl-trp	33.61 ± 7.08	281.67 ± 57.62
	po1f/peCASyl-trp	62.78 ± 5.20	211.33 ± 69.94
	po1fe3-1/pCEN1-Leu-gTrp	95.19 ± 1.70	189.00 ± 50.04

Note: Results are presented as mean ± SD for three independent experiments.

It is widely recognized that an increase in gene editing efficiency typically leads to a decrease in the survival rate of edited cells, as more cells are susceptible to genomic damage [19]. As demonstrated in Table 3, we attained comparable transformation efficiency between the po1fe3-1 and peCASyl systems while significantly enhancing gene editing efficiency. We attribute this difference to the plasmid size, as the po1fe3-1 system requires a smaller plasmid than the peCASyl system, which does not require transfection of the *eSpCas9* gene. This smaller plasmid size allows more cells to be transfected with the plasmid under the same mass of plasmid, yielding more transformants.

Additionally, to further validate the po1fe strains, we expanded the testing gene to include the *LIP2* gene. The *LIP2* gene encodes extracellular lipase, and in its absence, the strains exhibit a reduced halo of triglyceride hydrolysis when cultured on medium containing tributyrin (Figure 1B) [30]. The results show a substantial enhancement in gene editing efficiency for the *LIP2* gene, increasing from 33.61% ± 7.08 for the pCASyl system to 95.19% ± 1.7 for the po1fe3-1 system (Table 3). The trend of transformation efficiency among the three systems is consistent with that of the *TRP1* gene. Additionally, the transformation efficiency in *LIP2* exhibited a specific reduction compared to *TRP1*, which can be attributed to the higher gene editing efficiency in *LIP2*, making more cells vulnerable to genomic damage.

3.3. Effect of Enhancing Gene Expression Level of Integrated eSpCas9 on Gene Editing Efficiency

Previous studies have shown that increasing the expression of Cas9 nuclease can enhance the indel rate in mammalian cells [31]. While we achieved high gene editing efficiency with po1fe3-1, there is potential for further improvement by regulating the *eSpCas9* expression level. Therefore, we enhanced the original promoter TEFin by adding 16 tandem copies of an upstream activation sequence (UAS) from P_{TEF} [32,33]. Subsequently, we integrated the eSpCas9 expression cassette containing the enhanced promoter into yeast cells to create a recombinant strain named po1fe4 in the same way as po1fe3, choosing po1fe4-2 as the tested strain based on relative copy number data. Additionally, according to the experience in previous studies, the expression level of heterologous proteins in *Y. lipolytica* varies with the growth phases [9]. This suggests that the growth phases may potentially influence the intracellular eSpCas9 protein content before transforming the plasmids containing the sgRNA cassette, which could consequently impact gene editing efficiency. Therefore, to confirm this hypothesis, we tested the gene editing efficiency of the selected genes after enhancing the promoter and preparing receptive cells at different growth cycles.

At first, we assessed the relative expression levels among the above strains using real-time quantitative PCR. As shown in Figure 4A, the enhanced promoter led to a nearly 12-fold increase in the eSpCas9 expression level. While the influence of the growth cycle on eSpCas9 expression was not as significant as the influence of the enhanced promoter, the most significant difference in eSpCas9 expression between the early log-phase and mid-stationary phase was less than two-fold (Figure 4B). However, as depicted in Figure 5, whether the *eSpCas9* expression level was significantly increased by enhancing the promoter or slightly adjusted by transforming at different growth phases, there were no significant variances in gene editing efficiency of *TRP1* analyzed with an unpaired Student's t-test, showing an efficiency of approximately 90%. We speculated that the expression of *eSpCas9* reached a sufficient level to achieve the highest editing efficiency under the regulation of promoter TEFin, indicating no need to pursue a higher expression level of the nuclease. Furthermore, the limitation of sgRNA expression may also be the reason for the lack of further improvement in editing efficiency, as the level of the sgRNA-Cas9 complex failed to increase with the level of eSpCas9 nuclease [34].

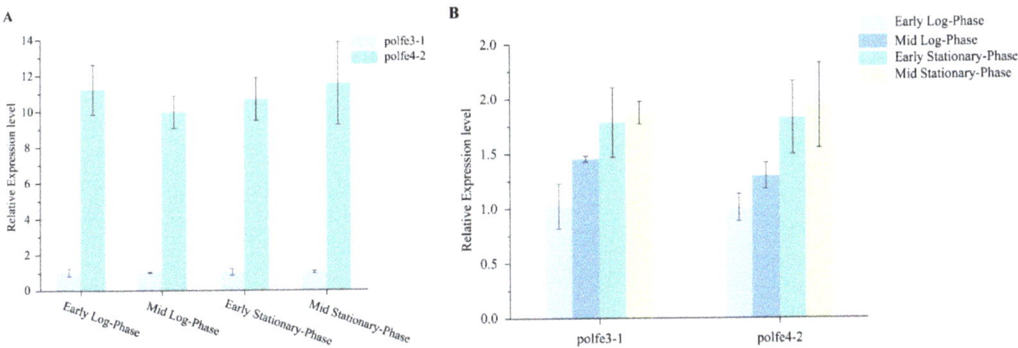

Figure 4. The relative expression level of eSpCas9 in po1fe3-1 and po1fe4-2. (**A**) Relative expression level with po1fe 3-1 as control. (**B**) Relative expression level with early log phase as control. Values and error bars represent the average readings and standard deviations for three independent experiments.

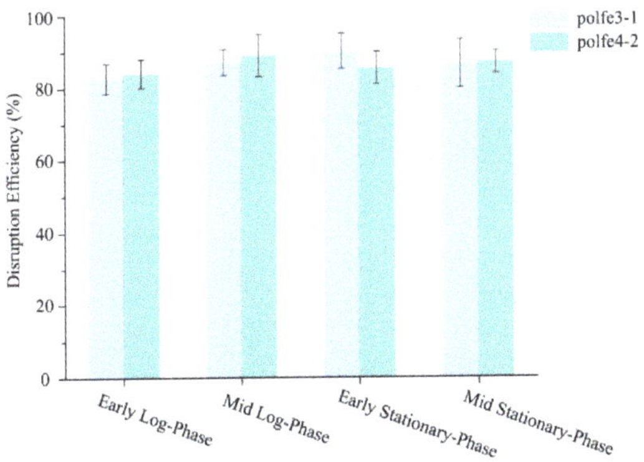

Figure 5. The disruption efficiency of *TRP1* under different *eSpCas9* expression levels. po1fe3-1 and po1fe4-2 shows that *eSpCas9* expression was controlled under promoter *pTEFin* and *pUAS16TEFin*; the *x* axis shows the preparation of receptive cells in different growth cycles. Values and error bars represent the average readings and standard deviations for three independent experiments.

3.4. Effect of Integrated eSpCas9 on Growth of Y. lipolytica

Cas9 nucleases are recognized for their toxicity to various species, especially when expressed at high levels in host cells [35]. Therefore, we investigated the impact of *eSpCas9* on the growth of *Y. lipolytica* at varying expression levels using growth curves. Initially, we transformed pINA1312 plasmids into po1f to restore the *ura* auxotroph, constructing po1f-ura+ as a control strain. Subsequently, we prepared seed cultures of the strains po1f-ura+, po1fe3-1, po1fe4-2, and inoculated them into fresh YPD medium for cultivation at 28 °C, 200 rpm, measuring the OD_{600} every 4 h as the data of growth curves. As depicted in Figure 6A, the growth of po1fe3-1 closely resembled that of the control strain po1f-ura+. Despite significantly increased expression levels using a UAS16-TEFin promoter, the growth of po1fe4-2 only suffered slight suppression (Figure 6A). In order to further explore the effect of Cas9 on strains' growth, a dilution assay was performed [36]. The cultures of po1f-ura+, po1fe3-1 and po1fe4-2 were adjusted to an OD_{600} of 2.5, followed by serial dilutions, and then plating of 5 µL spots on YNBD-Leu plates. As shown in Figure 6B, after cultivation for 12 h and 24 h, the growth of strains integrated with *eSpCas9*, po1fe3-1 and po1fe4-2, exhibited slightly impaired growth compared to the control strain po1f-ura+. However, similar delayed growth between po1fe3-1 and po1fe4-2 was not readily observed, indicating a high tolerance of *Y. lipolytica* to heterologous protein to some extent. In summary, control of eSpCas9 expression using strong inducible promoters may be a potential avenue for future research.

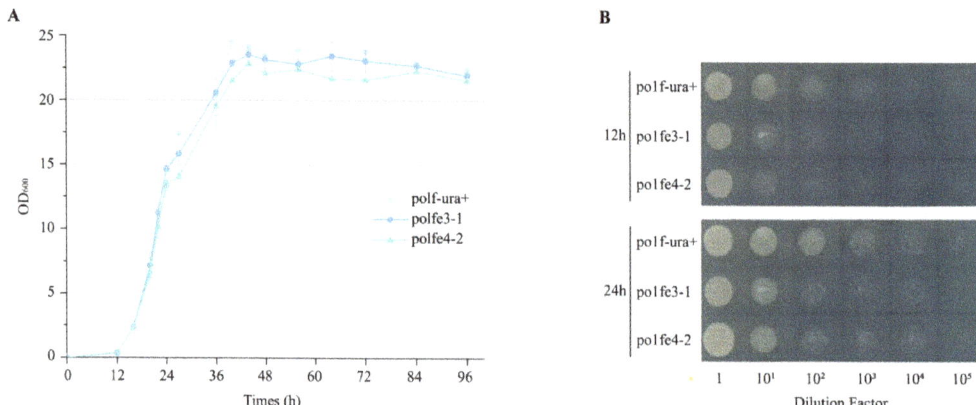

Figure 6. The effect of eSpCas9 protein on growth of *Y. lipolytica*. Strain po1f-ura+ was set as control. (**A**) Growth curves of po1f-ura+, po1fe3-1, po1fe4-2. Values and error bars represent the average readings and standard deviations for three independent experiments. (**B**) Dilution assay of po1f-ura+, po1fe3-1, po1fe4-2, with a consistent initial OD_{600} of 2.5.

4. Conclusions

In this study, we developed an enhanced gene editing system for *Y. lipolytica* based on the CRISPR/Cas9 system. The high-fidelity eSpCas9 protein was incorporated into the system to improve gene editing efficiency. Subsequently, *eSpCas9* was integrated into the genome of *Y. lipolytica* under the control of the TEFin promoter, resulting in significantly improved gene editing efficiency without the need for outgrowth. To further maximize efficiency, the TEFin promoter was replaced by the stronger UAS16-TEFin promoter. However, the potential improvement in gene editing was constrained by the original high efficiency and expression level of sgRNA. Additionally, we confirmed that the consistent expression of eSpCas9 protein slightly suppressed the growth of *Y. lipolytica*, revealing that the strong inducible promoters may be a favorable approach for future research. This approach simplifies the gene editing process in *Y. lipolytica*, which has the potential to broaden the synthetic capabilities of *Y. lipolytica* and offer valuable insights for other comparable microorganisms.

Supplementary Materials: The following supporting information can be downloaded at: https://www.mdpi.com/article/10.3390/jof10010063/s1, Figure S1: Maps of representative plasmids. (A) Maps of plasmids pCASyl, peCASyl, and pCEN1-Leu-sgRNA; (B) Maps of pINA1312, pINA1312-TEFin-eCAS9 and pINA1312-UAS16TEFin-eCAS9; Table S1: Codon-optimized sequence of the *eSpCas9* gene.

Author Contributions: Conceptualization, J.C. and B.Z.; methodology, J.C. and B.Z.; software, J.C.; validation, J.C.; formal analysis, J.C.; investigation, J.C.; resources, J.C.; data curation, J.C.; writing—original draft preparation, J.C.; writing—review and editing, J.C.; visualization, J.C.; supervision, B.Z.; project administration, B.Z. All authors have read and agreed to the published version of the manuscript.

Funding: This research received no external funding.

Institutional Review Board Statement: Not applicable.

Informed Consent Statement: Not applicable.

Data Availability Statement: The data supporting the conclusions of this study are available and included within the article.

Acknowledgments: The authors thank the National Engineering Research Center of Functional Food (Jiangnan University) for access to their facilities and technical assistance. The authors would like to thank Catherine Madzak (Paris-Saclay University, INRA, AgroParisTech, UMR SayFood, Thiverval-Grignon, France) for supplying the *Y. lipolytica* strains and expression system. The authors thank Sheng Yang (CAS Center for Excellence in Molecular Plant Science/Institute of Plant Physiology and Ecology, Chinese Academy of Sciences, Shanghai, China) for providing the CRISPR plasmids.

Conflicts of Interest: The authors declare no conflict of interest.

References

1. Raschmanova, H.; Weninger, A.; Glieder, A.; Kovar, K.; Vogl, T. Implementing CRISPR-Cas technologies in conventional and non-conventional yeasts: Current state and future prospects. *Biotechnol. Adv.* **2018**, *36*, 641–665. [CrossRef]
2. Naseri, G.; Koffas, M.A.G. Application of combinatorial optimization strategies in synthetic biology. *Nat. Commun.* **2020**, *11*, 2446. [CrossRef] [PubMed]
3. Brooks, S.M.; Alper, H.S. Applications, challenges, and needs for employing synthetic biology beyond the lab. *Nat. Commun.* **2021**, *12*, 1390. [CrossRef] [PubMed]
4. Madzak, C. *Yarrowia lipolytica* strains and their biotechnological applications: How natural biodiversity and metabolic engineering could contribute to cell factories improvement. *J. Fungi* **2021**, *7*, 548. [CrossRef]
5. Miller, K.K.; Alper, H.S. *Yarrowia lipolytica*: More than an oleaginous workhorse. *Appl. Microbiol. Biotechnol.* **2019**, *103*, 9251–9262. [CrossRef] [PubMed]
6. Fickers, P.; Cheng, H.; Sze Ki Lin, C. Sugar alcohols and organic acids synthesis in *Yarrowia lipolytica*: Where are we? *Microorganisms* **2020**, *8*, 574. [CrossRef] [PubMed]
7. Egermeier, M.; Russmayer, H.; Sauer, M.; Marx, H. Metabolic flexibility of *Yarrowia lipolytica* growing on glycerol. *Front. Microbiol.* **2017**, *8*, 49. [CrossRef] [PubMed]
8. Xue, Z.; Sharpe, P.L.; Hong, S.P.; Yadav, N.S.; Xie, D.; Short, D.R.; Damude, H.G.; Rupert, R.A.; Seip, J.E.; Wang, J.; et al. Production of omega-3 eicosapentaenoic acid by metabolic engineering of *Yarrowia lipolytica*. *Nat. Biotechnol.* **2013**, *31*, 734–740. [CrossRef]
9. Zhang, B.; Chen, H.; Li, M.; Gu, Z.; Song, Y.; Ratledge, C.; Chen, Y.Q.; Zhang, H.; Chen, W. Genetic engineering of *Yarrowia lipolytica* for enhanced production of *trans*-10, *cis*-12 conjugated linoleic acid. *Microb. Cell Factories* **2013**, *12*, 70. [CrossRef]
10. Obsen, T.; Faergeman, N.J.; Chung, S.; Martinez, K.; Gobern, S.; Loreau, O.; Wabitsch, M.; Mandrup, S.; McIntosh, M. *Trans*-10, *cis*-12 conjugated linoleic acid decreases de novo lipid synthesis in human adipocytes. *J. Nutr. Biochem.* **2012**, *23*, 580–590. [CrossRef]
11. Liu, Y.; Jiang, X.; Cui, Z.; Wang, Z.; Qi, Q.; Hou, J. Engineering the oleaginous yeast *Yarrowia lipolytica* for production of alpha-farnesene. *Biotechnol. Biofuels* **2019**, *12*, 296. [CrossRef] [PubMed]
12. Wei, W.; Zhang, P.; Shang, Y.; Zhou, Y.; Ye, B.C. Metabolically engineering of *Yarrowia lipolytica* for the biosynthesis of naringenin from a mixture of glucose and xylose. *Bioresour. Technol.* **2020**, *314*, 123726. [CrossRef] [PubMed]
13. Park, Y.K.; Ledesma-Amaro, R. What makes *Yarrowia lipolytica* well suited for industry? *Trends Biotechnol.* **2023**, *41*, 242–254. [CrossRef] [PubMed]
14. Mir, A.; Edraki, A.; Lee, J.; Sontheimer, E.J. Type II-C CRISPR-Cas9 biology, mechanism, and application. *ACS Chem. Biol.* **2018**, *13*, 357–365. [CrossRef] [PubMed]
15. Ran, F.A.; Hsu, P.D.; Wright, J.; Agarwala, V.; Scott, D.A.; Zhang, F. Genome engineering using the CRISPR-Cas9 system. *Nat. Protoc.* **2013**, *8*, 2281–2308. [CrossRef] [PubMed]
16. Asmamaw, M.; Zawdie, B. Mechanism and applications of CRISPR/Cas-9-mediated genome editing. *Biologics* **2021**, *15*, 353–361. [CrossRef]
17. Shi, T.Q.; Huang, H.; Kerkhoven, E.J.; Ji, X.J. Advancing metabolic engineering of Yarrowia lipolytica using the CRISPR/Cas system. *Appl. Microbiol. Biotechnol.* **2018**, *102*, 9541–9548. [CrossRef]
18. Schwartz, C.M.; Hussain, M.S.; Blenner, M.; Wheeldon, I. Synthetic RNA polymerase III promoters facilitate high-efficiency CRISPR-Cas9-mediated genome editing in *Yarrowia lipolytica*. *ACS Synth. Biol.* **2016**, *5*, 356–359. [CrossRef]

19. Gao, S.; Tong, Y.; Wen, Z.; Zhu, L.; Ge, M.; Chen, D.; Jiang, Y.; Yang, S. Multiplex gene editing of the *Yarrowia lipolytica* genome using the CRISPR-Cas9 system. *J. Ind. Microbiol. Biotechnol.* **2016**, *43*, 1085–1093. [CrossRef]
20. Sander, J.D.; Joung, J.K. CRISPR-Cas systems for editing, regulating and targeting genomes. *Nat. Biotechnol.* **2014**, *32*, 347–355. [CrossRef]
21. Guo, C.; Ma, X.; Gao, F.; Guo, Y. Off-target effects in CRISPR/Cas9 gene editing. *Front. Bioeng. Biotechnol.* **2023**, *11*, 1143157. [CrossRef]
22. Wilding-Steele, T.; Ramette, Q.; Jacottin, P.; Soucaille, P. Improved CRISPR/Cas9 tools for the rapid metabolic engineering of *Clostridium acetobutylicum. Int. J. Mol. Sci.* **2021**, *22*, 3704. [CrossRef] [PubMed]
23. Slaymaker, I.M.; Gao, L.; Zetsche, B.; Scott, D.A.; Yan, W.X.; Zhang, F. Rationally engineered Cas9 nucleases with improved specificity. *Science* **2016**, *351*, 84–88. [CrossRef] [PubMed]
24. Madzak, C.; Treton, B.; Blanchin-Roland, S. Strong hybrid promoters and integrative expression/secretion vectors for quasi-constitutive expression of heterologous proteins in the yeast *Yarrowia lipolytica. J. Mol. Microbiol. Biotechnol.* **2000**, *2*, 207–216.
25. Nicaud, J.M.; Madzak, C.; van den Broek, P.; Gysler, C.; Duboc, P.; Niederberger, P.; Gaillardin, C. Protein expression and secretion in the yeast *Yarrowia lipolytica. FEMS Yeast Res.* **2002**, *2*, 371–379. [CrossRef]
26. Hsu, P.D.; Scott, D.A.; Weinstein, J.A.; Ran, F.A.; Konermann, S.; Agarwala, V.; Li, Y.; Fine, E.J.; Wu, X.; Shalem, O.; et al. DNA targeting specificity of RNA-guided Cas9 nucleases. *Nat. Biotechnol.* **2013**, *31*, 827–832. [CrossRef]
27. Koonin, E.V.; Gootenberg, J.S.; Abudayyeh, O.O. Discovery of diverse CRISPR-Cas systems and expansion of the genome engineering toolbox. *Biochemistry* **2023**, *62*, 3465–3487. [CrossRef] [PubMed]
28. Gong, G.; Zhang, Y.; Wang, Z.; Liu, L.; Shi, S.; Siewers, V.; Yuan, Q.; Nielsen, J.; Zhang, X.; Liu, Z. GTR 2.0: gRNA-tRNA array and cas9-ng based genome disruption and single-nucleotide conversion in *Saccharomyces cerevisiae. ACS Synth. Biol.* **2021**, *10*, 1328–1337. [CrossRef] [PubMed]
29. Mans, R.; van Rossum, H.M.; Wijsman, M.; Backx, A.; Kuijpers, N.G.; van den Broek, M.; Daran-Lapujade, P.; Pronk, J.T.; van Maris, A.J.; Daran, J.M. CRISPR/Cas9: A molecular Swiss army knife for simultaneous introduction of multiple genetic modifications in *Saccharomyces cerevisiae. FEMS Yeast Res.* **2015**, *15*, fov004. [CrossRef]
30. Pignede, G.; Wang, H.J.; Fudalej, F.; Seman, M.; Gaillardin, C.; Nicaud, J.M. Autocloning and amplification of LIP2 in *Yarrowia lipolytica. Appl. Environ. Microbiol.* **2000**, *66*, 3283–3289. [CrossRef]
31. Duda, K.; Lonowski, L.A.; Kofoed-Nielsen, M.; Ibarra, A.; Delay, C.M.; Kang, Q.; Yang, Z.; Pruett-Miller, S.M.; Bennett, E.P.; Wandall, H.H.; et al. High-efficiency genome editing via 2A-coupled co-expression of fluorescent proteins and zinc finger nucleases or CRISPR/Cas9 nickase pairs. *Nucleic Acids Res.* **2014**, *42*, e84. [CrossRef] [PubMed]
32. Tai, M.; Stephanopoulos, G. Engineering the push and pull of lipid biosynthesis in oleaginous yeast *Yarrowia lipolytica* for biofuel production. *Metab. Eng.* **2013**, *15*, 1–9. [CrossRef] [PubMed]
33. Blazeck, J.; Liu, L.; Redden, H.; Alper, H. Tuning gene expression in *Yarrowia lipolytica* by a hybrid promoter approach. *Appl. Environ. Microbiol.* **2011**, *77*, 7905–7914. [CrossRef] [PubMed]
34. Shin, J.; Lee, N.; Song, Y.; Park, J.; Kang, T.J.; Kim, S.C.; Lee, G.M.; Cho, B.K. Efficient CRISPR/Cas9-mediated multiplex genome editing in CHO cells high-level sgRNA-Cas9 complex. *Biotechnol. Bioprocess Eng.* **2015**, *20*, 825–833. [CrossRef]
35. Jacobs, J.Z.; Ciccaglione, K.M.; Tournier, V.; Zaratiegui, M. Implementation of the CRISPR-Cas9 system in fission yeast. *Nat. Commun.* **2014**, *5*, 5344. [CrossRef] [PubMed]
36. Cui, L.; Vigouroux, A.; Rousset, F.; Varet, H.; Khanna, V.; Bikard, D. A CRISPRi screen in *E. coli* reveals sequence-specific toxicity of dCas9. *Nat. Commun.* **2018**, *9*, 1912. [CrossRef]

Journal of
Fungi

Article

Production of Xylanase by *Trichoderma* Species Growing on Olive Mill Pomace and Barley Bran in a Packed-Bed Bioreactor

Kholoud M. Alananbeh [1,*,†], **Rana Alkfoof** [2,†], **Riyadh Muhaidat** [2] **and Muhannad Massadeh** [3,*]

[1] Department of Plant Protection, School of Agriculture, The University of Jordan, Amman 11942, Jordan
[2] Department of Biological Sciences, Faculty of Science, Yarmouk University, Irbid P.O. Box 21163, Jordan; kfoofrana@yahoo.com (R.A.); muhaidat@yu.edu.jo (R.M.)
[3] Department of Biology and Biotechnology, Faculty of Science, The Hashemite University, Zarqa P.O. Box 11315, Jordan
* Correspondence: k.alananbeh@ju.edu.jo (K.M.A.); massadeh@hu.edu.jo (M.M.); Tel.: +962775321023 (K.M.A.); +962798518882 (M.M.)
† These authors contributed equally to this work.

Abstract: Xylanases are hydrolytic enzymes that have tremendous applications in different sectors of life, but the high cost of their production has limited their use. One solution to reduce costs and enhance xylanase production is the use of agro-wastes as a substrate in fungal cultures. In this study, olive mill pomace (OMP) and barley bran (BB) were used as carbon sources and possible inducers of xylanase production by three species of *Trichoderma* (*atroviride*, *harzianum*, and *longibrachiatum*), one major xylanase producer. The experiments were conducted under a solid-state fermentation system (SSF) in flask cultures and a packed-bed bioreactor. Cultures of OMP and BB were optimized by examining different ratios of OMP and BB, varied particle sizes, and inoculum size for the three species of *Trichoderma*. The ratio of 8:2 OMP and BB yielded the highest xylanase activity, with a particle size of 1 mm at 29 °C and an inoculum size of 1×10^7 spores/mL. Studying the time profile of the process revealed that xylanase activity was highest after seven days of incubation in flask SSF cultures (1.779 U/mL) and after three days in a packed-bed bioreactor (1.828 U/mL). The maximum percentage of OMP degradation recorded was about 15% in the cultures of *T. harzianum* flask SSF cultures, compared to about 11% in *T. longibrachiatum* bioreactor cultures. Ammonium sulfate precipitation and dialysis experiments showed that Xylane enzyme activity ranged from 0.274 U/mL in *T. harzianum* to 0.837 U/mL in *T. atroviride* when crude extract was used, with the highest activity (0.628 U/mL) at 60% saturation. Xylose was the main sugar released in all purified fractions, with the G-50 and G-75 fractions showing the maximum units of xylanase.

Keywords: barley bran; olive mill pomace; packed-bed bioreactor; partial purification; solid-state fermentation; *Trichoderma* spp.; xylanase

Citation: Alananbeh, K.M.; Alkfoof, R.; Muhaidat, R.; Massadeh, M. Production of Xylanase by *Trichoderma* Species Growing on Olive Mill Pomace and Barley Bran in a Packed-Bed Bioreactor. *J. Fungi* **2024**, *10*, 49. https://doi.org/10.3390/jof10010049

Academic Editor: Baojun Xu

Received: 11 November 2023
Revised: 27 December 2023
Accepted: 29 December 2023
Published: 5 January 2024

1. Introduction

There is a growing interest in screening natural sources to find bioactive compounds with nutraceutical and industrial benefits [1]. Increasing synthetic costs and environmental concerns rationalize the search for natural, high-value products to produce different commodities [2]. Therefore, attention has been drawn to plant biomass and agricultural wastes (agro-industrial residues) because they are cheap and abundant around the world, and major research efforts are underway to use them from different perspectives [3]. The agro-residues are primary reservoirs of lignocellulose, rendering them particularly appealing for many products essential to human society [4].

Xylan, a major hemicellulosic constituent of lignocellulosic materials, is markedly valued from industrial and biomedical perspectives and for the production of value-added products [5,6]. It is the most abundant hemicellulose in plant cell walls, accounting for more than 30% of the plant biomass [4,7,8]. Appreciable proportions of xylan exist in hardwoods and softwoods, as well as herbaceous plants, and form up to 35% of the lignocellulosic materials [9].

Xylan has a backbone of β-1,4-D-xylose residues with different α-glycosidically linked substitutions in side chains, primarily including L-arabinofuranose, D-galactose, and D-glucuronic or 4-O-methyl-d-glucuronic acid [10,11]. Substituents such as acetyl, arabinosyl, glucuronysyl, coumaroyl, and ferulic acid esters can also be found. The proportion of these substituents and the degree of branching are variable, depending on the plant source [12]. Complete degradation of xylan is achieved by a set of hydrolytic enzymes collectively called xylanases or xylanolytic enzymes [13]. These enzymes have been the target of many research endeavors; yet they are expensive and required in large amounts to be recruited in food, industrial, and medical applications. For this reason, attention has been paid to the natural producers of xylanases and agro-industrial wastes to develop nutrient-rich media for xylanase-producing organisms.

Xylanases find extensive applications across various industries, including food and feed, paper and pulp, textiles, pharmaceuticals, and lignocellulosic biorefinery [9]. Xylanases are ubiquitous and produced by a variety of organisms, including bacteria, protozoa, algae, fungi, snails, crustaceans, and insects [9,10,14]. Among these, multicellular fungi have been highly acknowledged as tremendous producers of xylanases [9,15]. Their filamentous structure provides large surface areas for absorptive nutrition and enhanced production of xylanases [2,16]. Species of the genus *Trichoderma* such as *T. reesi*, *T. viride*, and *T. harzianum* have held a key position in research studies aiming at xylanase production and purification [9,17].

Olive mill pomace (OMP) is one of the most cost-effective and nutrient-rich substrates used for cultivating fungi. Large quantities of OMP are produced in the Mediterranean region. The extraction of olive oil yields ca. 15,000 tons of olive oil and 80,000 tons of OMP a year [18]. In recent years, Jordan has been the eighth largest producer of olive oil worldwide, with 24,000 tons of oil and 60,000 tons of OMP per year [19]. Traditionally, OMP is used as firewood, but this causes air pollution. Consequently, the authorities placed restrictions on olive oil extraction factories in terms of waste accumulation and re-use. One solution is to make use of OMP as a growth medium for many microorganisms in attempts to produce commercial and high-value products (i.e., enzymes, bio-control agents, fertilizers, and biofuel), and interest in this is increasing [20]. However, few studies exploited OMP to produce enzymes and other value-added products.

The present study aims to examine the capacity of selected *Trichoderma* species isolated from plant sources for xylanase production using solid-state fermentation (SSF) and varied combinations of OMP and barely bran (BB) substrates, to determine optimal growth conditions for xylanase production, and to extract and partially purify xylanase. Despite the many studies on xylanase production from *Trichoderma* spp. around the world, this study is, to our knowledge, the first on Jordanian isolates of *Trichoderma*. The process of SSF has been accredited for the production of value-added microbial enzymes using agro-industrial origins. It is highly productive, with higher product stability and lower processing costs compared to the submerged fermentation process [21].

2. Materials and Methods

2.1. Fungal Species Isolates and Inoculum Preparation

Isolates of three distinct *Trichoderma* species (*T. atroviride* [accession MT626716], *T. harzianum* [accession MT626717], and *T. longibrachiatum* [accession MT626720]) were examined for xylanase production. The isolates were sub-cultured on potato dextrose agar (PDA) growth medium, incubated for 5 days at 28 °C, and used for inoculum preparation.

The inoculum of each isolate was prepared by adding 1 mL of sterile distilled water to a fresh culture of PDA plate, swept with a loop, and filtered to produce a slurry. After that, the suspension was transferred to 99 mL of sterile distilled water. The spores were counted using a hemocytometer to obtain a spore concentration of (10^7 spores/mL), and the suspension was stored in the refrigerator at 4 °C until use [22].

2.2. Preparation of OMP

Freshly harvested OMP was collected from olive mills located in Zarqa city (Jordan) during the period of olive oil extraction (October–December 2019). The collected pomace was dried in the greenhouse and used as the main substrate in culturing and fermentation processes.

2.3. Flask Solid State Fermentation for Xylanase Production

This part of the study was carried out following [8,23]. For each *Trichoderma* sp., a flask containing a 10 g mixture of OMP and barley bran (BB) was autoclaved at 121 °C for 15 min. Then, the mixture was moistened with 5 mL of sterile distilled water and inoculated with a spore suspension of each *Trichoderma* species (10^7 spores/mL). All flasks were incubated for seven days at 29 °C until further use in further biochemical analyses.

2.4. Xylanase Assay

The extraction and assaying of xylanase were carried out according to Al Sheikh [24], Assamoi et al. [25], and Mardawati et al. [8]. The contents of each prepared flask were collected in new flasks, and 50 mM sodium citrate solution (pH 5) was added. Then, the flasks were rotated at 150 rpm at room temperature for 30 min on an orbital shaker (Human Lab, Seoul, Republic of Korea) and filtered through Whatman No. 1 filter paper. The filtrate was centrifuged at 6000 rpm (Wagtek, Thatcham, UK) for 15 min at 4 °C, and the supernatant was used for the xylanase assay. Xylanase activity was measured by adding 0.2 mL of the sample supernatant to 0.5 mL of 1% oat-spelt xylan solution in 0.3 mL of 50 mM citrate buffer (pH 5) at 45 °C for 30 min. Reducing sugars released by xylanase were determined using 3 mL of 3.5-dinitrosalicylic acid (DNS) at 100 °C for 5 min according to the DNS method [15,26,27]. The amount of reducing sugars released in the reactions was measured spectrophotometrically at 540 nm. Xylanase activity was calculated based on a standard curve of serial concentrations of xylose. One unit (U) of xylanase activity was defined as the amount of enzyme releasing one μmole of reducing sugar (xylose equivalent) per min under the assay conditions [8,24].

2.5. Assessment of Reducing Sugars

Reducing sugar released by each culture was assessed using the Nelson–Somogyi methodology [28]. One milliliter of Somogyi reagent was added to 1 mL of the sample supernatant. The mixture was vortexed and boiled for 15 min at 100 °C. Next, 1 mL of Nelson reagent was added, and the absorbance was measured at 520 nm. Concentrations of reducing sugars were determined based on a standard curve of xylose concentrations and following the methodology of [29].

2.6. Total Carbohydrates

The total carbohydrates of each culture were analyzed by the phenol–sulfuric acid method and detected photometrically [30]. Fifty microliters of 80% phenol solution was added to 50 μL of sample supernatant, then 2 mL of sulfuric acid were added. The absorbance was read at 490 nm after 10 min of reaction progress. Concentrations of the total carbohydrates for each culture were estimated using a standard curve.

2.7. Optimization of Xylanase Production by Trichoderma sp.

Different ratios of OMP and BB (9.5:0.5, 9:1, and 8:2) and of different supplement particle sizes of OMP and BB (5 mm, 2 mm, and 1 mm) were prepared to determine the best combination that enhances growth of *Trichoderma* species and xylanase production. Test trials were conducted in flasks containing trace elements of a nitrogen source (ammonium sulfate or peptone) and a carbon source (xylose or xylan) at a concentration of 1% of the culture medium [31]. A culture containing only OMP was included as a control. The mixture of 8:2 OMP and BB (8 and 2 g/10 mL of distilled water) with particle sizes of 1 mm supplied with xylan (1% of medium) and peptone (1% of medium) yielded the highest growth and so was chosen as the culture medium in the following packed-bed column SSF.

2.8. Packed-Bed Column Solid State Fermentation of OMP and BB for Xylanase Production

This experiment was conducted following the method of [32]. Cultures yielding optimal growth from the flask fermentation were transferred into a sterile packed-bed bioreactor composed of a double-jacketed glass column (50 × 5 cm) connected to an air filter pump via silica tubes supplied through a 0.22 μm pore size filter at the bottom with a 20–25 air bubble flow rate that is controlled by a flow meter. The bioreactor was packed with multilayers of sterile stainless-steel mesh, and the culture medium was moistened with sterile distilled water and mixed with a spore suspension of *Trichoderma* species (10^7 spores/mL). The column was then incubated at 29 °C for 3 days and supplied with air at 1 vvm. Finally, the fermented substrate was used for the different biochemical analyses employed in the abovementioned flask SSF. This experiment was repeated three times.

2.9. Percentage of Degradation

To determine the percentage of the substrate consumed by *Trichoderma* sp., the fermented culture medium was weighed (initial weight) and then dried at 50 °C for 24 h and weighed (final weight). The degradation percentage was calculated using the following equation:

$$(\text{Initial weight} - \text{Final weight})/\text{Initial weight} \times 100\%$$

2.10. Times Profile of the Process

To determine the optimal incubation time for the cultures of the three *Trichoderma* isolates, two flask SSF replicates of each species were mixed with its spore suspension (10^7 spores/mL) and incubated at 29 °C for five days. Afterwards, they were filtered daily, and the filtrate was used for biochemical assays of xylanase activity, total carbohydrates, reducing sugars, and percentages of degradation.

2.11. Partial Purification of Xylanase

2.11.1. Ammonium Sulfate Precipitation and Dialysis

Crude xylanase extract was subjected to a graded ammonium sulfate saturation (20 to 70%) with continuous stirring on ice. Samples were then centrifuged at 5000 rpm for 15 min. For each percentage, the pellet formed was dissolved in a minimum amount of 50 Mm citrate buffer (pH 5), and enzyme activity and protein concentration were measured. The sample with the highest xylanase activity was further processed using dialysis bags in citrate buffer for 24 h at 4 °C. Afterwards, xylanase activities and protein concentrations were re-measured [33,34].

2.11.2. Size-Exclusion Chromatography

After ammonium sulfate precipitation and desalting of proteins, the highest protein concentration of dialyzed samples was subjected to a column (diameter: 1.5 cm, length: 30 cm, Bio-Rad, Hercules, CA, USA) containing Sephadex G-75 as the stationary phase, and in another trial, the column contained Sephadex G-50. The column was eluted with citrate buffer, pH 5. The eluted solution was collected in Cuvette tubes as fractions, each of 1 mL in size. Each fraction was used to measure dissolved protein concentration at 280 nm, enzyme activity in U/mL, and specific activity in U/mg protein as previously described [35,36].

2.12. Statistical Analysis

Standard curves of xylanase enzyme activity, reducing sugar, total carbohydrates, and protein concentration were drawn using Microsoft Excel 2010. One-way analysis of variance (ANOVA) was used for different testing among the experimental groups. Data were calculated as the mean ± standard error of the mean (SE) for duplicate independent experiments, and $p \leq 0.05$ was considered a significant difference.

3. Results

3.1. Flask Solid State Fermentation (SSF) for Xylanase Production by Trichoderma sp.

Three *Trichoderma* species (*T. atroviride* MT626716, *T. harzianum* MT626717, and *T. longibrachiatum* MT626720) were tested for xylanase production in flask SSF. Their optimum growth temperature was examined using PDA agar plates. The optimum temperature for their growth was 29–30 °C, where maximum biomass concentration was observed. The three species entered the stationary phase, as indicated by spore formation, after five days of incubation.

To perform SSF for enzyme production, the three *Trichoderma* spp. were grown on OMP with or without BB under static conditions without any supplements. It was noticed that fungal mycelia penetrated through the substrate particles of the culture after three days of incubation, and sporulation started to appear afterward. Xylanase activity, reducing sugar, and total carbohydrate concentrations were measured for seven-day-old *Trichoderma* spp. cultures (Figure 1). The data illustrate that supplementing OMP with BB enhanced enzyme production and fungal growth as well. The three fungal species showed almost the same behavior in producing the xylanase enzyme (Figure 1).

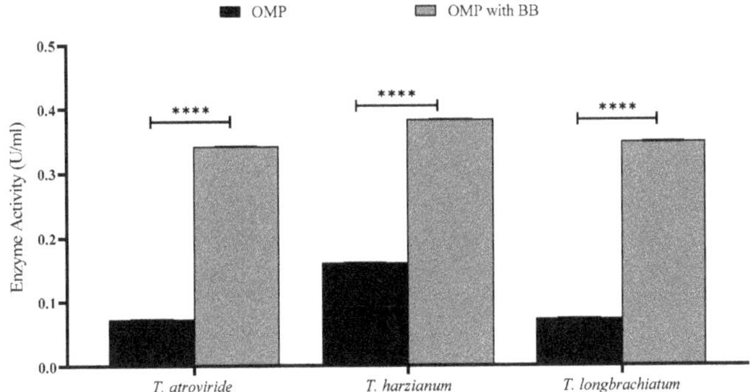

Figure 1. Xylanase activity of *T. atroviride*, *T. harzianum*, and *T. longibrachiatum* cultured on OMP with and without BB after seven days of fermentation. Bars represent the means ($\pm$SE) of two replicates for each culture. Asterisks indicate statistically significant differences at $p \leq 0.05$. **** $p < 0.0001$.

T. atroviride was able to liberate 43.725 µg/mL of reducing sugars compared to 22.835 mg/mL in the culture of *T. longibrachiatum*. On the other hand, the maximum total carbohydrate concentration obtained was from the cultures of *T. longibrachiatum* (1.92 µg/mL). Based on these results, combination cultures of OMP and BB were used for further studies.

3.2. Optimization of SSF for Xylanase Production

The effect of OMP and BB mixing ratios and sizes, in addition to the supplementation of cultures with trace elements, was studied. The results revealed a significant difference in the activity of xylanase and concentrations of reducing sugars and total carbohydrates in the three cultures of *Trichoderma* sp. compared to that of the control ($p < 0.001$; Figures 2 and 3).

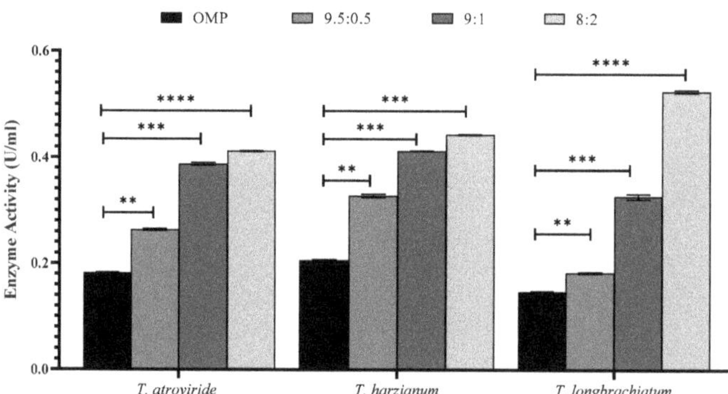

Figure 2. The effect of the culture ratio between OMP and BB on xylanase activity in the three studied *Trichoderma* species after seven days of fermentation. Bars represent the means ($\pm$SE) of two replicates for each culture. Asterisks indicate statistically significant differences at $p \leq 0.05$. ** $p < 0.01$, *** $p < 0.001$, and **** $p < 0.0001$.

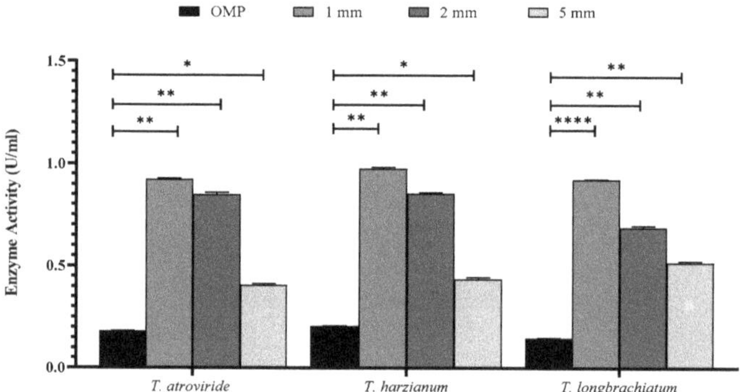

Figure 3. The effect of particle size of OMP and BB on xylanase activity in the three studied *Trichoderma* species after seven days of fermentation. Bars represent the means ($\pm$SE) of two replicates for each culture. Asterisks indicate statistically significant differences at $p \leq 0.05$. * $p < 0.05$, ** $p < 0.01$, and **** $p < 0.0001$.

The activity of xylanase was significantly higher in all ratios and sizes tested compared to that of the control. It was noticed that there were variations in xylanase activity within each *Trichoderma* sp. culture (Figures 2 and 3). The ratios of 8:2 and 9.5:0.5 of OMP and BB gave the highest and lowest xylanase activity, respectively (Figure 2). In Figure 3, the particle size of 1 mm was found to be the most effective in xylanase production compared to 2 mm and 5 mm in all *Trichoderma* species studied. Similarly, the inclusion of trace elements in growth cultures exerted a significant effect on enhancing *Trichoderma* to produce more active xylanase (Figure 3).

However, the effect was dependent on the type of supplement included in the culture. In all species, supplementation with xylan, peptone, or a combination of both yielded higher xylanase activity compared to all other types of supplements and the controls (Figure 4). Notably, peptone alone was found to be the most effective in xylanase production in all *Trichoderma* species, irrespective of the presence or absence of xylan and despite the overall variations observed (Figure 4).

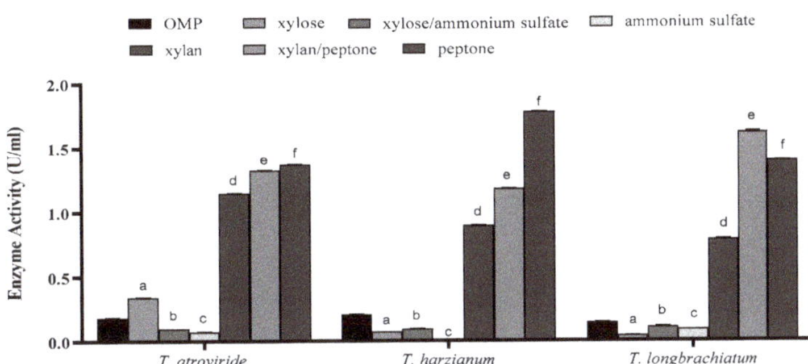

Figure 4. The effect of supplementation of growth cultures with trace elements on xylanase activity for the three studied *Trichoderma* species after seven days of fermentation. Bars represent the means (±SE) of two replicates for each culture. Different letters indicate statistically significant differences at $p \leq 0.05$.

3.3. Time Profile of the Process

The time profile for the production process was studied in the three tested cultures of *Trichoderma* sp. at the optimum conditions determined previously. The xylanase activity, reducing sugar, and total carbohydrate concentrations were measured independently every day over the incubation period (Figure 5). Over the incubation period, the production of the xylanase enzyme for the three tested species of *Trichoderma* reached its maximum on day three. Therefore, three-day-old cultures were chosen for further studies.

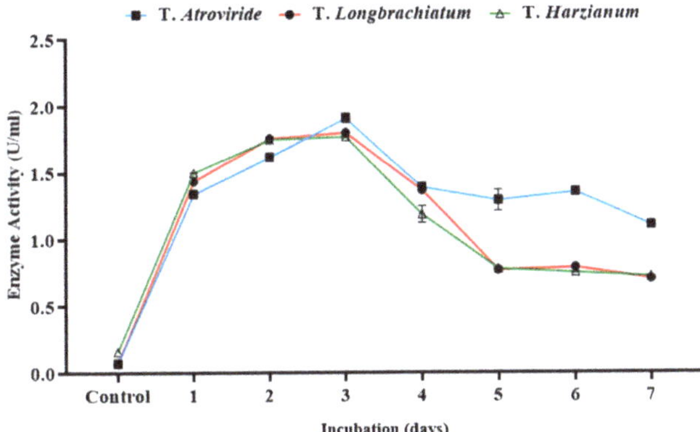

Figure 5. Xylanase activity over a seven-day incubation period in three studied *Trichoderma* species incubated at 29 °C.

3.4. Packed-Bed Column Solid State Fermentation

The data of the flask SSF process for xylanase enzyme production were translated into a column bioreactor packed with OMP and BB under SSF conditions. After three days of fermentation, all measures were significantly higher compared to those achieved from flask SSF cultures (Figures 6 and 7). For instance, the maximum xylanase activity achieved was 1.998 U/mL for *T. harizianum* (Figure 7). The maximum total reducing sugars (867.08 mg/mL) and total carbohydrate (1.667 mg/mL) were the highest in *T. atroviride* compared to the other two species.

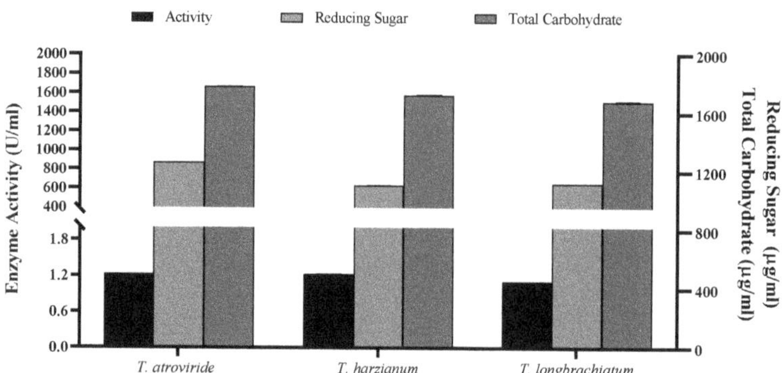

Figure 6. Xylanase activity and concentrations of reducing sugars and total carbohydrate of the three *Trichoderma* cultured species using packed-bed bioreactor.

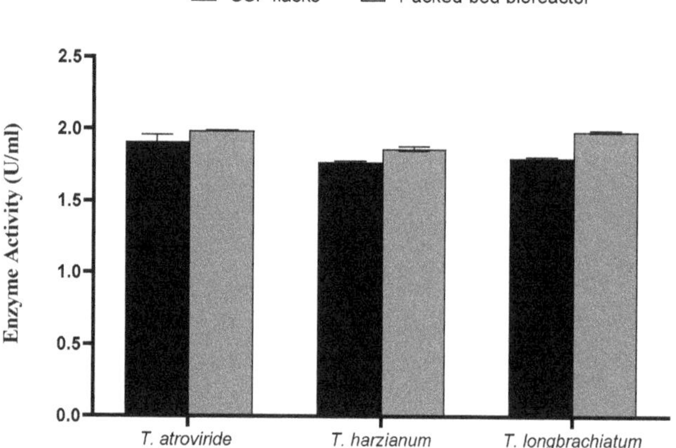

Figure 7. Xylanase activity using SSF and packed-bed bioreactor.

3.5. Percentage of OMP Degradation

The percentage of substrate degradation varied between SSF in flasks and in the packed-bed bioreactor, with the former being generally greater than the latter except for *T. longibrachiatum*. Using the SSF, the percentage varied in the following order: *T. harzianum* > *T. longibrachiatum* > *T. atroviride* (Figure 8). The maximum percentage of degradation recorded was about 15% in the cultures of *T. harzianum* flask SSF cultures, compared to about 11% in *T. longibrachiatum* bioreactor cultures.

3.6. Partial Purification of Xylanase

The crude extract of each *Trichoderma* species that gave the highest xylanase activity using a packed-bed bioreactor was partially purified through a series of ammonium sulfate precipitation reactions and dialysis. The highest dissolved protein concentration was obtained using a 60% ammonium sulfate saturation, and this percentage was chosen for further purification of xylanase using size-exclusion chromatography.

The experiment of size-exclusion chromatography yielded 25 fractions per *Trichoderma* species tested (Figure 9). Dissolved proteins and xylanase activity were highest in fractions 9 to 12 for *T. atroviride*, 4 to 7 for *T. harzianum*, and 9 to 14 for *T. longibrachiatum* (Figure 9).

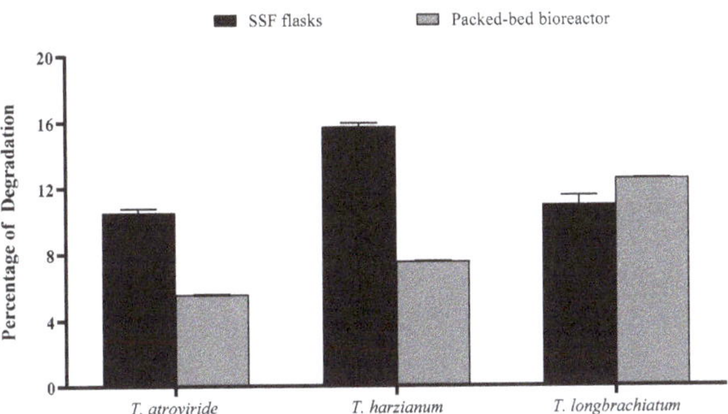

Figure 8. Consumed and percentage of culture media after seven days of incubation using SSF and packed-bed bioreactor by the three *Trichoderma* species studied. Bars represent the means (±SE) of four replicates per species.

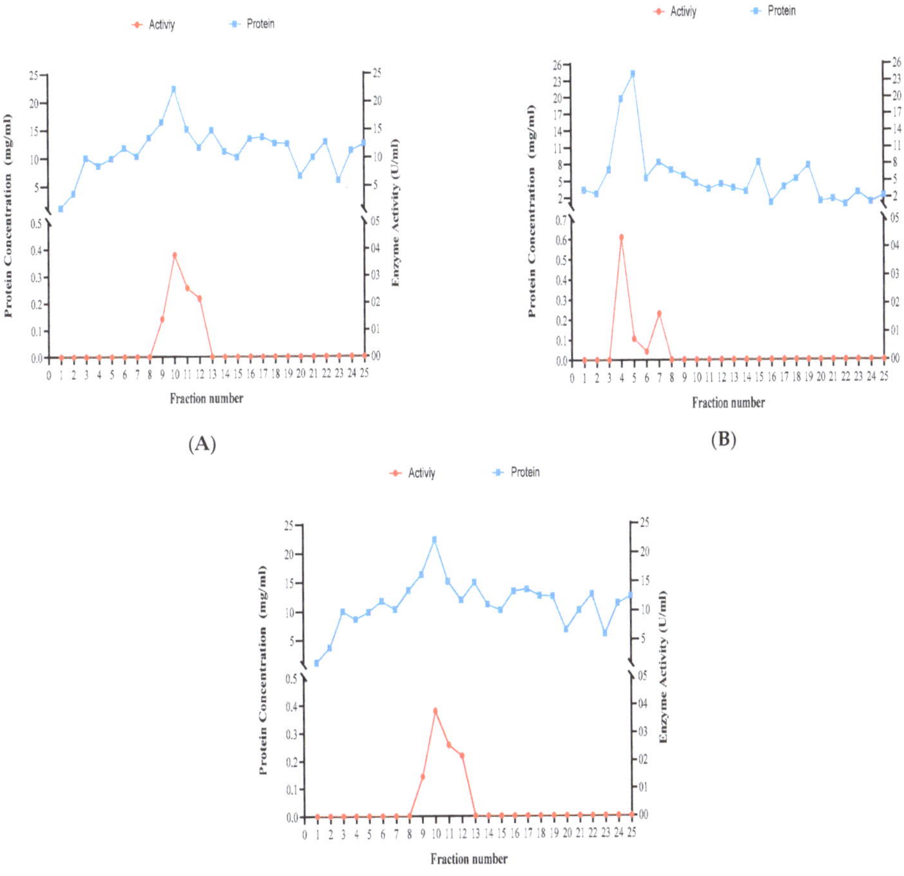

(A)

(B)

(C)

Figure 9. Concentrations of dissolved proteins and xylanase activity in the fractions obtained using size-exclusion chromatography for the three studied *Trichoderma* species. (**A**) *T. atroviride*, (**B**) *T. harzianum*, and (**C**) *T. longibrachiatum*.

Xylanase enzymatic activity ranged from 0.274 U/mL in *T. harzianum* to 0.837 U/mL in *T. atroviride* when crude extract was used. In general, xylanase activity was the highest in *T. atroviride* compared to the other two species across all the ammonium sulfate precipitation concentrations (20–70%). In addition, the specific activity of xylanase was measured for the three tested species of *Trichoderma*, and data ranging from 0.001 to 0.064 U/mg for *T. atroviride*, 0.004 to 0.031 mg/U for *T. harzianum*, and from 0.003 to 0.057 for *T. longibrachiatum* were obtained (Appendix A).

4. Discussion

In this study, a combination of olive mill pomace (OMP) and barley bran (BB) was studied as substrates for the production of the xylanase enzyme by *Trichoderma* sp. Different lignocellulosic materials were used as substrates to produce xylanase, and several reports illustrated the ability of *Trichoderma* species to use lignocellulosic materials as nutrient sources [37]. The agro-industrial residues of OMP have been well recognized in the literature as a rich source of organic matter, comprising a large amount of cellulose, hemicellulose, lignin, lipids, carbohydrates, and phenols [38]. This complex structure makes it suitable for the growth and metabolism of many microorganisms, specifically *Trichoderma* [39]. However, there are limited studies in the literature on using a combination of OMP and BB as an enzyme inducer to produce xylanase.

The activity of xylanase was studied with and without BB throughout the incubation period. The highest activity of xylanase was recorded with BB, and the incubation period was seven days. This indicates that the BB used in this study supports the growth of *Trichoderma* spp. and enhances their capacity to produce xylanase, an inducible enzyme. This is consistent with the findings of Soliman [40], who underscored BB as an excellent nutritive material that stimulates *Trichoderma* to produce xylanases because its hemicellulosic content is composed largely of xylan [40]. The substrates OMP and BB are thus vital for enhancing gene expression and protein synthesis, and, in consequence, the activity of cells to produce xylanase is augmented. It is well accepted that when cells are in the stationary phase of growth, their priority to express a specific gene varies depending on their needs and the signals they receive from their environment.

Xylanase activity increased after the inclusion of OMP and BB at a ratio of 8:2 and a particle size of 1 mm. It was necessary to reduce the substrate particle size, which may provide a favorable surface area for fungal growth, uptake of gases, accessibility, and hence improved production of xylanase. Lakshmi [41] affirmed that a smaller substrate particle size provides more efficient nutrient uptake, better support for microbial attachment, and better transport of substrate components compared to larger particle sizes.

To further enhance fungal growth and enzyme production, 1% xylan and peptone were added to the cultures as additional carbon and nitrogen sources, respectively. This supplement increases xylanase activity. Xylan was added as a simple inducer of gene expression and xylanase synthesis, whereas peptone was utilized by *Trichoderma* as a protein source. Once a high fungal biomass is established, *Trichoderma* can achieve complete hydrolysis of xylan. Therefore, the addition of 1% of these constituents to the fermentation media helps the microorganism attain its maximum xylanase production capability in a shorter period of time owing to the easily utilized carbon and nitrogen sources in the media. Our results support those of Gauterio et al. [42], who maximized xylanase production by *Aureobasidium pullulans* using a by-product of rice grain supplied with xylan, peptone, and other nutritional sources. Moreover, a considerable increase in xylanase activity by *T. harzianum* 1073 D3 was reported when 1% xylose was included, surpassing the effects observed using alternative sugars such as glucose, galactose, fructose, lactose, and sucrose [43]. In a study by El-Gendi et al. [44] on xylanase production by *Bacillus subtilis* using varying peptone concentrations, a direct correlation between peptone concentration and xylanase production and strain growth was reported. The highest growth and productivity were found at a peptone concentration of 1.4%. Taking these findings as a baseline, peptone emerged as a primary organic nitrogen source required to enhance xylanase production and activity by bacterial and fungal species [43].

In this study, examining the temporal profile of the production process reveals that xylanase activity peaked after three days of incubation, followed by a sudden drop. This indicates that the fungal ability to generate the enzyme was strikingly restricted to the early phases of growth, allowing them to access available substrates necessary for their survival and reproduction. Subsequently, xylanase generation decreased during the transition between log and stationary phases of their normal growth cycles. From a commercial perspective, repeating solid-state fermentation (SSF) beyond a three-day incubation period would not yield significant results for xylanase production. The decline in nutrient availability and the build-up of metabolic by-products adversely impact fungal growth, pushing them into the stationary phase, where spore formation is induced. A similar trend of xylanase activity by *Trichoderma* using SSF was previously reported by Pandey [45].

To study xylanase enzyme properties, it was necessary to produce the enzyme in a larger quantity. For that reason, a packed-bed bioreactor was employed. Xylanase enzyme production was enhanced in the packed-bed bioreactor compared to the SSF flask due to the efficient aeration pattern in the packed-bed bioreactor, which provides air circulation in a closed system [45]. Also, a packed-bed bioreactor maintains adequate and homogeneous levels of temperature and moisture, allowing efficient transfer of nutrients and metabolites [46]. In a recent study by Massadeh et al. [47], the entrapment/immobilized fermentation method has been proposed as an alternative to classical free-cell fermentation owing to its prolonged periods of growth, enhanced metabolic activities, and repeated use of cells. Cell entrapment is a common methodology for whole-cell immobilization carried out in a polymer matrix, carrageenan and alginate, and synthetic fibers. As a result, fungal growth was greatly improved, xylanase activity increased, and the maximum percentage of substrate degradation was accomplished.

Partial purification of the xylanase enzyme was performed after protein precipitation with ammonium sulfate. The highest xylanase activity was achieved at 60% saturation of ammonium sulfate. The crude enzyme precipitate was dialyzed against solid sodium citrate, which caused a reduction in enzyme activity compared to the crude enzyme extract, which might be due to the high polarity of the protein that forced them to remain in their physical medium. This indicates that it is possible to separate proteins from a mixture on the basis of their relative hydrophilicity by gradually increasing the concentration of ammonium sulphate [42]. Gauterio et al. [42] reported a reduction in xylanase and cellulase activities when employing ammonium sulfate precipitation and dialysis against different buffer solutions.

The dialyzed samples with the highest activity were subjected to gel filtration chromatography using Sephadex G-50 and G-75. The results revealed that xylanase was partially purified, and its activity increased. This improvement could be due to the removal of other compounds affecting xylanase activity, as Sephadex G-50 and G-75 allowed the adhesion of proteins of specific sizes. Our results are in agreement with Silva et al. [48], who purified xylanase from *Trichoderma* sp. using Sephadex G-75 and claimed that the xylanase enzyme was purified [48].

Fractions showing the highest activity from the gel filtration chromatography were collected, and a sample of each fraction was spotted on silica gel plates to perform TLC. It was shown that xylose and glucose were the main sugars present in the enzyme extracts when compared to the standard employed (R. Alkfoof, K., Alananbeh, M. Massadeh and R. Muhaidat, unpublished results). Two distinct materials, namely OMP and BB, were employed as substrates for *Trichoderma* culture. The induction process triggered the enzymatic activity, leading to enzyme-mediated transformations. Upon conducting thin layer chromatography (TLC) post-enzyme purification, it was observed that OMP and BB exhibited high heterogeneity and intricate structural complexity. Their native structures posed challenges in comprehension. The enzymatic action on these materials resulted in not only xylose but also the formation of disaccharides and oligosaccharides. Given the endoxylanase nature of the enzyme, it acts on internal bonds within these structurally complex substrates. Unfortunately, due to the limited availability of disaccharides and oligosaccharides in the laboratory, a comprehensive experiment elucidating the complete spectrum of these materials was not conducted. Nevertheless, our analysis confirmed the presence of xylose in the reaction mixture. Our results agree with those of Silva et al. [48], who claimed that complete degradation of xylan releases xylose as the final product.

The xylanase enzyme has tremendous applications in industry, and it can be used in the synthesis of biofuels, foodstuffs, textiles, organic acids, and many value-added products. The agro-residues of OMP are cheap and available all through the year, and hence it is recommended to study its supplementation with different cost-effective elements to support the growth and metabolism of *Trichoderma*. To elucidate the fungal behavior in producing an active xylanase enzyme, it is recommended to investigate the type of xylanase produced by using SDS-PAGE, native PAGE electrophoresis, and HPLC for detecting the reaction products. In doing so, additional species of *Trichoderma* should be studied.

5. Conclusions

This study showed the ability of three *Trichoderma* species to produce xylanase by using OMP and BB as substrates in solid-state fermentation. After optimizing the growth of the three *Trichoderma* sp., xylanase enzyme production was assessed. It was found that BB supplementation was necessary to induce xylanase production. The optimum ratio of OMP and BB was 8:2, corresponding with the highest activity of xylanase recorded (0.527 U/mL) from the culture of *T. longibrachiatum*. Furthermore, using a 1 mm particle size of both substrates enhanced the activity of xylanase (0.983 U/mL) from the culture of *T. harzianum*. Supplementing the culture with xylan and peptone showed a positive impact on fungal growth and xylanase production in all *Trichoderma* sp. cultures. The maximum xylanase activity (1.944 U/mL) was obtained in the culture of *T. atroviride* after 3 days in SSF flask culture and in the packed-bed SSF bioreactor (1.998 U/mL). To extract the enzyme and partially purify it, the crude extract was collected from the packed-bed bioreactor culture of the three *Trichoderma* sp. and subjected to 60% ammonium sulfate precipitation and dialysis against solid sodium citrate. The dialyzed enzyme samples were introduced to size-exclusion chromatography using Sephadex G-75 and G-50. It was noticed that xylanase activity was concentrated, and to confirm its purity, the fractions were subjected to a TLC reaction. The results revealed that the selected fractions collected from the three *Trichoderma* sp. were highly active, liberating xylose as the main sugar observed.

Author Contributions: Conceptualization, R.A., R.M., M.M. and K.M.A.; methodology, R.A., R.M., M.M. and K.M.A.; investigation, R.A. and R.M.; resources, R.A., R.M., M.M. and K.M.A.; project administration. R.M., M.M. and K.M.A.; data curation, R.A. and R.M.; writing—original draft preparation, R.A., R.M., M.M. and K.M.A.; writing—review and editing, R.M., M.M. and K.M.A.; supervision, R.M., M.M. and K.M.A.; funding acquisition R.M., M.M. and K.M.A. All authors have read and agreed to the published version of the manuscript.

Funding: This research was funded by research grant number 1600204/9 to M.M. from the Deanship of Research of the Hashemite University, and the research lab facilities of the Department of Biological Sciences of Yarmouk University.

Institutional Review Board Statement: Not applicable.

Informed Consent Statement: Not applicable.

Data Availability Statement: The data presented in this study are available on request from the corresponding author.

Conflicts of Interest: The authors declare no conflicts of interest.

Appendix A

Xylanase specific activity at different fractions for the three tested *Trichoderma* spp.

| Fractions | **Trichoderma species** | | | | | | | | | | | |
| | *T. atroviride* | | | | *T. harzianum* | | | | *T. longibrachiatum* | | | |
	OD_{280}	Protein mg/mL	Activity U/mL	Specific Enzyme Activity U/mg	OD_{280}	Protein mg/mL	Activity U/mL	Specific Enzyme Activity U/mg	OD_{280}	Protein mg/mL	Activity U/mL	Specific Enzyme Activity U/mg
1	-0.035	-1.750	0.000	0.000	0.068	3.410	0.000	0.000	0.006	0.300	0.000	0.000
2	-0.030	-1.510	0.000	0.000	0.055	2.735	0.000	0.000	0.009	0.450	0.000	0.000
3	-0.001	-0.050	0.000	0.000	0.140	7.015	0.000	0.000	0.065	3.250	0.000	0.000
4	0.005	0.260	0.000	0.000	0.396	19.820	0.610	**0.031**	0.149	7.440	0.000	0.000
5	0.013	0.640	0.000	0.000	0.487	24.325	0.106	**0.004**	0.212	10.595	0.000	0.000
6	0.070	3.486	0.000	0.000	0.111	5.550	0.042	**0.007**	0.213	10.670	0.000	0.000
7	0.016	0.796	0.000	0.000	0.168	8.415	0.230	**0.027**	0.300	15.005	0.000	0.000
8	0.035	1.750	0.000	0.000	0.140	6.980	0.000	0.000	0.309	15.440	0.000	0.000
9	0.159	7.950	0.511	**0.064**	0.121	6.040	0.000	0.000	0.397	19.840	0.059	**0.003**
10	0.200	10.005	0.093	**0.009**	0.093	4.635	0.000	0.000	0.485	24.235	0.469	**0.019**
11	0.350	17.500	0.106	**0.006**	0.072	3.600	0.000	0.000	0.391	19.540	0.140	**0.007**
12	0.456	22.780	0.041	**0.002**	0.090	4.475	0.000	0.000	0.286	14.315	0.250	**0.017**
13	0.388	19.400	0.000	0.000	0.076	3.775	0.000	0.000	0.209	10.425	0.597	**0.057**
14	0.300	15.000	0.000	0.000	0.062	3.120	0.000	0.000	0.244	12.220	0.050	**0.004**
15	0.298	14.900	0.000	0.000	0.167	8.345	0.000	0.000	0.304	15.205	0.000	0.000
16	0.266	13.300	0.000	0.000	0.022	1.120	0.000	0.000	0.300	15.005	0.000	0.000
17	0.255	12.750	0.000	0.000	0.079	3.935	0.000	0.000	0.223	11.130	0.000	0.000
18	0.250	12.500	0.000	0.000	0.108	5.390	0.000	0.000	0.270	13.480	0.000	0.000
19	0.203	10.165	0.000	0.000	0.155	7.745	0.000	0.000	0.257	12.830	0.000	0.000
20	0.100	5.005	0.000	0.000	0.007	0.345	0.000	0.000	0.270	13.500	0.000	0.000
21	0.125	6.250	0.000	0.000	0.034	1.715	0.000	0.000	0.200	10.005	0.000	0.000
22	0.200	10.000	0.000	0.000	0.015	0.765	0.000	0.000	0.190	9.480	0.000	0.000
23	0.100	5.010	0.000	0.000	0.058	2.885	0.000	0.000	0.156	7.775	0.000	0.000
24	0.109	5.450	0.000	0.000	0.023	1.170	0.000	0.000	0.098	4.910	0.000	0.000
25	0.100	5.000	0.000	0.000	0.047	2.335	0.000	0.000	0.085	4.260	0.000	0.000

References

1. Sorrenti, V.; Burò, I.; Consoli, V.; Vanella, L. Recent advances in health benefits of bioactive compounds from food wastes and by-products: Biochemical aspects. *Int. J. Mol. Sci.* **2023**, *24*, 2019. [CrossRef] [PubMed]

2. Knob, A.; Fortkamp, D.; Prolo, T.; Izidoro, S.C.; Almeida, J.M. Agro-residues as an alternative for xylanase production by filamentous fungi. *BioResources* **2014**, *9*, 5738–5773.

3. Gontard, N.; Sonesson, U.; Birkved, M.; Majone, M.; Bolzonella, D.; Celli, A.; Angellier, H.; Jang, H.; Verniquet, A.; Brosze, J.; et al. A Research challenge vision regarding management of agricultural waste in circular bio-based economy. *Crit. Rev. Environ. Sci. Technol.* **2018**, *48*, 614–654. [CrossRef]

4. Basit, A.; Jiang, W.; Rahim, K. Xylanase and its industrial applications. In *Biotechnological Applications of Biomass*; Basso, T.P., Basso, T.O., Basso, L.C., Eds.; Intechopen: London, UK, 2020; p. 638.

5. Ocreto, J.B.; Chen, W.H.; Rollon, A.P.; Ong, H.C.; Pétrissans, A.; Pétrissans, M.; De Luna, M.D.G. Ionic liquid dissolution utilized for biomass conversion into biofuels, value-added chemicals and advanced materials: A comprehensive review. *Chem. Eng. J.* **2022**, *445*, 136733. [CrossRef]

6. Rebouillat, S.; Pla, F. A Review: On smart materials based on some polysaccharides; within the contextual bigger data, insiders, "improvisation," and said artificial intelligence trends. *J. Biomater. Nanobiotech.* **2019**, *10*, 41–77. [CrossRef]

7. Kumar, A.; Naraian, R. Chapter 6—Differential expression of the microbial β-1,4-Xylanase and β-1,4-Endoglucanase genes. In *New and Future Developments in Microbial Biotechnology and Bioengineering*; Singh, H.B., Gupta, V.K., Jogaiah, S., Eds.; Elsevier: Amsterdam, The Netherlands, 2019; pp. 95–111.

8. Mardawati, E.; Sinurat, Y.; Yuliana, T. Production of crude xylanase from *Trichoderma* sp. using *Reutealis trisperma* exocarp substrate in solid state fermentation. In Proceedings of the International Conference of Sustainability Agriculture and Biosystem, Padang, Indonesia, 12–13 November 2019; IOP Publishing: Bristol, UK, 2020; p. 012024. [CrossRef]

9. Bhardwaj, N.; Kumar, B.; Verma, P. A Detailed overview of xylanases: An emerging biomolecule for current and future prospective. *Bioresour. Bioprocess.* **2019**, *6*, 40. [CrossRef]

10. Dutta, P.D.; Neog, B.; Goswami, T. Xylanase enzyme production from *Bacillus australimaris* P5 for prebleaching of bamboo (*Bambusa tulda*) pulp. *Mater. Chem. Phys.* **2020**, *243*, 122227. [CrossRef]

11. Fang, H.Y.; Chang, S.M.; Hsieh, M.C.; Fang, T.J. Production, Optimization growth conditions, and properties of the xylanase from *Aspergillus carneus* M34. *J. Mol. Catal. B Enzym.* **2007**, *49*, 36–42. [CrossRef]

12. Rojas, L.F.; Zapata, P.; Ruiz-Tirado, L. Agro-industrial waste enzymes: Perspectives in circular economy. *Curr. Opin. Green Sustain. Chem.* **2022**, *34*, 100585. [CrossRef]

13. Danso, B.; Ali, S.S.; Xie, R.; Sun, J. Valorization of wheat straw and bioethanol production by a novel xylanase- and cellulase-producing *Streptomyces* strain isolated from the wood-feeding termite, *Microcerotermes* Species. *Fuel* **2022**, *310*, 122333. [CrossRef]

14. Bajpai, P. Xylanases. In *Encyclopedia of Microbiology*, 3rd ed.; Schaechter, M., Lederberg, J., Eds.; Academic Press: San Diego, CA, USA, 2009; pp. 600–612.

15. Azzouz, Z.; Bettache, A.; Boucherba, N.; Amghar, Z.; Benallaoua, S. Optimization of xylanase production by newly isolated strain *Trichoderma afroharzianum* isolate AZ 12 in solid state fermentation using response surface methodology. *Cellul. Chem. Technol.* **2020**, *54*, 451–462. [CrossRef]

16. Haddadin, M.S.; Haddadin, J.; Arabiyat, O.I.; Hattar, B. Biological conversion of olive pomace into compost by using *Trichoderma harzianum* and *Phanerochaete chrysosporium*. *Bioresour. Technol.* **2009**, *100*, 4773–4782. [CrossRef] [PubMed]

17. Polizeli, M.; Rizzatti, A.; Monti, R.; Terenzi, H.; Jorge, J.A.; Amorim, D. Xylanases from fungi: Properties and industrial applications. *Appl. Microbiol. Biotechnol.* **2005**, *67*, 577–591. [CrossRef] [PubMed]

18. Abdelhadi, S.O.; Dosoretz, C.G.; Rytwo, G.; Gerchman, Y.; Azaizeh, H. Production of biochar from olive mill solid waste for heavy metal removal. *Bioresour. Technol.* **2017**, *244*, 759–767. [CrossRef]

19. Khdair, A.I.; Abu-Rumman, G.; Khdair, S.I. Pollution estimation from olive mills wastewater in Jordan. *Heliyon* **2019**, *5*, e02386. [CrossRef] [PubMed]

20. Prajapati, K.; Nayak, R.; Shukla, A.; Parmar, P.; Goswami, D.; Saraf, M. Polyhydroxyalkanoates: An exotic gleam in the gloomy tale of plastics. *J. Polym. Environ.* **2021**, *29*, 2013–2032. [CrossRef]

21. Abdollahi, F.; Jahadi, M.; Ghavami, M. Thermal stability of natural pigments produced by *Monascus purpureus* in submerged fermentation. *Food Sci. Nutr.* **2021**, *9*, 4855–4862. [CrossRef]

22. Singh, A.; Shahid, M.; Srivastava, M.; Pandey, S.; Sharma, A.; Kumar, V. Optimal physical parameters for growth of *Trichoderma* species at varying pH, temperature, and agitation. *Virol Mycol.* **2014**, *3*, 1–7.

23. Khanahmadi, M.; Arezi, I.; Amiri, M.-S.; Miranzadeh, M. Bioprocessing of agro-industrial residues for optimization of xylanase production by solid-state fermentation in flask and tray bioreactor. *Biocatal. Agric. Biotechnol.* **2018**, *13*, 272–282. [CrossRef]

24. Al Sheikh, A.H.M. Utilization of Dry Mill Residues (DOR) for the Production of Xylanase Enzyme by *Aspergillus terreus* in Solid State Fermentation. Master's Thesis, Hashemite University, Zarqa, Jordan, 2011.

25. Assamoi, A.A.; Destain, J.; Delvigne, F.; Lognay, G.; Thonart, P. Solid-state fermentation of xylanase from *Penicillium canescens* 10-10c in a multi-layer-packed bed reactor. *Appl. Biochem. Biotechnol.* **2007**, *145*, 87–98. [CrossRef]

26. Miller, G.L. Use of dinitrosalicylic acid reagent for determination of reducing sugar. *Anal. Chem.* **1959**, *31*, 426–428. [CrossRef]

27. Teixeira, R.S.; da Silva, A.S.; Ferreira-Leitao, V.S.; da Silva Bon, E.P. Amino acids interference on the quantification of reducing sugars by the 3,5-dinitrosalicylic acid assay mislead carbohydrase activity measurements. *Carbohydr. Res.* **2012**, *363*, 33–37. [CrossRef] [PubMed]

28. Ranjitha, P.; Karthy, E.; Mohankumar, A. Purification and characterization of the lipase from marine *Vibrio fischeri*. *Int. J. Biol.* **2009**, *1*, 48. [CrossRef]

29. Saryono, S.; Novianty, R.; Suraya, N.; Piska, F.; Devi, S.; Pratiwi, N.W.; Ardhi, A. Molecular identification of cellulase-producing thermophilic fungi isolated from Sungai Pinang hot spring, Riau Province, Indonesia. *Biodivers. J.* **2022**, *23*. [CrossRef]

30. Shallom, D.; Shoham, Y. Microbial hemicellulases. *Curr. Opin. Microbiol.* **2003**, *6*, 219–228. [CrossRef] [PubMed]

31. Pathania, S.; Sharma, N.; Verma, S.K. Optimization of cellulase-free xylanase produced by a potential thermoalkalophilic *Paenibacillus* sp. N1 isolated from hot springs of Northern Himalayas in India. *J. Microbiol. Biotechnol. Food Sci.* **2020**, *9*, 1–24.

32. Oliveira, F.; Salgado, J.M.; Abrunhosa, L.; Pérez-Rodríguez, N.; Domínguez, J.M.; Venâncio, A.; Belo, I. Optimization of lipase production by solid-state fermentation of olive pomace: From flask to laboratory-scale packed-bed bioreactor. *Bioprocess Biosyst. Eng.* **2017**, *40*, 1123–1132. [CrossRef]

33. Dhivahar, J.; Khusro, A.; Ahamad Paray, B.; Rehman, M.U.; Agastian, P. Production and partial purification of extracellular Xylanase from *Pseudomonas nitroreducens* using frugivorous bat (*Pteropus giganteus*) faeces as an ideal substrate and its role in poultry feed digestion. *J. King Saud Univ. Sci.* **2020**, *32*, 2474–2479. [CrossRef]

34. Siddiqui, M.A.H.; Biswas, M.; Faruk, M.O.; Roy, M.; Asaduzzaman, A.K.M.; Sharma, S.C.D.; Biswas, T.; Roy, N. Optimization, Isolation and characterization of cellulase–free thermostable xylanase from *Paenibacillus* sp. *Am. J. Life Sci.* **2016**, *4*, 93–98. [CrossRef]

35. Abusara, W.A. *Beta-Glucosidase Enzyme Production in Solid-State Fermentation Using Olive Mill Pomace (OMP) and Yeast Industry Wastewater (YIW)*; MSc. Hashemite University: Zarqa, Jordan, 2019.

36. Le Strat, Y.; Mandin, M.; Ruiz, N.; Robiou du Pont, T.; Ragueneau, E.; Barnett, A.; Déléris, P.; Dumay, J. Quantification of xylanolytic and cellulolytic activities of fungal strains isolated from *Palmaria palmata* to enhance R-phycoerythrin extraction of *Palmaria palmata*: From seaweed to seaweed. *Mar. Drugs* **2023**, *21*, 393. [CrossRef]

37. Naeimi, S.; Khosravi, V.; Varga, A.; Vágvölgyi, C.; Kredics, L. Screening of organic substrates for solid-state fermentation, viability and bioefficacy of *Trichoderma harzianum* AS12-2, a biocontrol strain against rice sheath blight disease. *J. Agron.* **2020**, *10*, 1258. [CrossRef]

38. Medouni-Haroune, L.; Zaidi, F.; Medouni-Adrar, S.; Kecha, M. Olive pomace: From an olive mill waste to a resource, an overview of the new treatments. *J. Crit. Rev.* **2018**, *5*, 1–6. [CrossRef]

39. Boutiche, M.; Sahir-Halouane, F.; Meziant, L.; Boulaouad, I.; Saci, F.; Fiala, S.; Bekrar, A.; Mesbah, R.; Abdessemed, A. Characterization and valorization of olive pomace for production of cellulase from *Trichoderma reesei* RUT C30 in solid-state fermentation. *Alger. J. Environ. Sci. Technol.* **2020**, *6*, 1381–1387.

40. Soliman, H.M.; Sherief, A.; EL-Tanash, A.B. Production of xylanase by *Aspergillus niger* and *Trichoderma viride* using some agricultural residues. *Int. J. Agric. Res.* **2012**, *7*, 46–57. [CrossRef]

41. Lakshmi, G.S.; Rao, C.S.; Rao, R.S.; Hobbs, P.J.; Prakasham, R.S. Enhanced production of xylanase by a newly isolated *Aspergillus terreus* under solid state fermentation using palm industrial waste: A statistical optimization. *Biochem. Eng. J.* **2009**, *48*, 51–57. [CrossRef]

42. Gautério, G.V.; da Silva, L.G.G.; Hübner, T.; da Rosa Ribeiro, T.; Kalil, S.J. Xylooligosaccharides production by crude and partially purified xylanase from *Aureobasidium pullulans*: Biochemical and thermodynamic properties of the enzymes and their application in xylan hydrolysis. *Process Biochem.* **2021**, *104*, 161–170. [CrossRef]

43. Seyis, I.; Aksoz, N. Xylanase production from *Trichoderma harzianum* 1073 D3 with alternative carbon and nitrogen sources. *Food Technol. Biotechnol.* **2005**, *43*, 37–40.

44. El-Gendi, H.; Badawy, A.S.; Bakhiet, E.K.; Rawway, M.; Ali, S.G. Valorization of lignocellulosic wastes for sustainable xylanase production from locally isolated *Bacillus subtilis* exploited for xylooligosaccharides' production with potential antimicrobial activity. *Arch. Microbiol.* **2023**, *205*, 315. [CrossRef]

45. Pandey, S.; Shahid, M.; Srivastava, M.; Sharma, A.; Singh, A.; Kumar, V.; Srivastava, Y. Isolation and optimized production of xylanase under solid state fermentation condition from *Trichoderma* sp. *Int. J. Adv. Res.* **2014**, *2*, 263–273.

46. De los Santos-Villalobos, S.; Hernández-Rodríguez, L.E.; Villaseñor-Ortega, F.; Peña-Cabriales, J.J. Production of *Trichoderma asperellum* T8a spores by a" home-made" solid-state fermentation of mango industrial wastes. *BioResources* **2012**, *7*, 4938–4951. [CrossRef]

47. Massadeh, M.I.; Fandi, K.; Al-Abeid, H.; Alsharafat, O.; Abu-Elteen, K. Production of citric acid by *Aspergillus niger* cultivated in olive mill wastewater using a two-stage packed column bioreactor. *Fermentation* **2022**, *8*, 153. [CrossRef]

48. Silva, L.d.O.; Terrasan, C.R.F.; Carmona, E.C. Purification and characterization of xylanases from *Trichoderma inhamatum*. *Electron. J. Biotechnol.* **2015**, *18*, 307–313. [CrossRef]

Journal of *Fungi*

Article

Ecdysteroid UDP-Glucosyltransferase Expression in *Beauveria bassiana* Increases Its Pathogenicity against Early Instar Silkworm Larvae

Xueqin Mao [1,†], Dongxu Xing [2,†], Die Liu [1], Haoran Xu [1], Luyu Hou [1], Ping Lin [1], Qingyou Xia [1], Ying Lin [1,*] and Guanwang Shen [1,*]

[1] Integrative Science Center of Germplasm Creation in Western China (Chongqing) Science City, Biological Science Research Center, Southwest University, Chongqing 400716, China
[2] Sericultural & Agri-Food Research Institute Guangdong Academy of Agricultural Sciences, /Key Laboratory of Functional Foods, Ministry of Agriculture and Rural Affairs, Guangdong Key Laboratory of Agricultural Products Processing, Guangzhou 510610, China
* Correspondence: ly908@swu.edu.cn (Y.L.); gwshen@swu.edu.cn (G.S.); Tel./Fax: +86-023-8919-8101 (Y.L. & G.S.)
† These authors contributed equally to this work.

Abstract: *Beauveria bassiana* (*B. bassiana*) is a broad-spectrum entomopathogenic fungus that can control pests in agriculture and forestry. In this study, encoding ecdysteroid uridine diphosphate glucosyltransferase gene (*egt*) was successfully screened in *B. bassiana* on the medium containing 500µg/mL G418 sulfate solution through the protoplast transformation method. This enzyme has the function of 20E (20-hydroxyecdysone) inactivation, thus increasing the mortality of the early instar larvae infected with *B. bassiana*. In this study, we transformed *B. bassiana* with the *egt* gene, which deactivates 20-hydroxyecdysone, a key hormone in insect development. The results showed that transgenic *B. bassiana* killed more silkworms of the 2nd instar larvae than the wild-type with a shorter LT50 time, which was reduced by approximately 20% (day 1 of the 2nd instar silkworm infection of *B. bassiana*) and 26.4% (day 2 of the 2nd instar silkworm infection of *B. bassiana*) compared to the wild-type, and also showed a higher mortality number before molting. The transgenic *B. bassiana* had a higher coverage of the body surface of silkworms compared to the wild type on the 3rd instar. In summary, improving entomopathogenic fungi using biological methods such as genetic engineering is feasible.

Keywords: fungal transgenic; 20E; deactivation; pathogenicity

Citation: Mao, X.; Xing, D.; Liu, D.; Xu, H.; Hou, L.; Lin, P.; Xia, Q.; Lin, Y.; Shen, G. Ecdysteroid UDP-Glucosyltransferase Expression in *Beauveria bassiana* Increases Its Pathogenicity against Early Instar Silkworm Larvae. *J. Fungi* 2023, 9, 987. https://doi.org/10.3390/jof9100987

Academic Editors: Baojun Xu and Jinkui Yang

Received: 4 August 2023
Revised: 29 September 2023
Accepted: 1 October 2023
Published: 4 October 2023

1. Introduction

The increasing environmental pollution caused by chemical pesticides, the higher resistance of pests to pesticides, and the potential harm to human health caused by pesticides have raised concerns regarding the use of chemical pesticides. Thus, biological control is receiving more attention. Entomopathogenic fungi is one of the major factors in controlling insect populations in nature and is an important tool in the biological control of pests. *B. bassiana* is the most common and widely used environmentally-friendly fungus in various insects in nature [1,2]. It is the most common and widely used fungal insecticide [3]. However, the pathogenicity of different strains in the host is different, and pathogenicity is a comprehensive effect determined by the multiple characteristics of the strain. Compared to chemical pesticides, its slow efficacy and inconsistent control results are the main limitations of its large-scale application due to the fungi's long growth and infection time [4,5]. Genetic modification of fungi using molecular biology techniques offers a faster and broader approach. Previous studies have demonstrated the increased expression of certain genes [6], such as protease Pr1A [4], CDEP1 [5], and chitinase CHIT1 [6] of *B. bassiana*, which are associated with the lethality of host infection. Some studies have also introduced exogenous toxin genes, such as the scorpion neurotoxin peptide AaIT1 [7,8] and the insecticidal proteins Vip3A [9] and Cry1Ac [10] from *Bacillus thuringiensis,* into *B. bassiana,* aiming to enhance its lethality to the host after infection.

Beauveria bassiana can infect more than 700 species of insects, including Lepidoptera, Hymenoptera, Coleoptera and 15 other orders, 149 families, 521 genera, and more than 10 species of ticks and mites in at least six families and seven genera [7]. As an important entomopathogenic fungus, *B. bassiana* is easy to cultivate and environmentally friendly to humans and animals. It has broad application prospects as an insecticide that biologically controls pests [1,8].

The fungal infection process in a host involves various steps, including host recognition, mechanical damage, nutrient competition, metabolic interference, toxin secretion, and tissue destruction in the host [9]. In particular, a certain amount of time is required for fungal conidia to adhere to the surface of the host insect, germinate, and penetrate the epidermis [10–12]. Most insects molt during their feeding period, and some insect instars have relatively short durations. Fungal infections may be eliminated before they invade the insect body, leading to potential failure [13]. Furthermore, although older instar insects have longer intervals between molts, the fungal infection must overcome the host's immune system [14]. The fungal infections grow mycelia in the insect hemocoel, proliferate extensively, and secrete toxins using nutrients in the hemolymph, ultimately leading to the host insect's death [15]. This process also requires a certain amount of time, during which the infected insects continue to feed. As a result, these pests can cause significant damage before their death following fungal infection. In conclusion, the fungus's relatively long infection and mortality times on the host insect contribute to the relatively poor efficacy of fungal biocontrol agents. This is currently a major factor limiting the large-scale application of fungal insecticides. The second instar silkworm, *Bombyx mori*, larvae infected with *B bassiana*, molts after about 48 h of growth and development [16]. Under normal circumstances, it is difficult for conventional strains of *B. bassiana* to infect the second instar larvae of silkworms on a large scale. In contrast, the last instar larvae of the silkworms have a relatively longer duration. However, from infection with *B. bassiana* to death, these larvae continue to feed for at least 5 days [17].

Ecdysteroid UDP glucosyltransferase (EGT) is a protein encoded by baculovirus [18,19]. EGT secreted by Autograph californica nucleopolyhedral virus (AcMNPV) can transfer UDP-glucoside to the hydroxyl group of the 22nd carbon atom of 20E in vitro, forming 20E-22-β-D-glucopyranoside without 20E activity [20–25]. When insects are infected by this virus, the secreted EGT can deactivate 20E, thereby hindering the normal molting of insects [20,26] We previously found that the expression of EGT in the last instar silkworm led to a decrease in the content of active 20E in the hemolymph of the last instar silkworm, which made the silkworm unable to pupate after spinning [27]. This study aimed to integrate *egt* into the genome of *B. bassiana*, which was selected as the host to explore the function of endowing *B. bassiana* to deactivate 20E, increase the mortality of young larvae infected with *B. bassiana*, and, to a certain extent, compensate for the application limitations of *B. bassiana* in pest biological control due to its long pathogenic cycle.

2. Materials and Methods

2.1. Screening of Antibiotics

The *B. bassiana* strain CMCC(B)G1-180910 was preserved and provided by our research group. Experimental silkworms were provided by the Biological Science Research Center of Southwest University (Chongqing, China). Three gradients of antibiotic solutions–Hygromycin B (Sheng gong, Shanghai, China), Zeocin (Sheng gong, Shanghai, China), and G418 sulfate solution (Sheng gong, Shanghai, China) were prepared at 50, 100, and 200 µg/mL concentrations, respectively. A potato dextrose agar (PDA) medium containing the corresponding antibiotic concentrations was prepared, and a control group without antibiotics was included. Using sterilized clean tweezers, we gently removed mycelium from preserved *B. bassiana* strains and inoculated them onto the PDA medium to allow the white strain to grow in medium free of bacterial or other fungal contamination. *B. bassiana* was purified by single-spore culture. First, a few fresh conidia and hyphae were dipped by a sterile cotton swab, added to sterilized ddH$_2$O, mixed well, and then passed through a 40 μm filter to remove mycelium. In this way, the conidia' suspension was prepared. Next, 1 μL was sucked for observation under a microscope and diluted with sterile water until the spore concentration was 1 μL, comprising only a single spore. This spore was inoculated and cultured in a petri dish. The purified *B. bassiana* strain was inoculated onto agar plates using sterile cotton swabs to ensure uniform distribution. It was then moderately dipped with sterile cotton swabs and evenly applied onto the PDA plate until no white conidia or mycelium were visible. The plates were inverted and cultured at 26 °C with a relative humidity (RH) of at least 90%. Suitable antibiotics and their concentrations were determined based on the growth of the *B. bassiana* colonies. If antibiotics had no inhibitory effect on the growth of *B. bassiana*, the PDA plate coated with the strain would be covered with white strain, and if they did have an inhibitory effect, it would not grow.

2.2. Construction of Transgenic Fungal Expression Vector

After optimizing the target gene containing EGT by using the preference codon of *B. bassiana* provided in the NCBI database, the Mcl1 signal peptide (SP, ATGCGTGAGCTCTC-CTCCGTTCTCGCTCTCTCCGGTCTCCTCGCTCTCGCTTCCGCT) and 6×His tag was added to the 5′ end of the *egt* gene. The linker (GGCGGCGGTGGTTCT) and nucleic acid sequence of green fluorescent protein (GFP) were added at the 3′ end. A nucleic acid sequence containing sp-6×His-EGT-linker-GFP with *Xho* I and *Kpn* I sites was synthesized by Takara Bio Corporation and cloned into the pMD19-T plasmid to generate pMD19-T [sp-6×His-EGT-linker-GFP]. pMD19-T [sp-6×His-EGT-linker-GFP] and the fungal expression vector pG418, containing the G418 resistance gene, were digested with *Xho* I and *Kpn* I, respectively. The recovered sp-6×His-EGT-linker-GFP was ligated into a transgenic vector. The *Xho* I-*Kpn* I fragment from pMD19-T [sp-6×His-EGT-linker-GFP] was excised and inserted into the *Xho* I-*Kpn* I site of pG418 to generate pG418 [sp-6×His-EGT-linker-GFP]. The pG418 fungal expression vector was kindly provided by Professor Yi Zou of the College of Pharmaceutical Sciences at Southwest University (Chongqing, China).

2.3. Construction and Selective Culture of Transgenic B. bassiana

The protoplast transformation method has been widely applied in the genetic transformation of fungi. Transgenic *B. bassiana* were generated as previously described [28], the details of which are following. *B. bassiana* colonies grown on PDA medium for 14 days were collected using a sterile 20% glycerol solution and the mycelia were filtered to obtain a spore suspension. The spore suspension was cultured at 25 °C, 220 *rpm* shake culture for 12 h, and obtained 70~80% of the spore germinated buds (the length of the buds was approximately 3~4 times the diameter of the conidia). The germinated spore buds in a 50 mL culture medium were collected and washed twice using 15 mL OM buffer (The OM buffer, including 1.2 3M MgSO$_4$·7H$_2$O and 10 mM Na–Phosphate buffer, was filter sterilized and stored at 4 °C. and the pH was adjusted to 5.8 using 1 M Na$_2$HPO$_4$) via centrifugation at 4 °C and 4000× *g* for 5 min. The 10 mM Na–Phosphate buffer was prepared from a 2 M NaPB stock solution containing 90.9 g of Na$_2$HPO$_4$ and 163.4 g of NaH$_2$PO$_4$ per liter, with a pH of 6.5. Next, 10 mL mixed enzyme solution of 0.03 g Lysing Enzymes from *Trichoderma harzianum* (Sigma-Aldrich, St. Louis, MO, USA) and 0.02 g Yatalase (Takara Bio, Otsu, Japan) dissolved in 10 mL OM buffer were added to digest the new buds. Protoplasts of *B. bassiana* were obtained via culturing at 28 °C and 80 *rpm* shake culture for 13 h. Then, 10 mL of trapping buffer (0.6 M sorbitol; 0.1 M Tris-HCl, pH 7.0; autoclaved and stored at 4 °C) were added and protoplast layers were collected by centrifugation at 4 °C and 4000× *g* for 20 min. Then, after being added to 10 mL STC buffer (1.2 M sorbitol; 10 mM CaCl$_2$·2H$_2$O; 10 mM Tris-HCl, pH 7.5; autoclaved and stored at 4 °C), the precipitate was obtained by centrifugation at 4 °C and 4000× *g* for 8 min and resuspended in 300 μL STC buffer. The 10 μL transgenic fungal expression vector pG418-sp-His-EGT-GFP ultrapure plasmid was mixed with 100 µL protoplasts and allowed to stand on ice for 40 min. Then, 600 µL PEG solution (60% PEG 4000; 50 mM $CaCl_2$; 50 mM Tris-HCl, pH 7.5; autoclaved and stored at 25 °C) was added and mixed evenly, standing at 25 °C for 30 min. The mixture was added to an SD-PDA (PDA with 21.86% sorbitol) plate containing 500 µg/mL G418 sulfate solution prepared in advance, and the mixture was manually rolled and spread. Under 26 °C, RH ≥ 90%, the mixture was cultured for 1 day and then inverted for 4 days. A white colony was selected and continuously cultured on PDA plate containing 500 µg/mL G418 sulfate solution, purer transgenic strains were obtained after 20 generations of single-spore cultured on PDA plates containing 500 µg/mL G418 sulfate solution.

2.4. Identification of Transgenic B. bassiana

The target strain was scraped with a sterile cotton swab and inoculated into a sterile test tube containing 8 mL of PDB medium. After shaking the culture at 26 °C and 200 *rpm* shake culture for 3 days, the precipitate was collected via centrifugation at 4 °C and 12,000× *g* for 3 min. After adding 700 µL CTAB buffer (2% CTAB; 8.182% NaCl; 0.1 M Tris-HCl, pH 8.0; 0.02 M EDTA, pH 8.0) and transferring to 1.5 mL sterile centrifuge tube, the mycelium fluid was quickly frozen with liquid nitrogen for 2 min, followed by a 65 °C water bath for 2 min, and repeatedly freeze-thawed five times. Then, 600 µL of the liquid was piped into a new 1.5 mL sterile centrifuge tube and added to 600 µL phenol: chloroform: isoamyl alcohol (25:24:1, *V/V/V*). The 500-µL supernate was obtained via centrifugation at 13,000× *g* for 10 min and then 500 µL phenol: chloroform: isoamyl alcohol (25:24:1, *V/V/V*) was added. After, 400 µL of supernate was obtained via centrifugation at 13,000× *g* for 10 min and added to 800 µL precooled isopropanol, then allowed to stand at −20 °C for 40 min. After centrifugation at 13,000× *g* for 5 min, the precipitate was washed using 600 µL 70% ethanol, dried at room temperature, and dissolved in sterile water sequentially. The total RNA was extracted using the RNA extraction kit SteadyPure Universal RNA Extraction Kit (Accurate Biology, Changsha, Hunan, China), then first-strand cDNA was generated using reverse transcription kit NovoScript® Plus All-in-one 1st Strand cDNA Synthesis SuperMix (gDNA Purge) (Novoprotein, Suzhou, China), according to the manufacturer's protocol.

Polymerase chain reaction (PCR) for *egt* fragment from *B. bassiana* genomic DNA. The PCR profile was used to amplify target sequences: 94 °C for 5 min; 30 cycles of 94 °C for 30 s, 60 °C for 30 s and 72 °C for 1 min; 72 °C for 10 min. PCR for the full-length sp-His-EGT-GFP from *B. bassiana* cDNA was used to amplify target sequences: 94 °C for 5 min; 30 cycles of 94 °C for 30 s, 58 °C for 30 s and 72 °C for 2 min 40 s; 72 °C for 10 min. 18S rRNA was used as the control. The PCR profile was 94 °C for 5 min; 30 cycles of 94 °C for 30 s, 58 °C for 30 s and 72 °C for 20 s; 72 °C for 10 min. Electrophoresis confirmed products on a 1.2% (*w/v*) agarose gel. The primers involved in PCR are shown in Table S1.

2.5. Validity Detection of Transgenic Beauveria bassiana

The wild-type and transgenic *B. bassiana* were inoculated in a PDA medium containing effective antibiotics, 26 °C, RH ≥ 90%, and cultured for 14 days. Conidia were scraped and dispersed in 0.05% Tween-80 solution to obtain a spore suspension with 1×10^7 conidia/mL concentration. Silkworm larvae were selected as the test insects on the first day of the second instar, the second day of the second instar, and the first day of the third instar. Each group comprised 300 silkworms of similar size. The spore suspension was evenly sprayed on small silkworms to observe the moisture on the body surface. A 0.05% Tween-80 solution without spore was used as a negative control. After 30 min of inoculation, mulberry leaves were fed to the silkworms, the temperature was raised to 26 °C, and the RH ≥ 90%. Silkworm excrement was cleaned, and fresh mulberry leaves were replaced daily. The death of each treatment was observed and recorded at 12 h intervals, and the median time to death of the test insects was counted prior to the death of wild-type and transgenic *B. bassiana*-infected silkworm. The transgenic *B. bassiana* treatment group was the experimental group, and the wild-type *B. bassiana* treatment group was the control group. Fresh *B. bassiana* with white villous hyphae and powdery spore on silkworm carcasses were selected, and the hyphae and conidia were collected. The genome and RNA were extracted and detected by PCR, as described above.

2.6. Data Collection and Visualization

Microsoft Excel was used to organize the raw data and graph the survival curves and mortality rates.

3. Results

3.1. Screening of Antibiotics and Determination of Culture Conditions

To screen and culture positive transgenic *B. bassiana*, sensitive antibiotics and the optimal concentration of *B. bassiana* were screened using the PDA medium. First, the activated *B. bassiana* conidia were inoculated on PDA solid medium with 50, 100, and 200 μg/mL of Hygromycin B, Zeocin, and G418 sulfate solution, respectively. After two days of culture, *B. bassiana* was grown on a medium containing 200 μg/mL Hygromycin B and Zeocin. The growth of *B. bassiana* was inhibited in the medium containing the G418 sulfate solution, and the higher the concentration, the more pronounced the inhibition (Figure 1A). Furthermore, *B. bassiana* was cultured on PDA solid medium containing 300 μg/mL, 400 μg/mL, and 500 μg/mL G418 sulfate solution for 4 days; the 500 μg/mL G418 sulfate solution could inhibit the growth of *B. bassiana* on PDA medium (Figure 1B). The above results showed that within 4 d, PDA medium containing 500 μg/mL G418 sulfate solution could be used as a screening medium for the genetic transformation of *B. bassiana* and positive transgenic strains of transformed offspring.

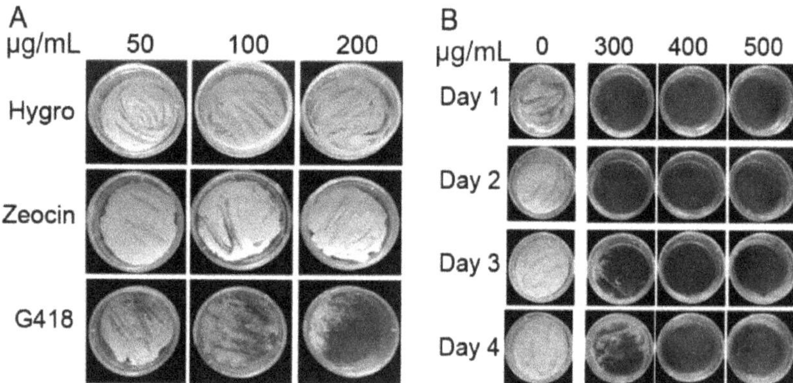

Figure 1. Screening of antibiotics and culture conditions of *Beauveria bassiana*. (**A**) The growth of *B. bassiana* on potato dextrose agar (PDA) solid medium with different concentrations of Hygromycin B, Zeocin, and G418 sulfate solution; (**B**) The growth of *B. bassiana* on PDA solid medium supplemented with different concentrations of G418 sulfate solution.

3.2. Generation of Transgenic Beauveria bassiana

The recombinant fungal expression vector pG418-sp-His-EGT-GFP driven by the gpdA constitutive promoter was digested with the restriction enzymes *Xho* I and *Kpn* I (Figure 2A). The size of the vector was 6365 bp, whereas the size of the target fragment was 2286 bp, consistent with the expected results (Figure 2B). This confirmed the successful construction of the transgenic fungal expression vector pG418-sp-His-EGT-GFP.

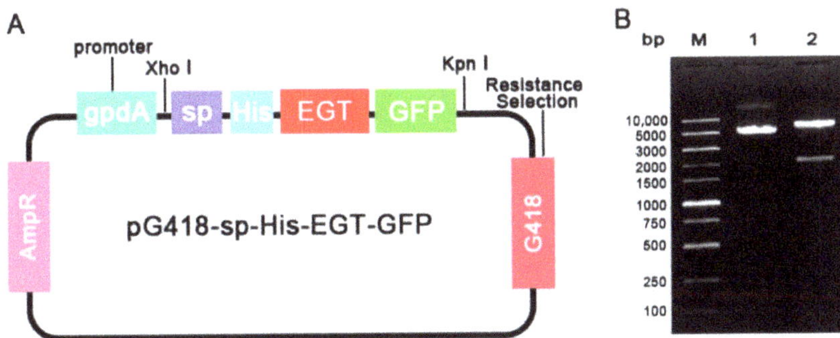

Figure 2. Construction of transgenic fungal expression vector. (**A**) Schematic illustration of the transgenic fungal expression vector pG418-sp-His-EGT-GFP; (**B**) The detection of recombinant fungal expression vector by double enzymes digestion.

Using lysing enzymes from *T. harzianum* and Yatalase, the newly germinated hyphae of *B. bassiana* were enzymatically digested to obtain protoplasts. Successfully digested protoplasts appeared as transparent circular structures (Figure 3A). The protoplasts were then transformed with the pG418-sp-His-EGT-GFP highly purified plasmid and uniformly spread on the corresponding culture media. After 4 days of cultivation, the culture media without antibiotic selection showed the growth of white mycelia (Figure 3(B1)), whereas no colony growth was observed in the culture media containing 500 µg/mL G418 sulfate solution without the plasmid (Figure 3(B2)). However, on G418 sulfate-containing PDA plates with plasmid transfer, single white fungal colonies emerged (Figure 3(B3)), indicating that the antibiotic effectively inhibited the growth of non-transgenic *B. bassiana*. Five randomly selected single fungal colonies were cultivated on G418 sulfate-containing PDA plates for 3 days. The transformed colonies showed powdery white growth with a radially symmetrical pattern (Figure 3B′). The genomic DNA of these five transformants were extracted, and PCR analysis revealed the amplification of an *egt* fragment of approximately 1000 bp in transformant B′3–5 (Figure 3C). Furthermore, RNA extracted from transformant B′3-5 confirmed the presence of 18s rRNA (Figure 3D) and the successful amplification of a target gene of approximately 2300 bp (Figure 3E). These results demonstrated the successful generation of transgenic *B. bassiana* containing EGT.

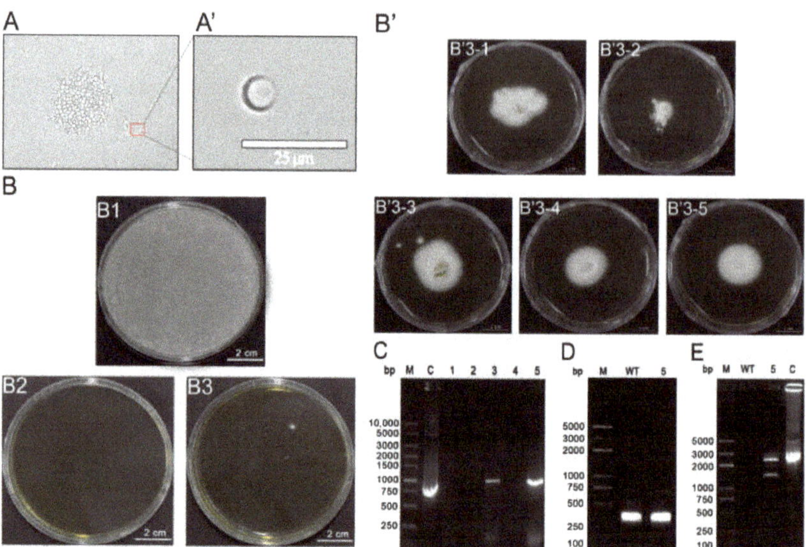

Figure 3. Generation and identification of transgenic *Beauveria bassiana*. (**A**) The protoplasts obtained from *B. bassiana*; (**A′**) Enlarged view of individual protoplast; (**B**) Screening of positive transgenic *B. bassiana*, (**B1**): the culture media without antibiotic selection, (**B2**): the culture media containing 500 µg/mL G418 sulfate solution without the plasmid, (**B3**): the culture media containing 500 µg/mL G418 sulfate solution with plasmid transfer; (**B′**) Cultivation of transformants, B′3 (1–5): five transformed colonies selected from B3, which is cultured for 3 days on PDA plates containing G418 sulphate; (**C**) 1–5: Gel electrophoresis of genomic PCR products of five transformants, C: PCR product of control plasmid pG418-sp-His-EGT-GFP; (**D**) Gel electrophoresis of 18S ribosomal RNA in transformant 5 RNA; (**E**) Gel electrophoresis of the target gene in transformant 5 RNA.

3.3. Validity Detection of Transgenic Beauveria bassiana

The median lethal time (LT50) of transgenic *B. bassiana* infecting silkworm larvae on the 1st and 2nd day of the second instar was shortened by 24 h compared to that of the wild-type (Figure 4A,B), while the LT50 of the transgenic *B. bassiana* was only 12 h shorter than that of the wild-type on the first day of the third instar larvae (Figure 4C). These results showed that the pathogenicity time of transgenic *B. bassiana* was shortened, and the effect on second-instar silkworms was more obvious. On the other hand, EGT has the function of hindering insect molting to take into consideration. We also collected the number of deaths before and after the molting of silkworms infected with *B. bassiana*. Comparing the inoculation of silkworms with *B. bassiana* on the first day with the second day of second instar larva, the number of silkworms that died before molting infected by transgenic *B. bassiana* was three times that of the wild-type (Figure 4(A1,A2)). On the second day of inoculation wild-type *B. bassiana*, all the silkworms died after molting, while there was still 20% of silkworms that died before molting infected by transgenic *B. bassiana* (Figure 4(B1,B2)). Even if inoculation has long period molting intervals for third instar silkworms, the number of silkworms infected by transgenic *B. bassiana* (96%) died before molting was still greater than those inoculated with wild-type *B. bassiana* (88%) (Figure 4(C1,C2)). This confirms that the transgenic *B. bassiana* increases in the availability of killed silkworms that die before molting.

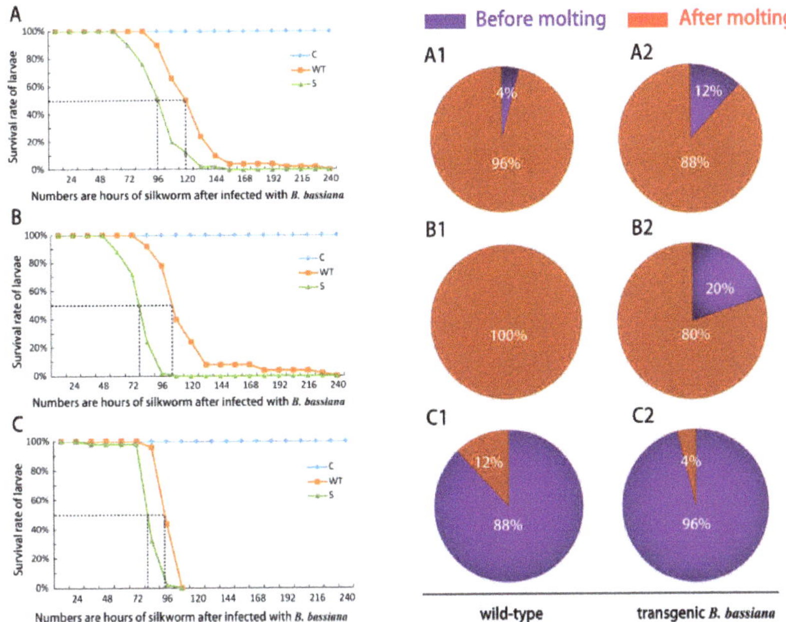

Figure 4. Validity detection of transgenic *Beauveria bassiana*. (**A**,**B**) The survival rate of silkworms inoculated with *B. bassiana* on the 1st/2nd day of the second instar, the ratio of death of silkworms before and after molting of the second instar inoculated with *B. bassiana* on the 1st day of the second instar of (**A1**,**A2**), and the ratio of death of silkworms before and after molting of the second instar of (**B1**,**B2**) inoculated with *B. bassiana* on the 2nd day of the second instar; (**C**) The survival rate of silkworms inoculated with *B. bassiana* on the first day of the third instar, the ratio of death of silkworms before and after molting of the third instar inoculated with *B. bassiana* on the first day of the third instar of (**C1**,**C2**). C: Negative control group treated with 0.05% Tween-80 solution; WT: control group treated with wild-type *B. bassiana*; S: Experimental group treated with transgenic *B. bassiana*.

Third-instar silkworms infected with *B. bassiana* died (108 h after inoculation), the phenotypes of 30 small silkworms were observed by photography. The proportion of small silkworms with mycelia on their body surface was approximately three times that of the control group. Some dead silkworms (Figure 5B) grew hyphae and conidia. Considering that the molting interval of the third instar silkworm (96 h) was longer than that of the second instar silkworm (48 h), *B. bassiana* could cause most silkworms to die before the third instar molting. Silkworms infected with transgenic *B. bassiana* exhibited greater ossification during the same period. The genome and RNA of *B. bassiana* from silkworm corpses were extracted for PCR detection. The results showed that the target gene fragment was successfully amplified in the experimental group with a size of approximately 1000 bp, consistent with the band size amplified from the control plasmid. In contrast, the control group was not amplified (Figure 5C). The difference observed in this study was due to the introduction of EGT into transgenic *B. bassiana*.

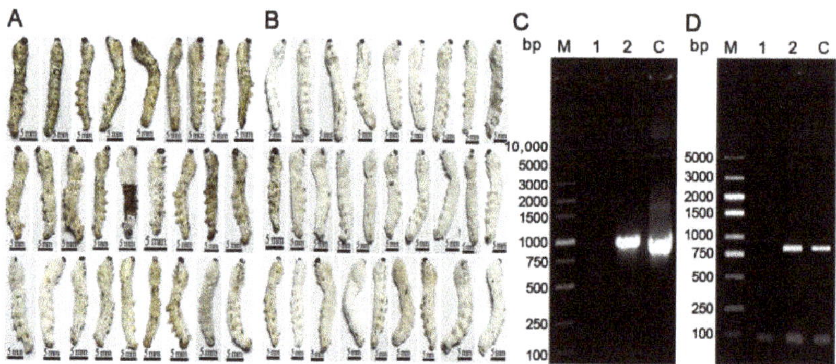

Figure 5. Pathogenicity analysis of transgenic *Beauveria bassiana*. (**A**) The phenotype of the third instar silkworms infected with wild-type *B. bassiana* after all death; (**B**) The phenotype of the third instar silkworms infected with transgenic *B. bassiana* at the same time; (**C**) Electrophoresis detection of PCR products of the target gene fragment in the genome of *B. bassiana*, 1: wild-type *B. bassiana* genome template, 2: transgenic *B. bassiana* genome template, C: control plasmid pG418-sp-His-EGT-GFP template; (**D**) Electrophoresis detection of PCR products of the target gene in *B. bassiana* RNA, 1: wild-type *B. bassiana* cDNA template, 2 transgenic *B. bassiana* cDNA template, C: control plasmid pG418-sp-His-EGT-GFP template.

4. Discussion

It should be noted that there are multiple methods for the genetic manipulation of fungal species [29]. Currently, there is no universal transformation method applicable to all fungi, and specific transformation protocols must be designed for different species. Therefore, it is of great practical significance to continuously explore and discover more efficient and suitable transformation methods for *B. bassiana* to improve the strains and facilitate future applications. In this study, we chose the protoplast transformation method because it is simple, effective, and does not require expensive equipment. Using protoplasts as recipient cells, this method produces many transformants and easily obtains homozygous transformants. However, this method inevitably has disadvantages, such as a low frequency of positive transformant regeneration and a long purification period. Additionally, the selection of positive strains using this method is commonly achieved through antibiotic resistance; however, different fungi have varying sensitivities to different antibiotics. To obtain positive strains efficiently and accurately, further exploration of plasmid resistance selection, antibiotic concentrations, and fungal culture methods is needed. In this study, we found that a solution of G418 sulfate at a concentration of 500 μg/mL completely inhibited the growth and reproduction of *B. bassiana* within 4 days. Therefore, it was used as a screening marker for the genetic transformation of *B. bassiana* and as a selective pressure for offspring after transformation. These findings provide a reference for selecting *B. bassiana* strains carrying other genes. However, it is necessary to explore further whether other antibiotics can make *B. bassiana* more sensitive.

Many genes have been introduced into *B. bassiana* to enhance its pathogenicity. Fungal infections of insects first penetrate the exoskeleton of insects, which is mainly composed of chitin and proteins. During the process of fungal infection of insect body surfaces, enzymes such as protease (Pr1A) [30] and Pr1A like proteases CDEP1 [31], and chitinase CHIT1 [32], are synthesized to dissolve insect body walls and aid in fungal penetration [2,12]. Therefore, increasing the expression of such proteins in fungi through transgenic transformation is one of the means to enhance their pathogenicity. In addition, some non-fungal endogenous proteins that have toxic effects on insects can be used to enhance fungal virulence such as the crystal insecticidal proteins (Vip3A) and Cry1Ac of *B. thuringiensis*, which can dissolve insect midgut epithelial cells, leading to cell lysis [33]. Scorpion neurotoxic peptide (ATM1) can cause instant spastic paralysis in insects [34]. The main goal of this study is to address the issue of short interval between molting of the early instar insect larvae, which leads to shedding of epidermis of *B. bassiana*-infected larvae after molting. Shedding of epidermis significantly affects the success of *B. bassiana* infection. Shedding of epidermis significantly affects the success of *B. bassiana* infection. Therefore, we utilized the characteristic of EGT to deactivate 20E [21,25,27] and expressed EGT in *B. bassiana* to inhibit the molting hormone in insects, thus extending the interval between molting of early instar larvae to improve the success rate of *B. bassiana* infection. Due to cost and limitations in experimental facilities, the conclusions of this study have only been validated in the laboratory. To confirm the effectiveness of larger-scale applications or the establishment of a relevant technological systems, further verification is needed by expanding the sample population in different or the same varieties of silkworms.

Lepidoptera is the second largest order in the class Insecta, with the most agricultural and forestry pests. Silkworms are commonly used as model insects for research on lepidopteran [35,36] and fungal infections [37]. In this study, although the improved genetically modified *B. bassiana* in this study theoretically has broad-spectrum biological control against Lepidoptera pests, its true effect still needs to be verified in other insects. In addition, it is noted that although the molting of larvae from almost all insects requires 20E to participate [38], the molting interval of late-stage larvae is inherently long (around 10 days for the last instar silkworms). During this time, the advantages of our genetically modified *B. bassiana* strain compared to the wild-type strain may weaken or even disappear. Therefore, regardless of whether it is the endogenous protein of the fungus or the exogenous protein, the extent of improvement is limited by the expression level of the target protein and the duration of application. In the future, genes that can enhance fungal infection efficiency (such as CDEP1 and CHIT1), genes toxic to insects (such as Vip3A and Cry1Ac proteins), and genes affecting insect molting (EGT) can be simultaneously expressed in *B. bassiana* or mixed in different strains of *B. bassiana* containing these proteins to enhance the insecticidal activity of *B. bassiana* against harmful insects and ultimately expand the application of *B. bassiana* in biological control.

5. Conclusions

Based on the characteristic of insect molting, we focused on "giving" *B. bassiana* a new function—deactivating 20E. This successfully caused the death of second instar silkworm larvae, which are usually difficult to infect with *B. bassiana* because of their short molting period. This approach partially compensates for the limitations of the long pathogenic cycle of *B. bassiana* in the application of pest biological control and provides insights into the improvement of biocontrol fungi and their efficacy against other insects.

Supplementary Materials: The following supporting information can be downloaded at: https://www.mdpi.com/article/10.3390/jof9100987/s1, Table S1: Primers used in the study.

Author Contributions: Conceptualization, G.S.; Data curation, P.L., Q.X., Y.L. and G.S.; Formal analysis, D.X., Y.L. and G.S.; Funding acquisition, P.L. and G.S.; Investigation, X.M., D.X. and D.L.; Methodology, Q.X. and G.S.; Project administration, Q.X.; Resources, Q.X. and Y.L.; Software, D.L., H.X. and L.H.; Validation, H.X. and L.H.; Writing—original draft, X.M., D.X., H.X., L.H. and P.L.; Writing—review & editing, X.M., D.L., P.L., Y.L. and G.S. All authors have read and agreed to the published version of the manuscript.

Funding: This research was funded by the Doctoral Fund Project of Southwest University, China, Grant No. SWU7110300034, the Fundamental Research Funds for the Central Universities, Southwest University, China, grant no. SWU-KR22013.

Acknowledgments: Thanks to Zou Yi, College of Pharmaceutical Sciences, Southwest University, and Fan Yanhua, Biotechnology Research Center, Southwest University, for their guidance and help in the cultivation and transformation of *Beauveria bassiana*.

Conflicts of Interest: The authors declare no conflict of interest.

References

1. Wang, C.; Fan, M.; Li, Z.; Butt, T.M. Molecular monitoring and evaluation of the application of the insect-pathogenic fungus *Beauveria bassiana* in southeast China. *J. Appl. Microbiol.* **2004**, *96*, 861–870. [CrossRef]
2. Clarkson, J.M.; Charnley, A.K. New insights into the mechanisms of fungal pathogenesis in insects. *Trends Microbiol.* **1996**, *4*, 197–203. [CrossRef] [PubMed]
3. Pell, J.K.; Hannam, J.J.; Steinkraus, D.C. Conservation biological control using fungal entomopathogens. *Biocontrol* **2010**, *55*, 187–198. [CrossRef]
4. Amsellem, Z.; Cohen, B.A.; Gressel, J. Engineering hypervirulence in a mycoherbicidal fungus for efficient weed control. *Nat. Biotechnol.* **2002**, *20*, 1035–1039. [CrossRef]
5. St Leger, R.; Joshi, L.; Bidochka, M.J.; Roberts, D.W. Construction of an improved mycoinsecticide overexpressing a toxic protease. *Proc. Natl. Acad. Sci. USA* **1996**, *93*, 6349–6354. [CrossRef] [PubMed]
6. Valero-Jimenez, C.A.; Wiegers, H.; Zwaan, B.J.; Koenraadt, C.J.; van Kan, J.A. Genes involved in virulence of the entomopathogenic fungus *Beauveria bassiana*. *J. Invertebr. Pathol.* **2016**, *133*, 41–49. [CrossRef]
7. Feng, M.G.; Poprawski, T.J.; Khachatourians, G.G. Production, pormulation and application of the entomopathogenic fungus *Beauveria Bassiana* for insect control—current status. *Biocontrol Sci. Technol.* **1994**, *4*, 3–34. [CrossRef]
8. Lecuona, R.E.; Turica, M.; Tarocco, F.; Crespo, D.C. Microbial control of *Musca domestica* (Diptera: Muscidae) with selected strains of *Beauveria bassiana*. *J. Med. Entomol.* **2005**, *42*, 332–336. [CrossRef]
9. Osherov, N.; May, G.S. The molecular mechanisms of conidial germination. *Fems Microbiol. Lett.* **2001**, *199*, 153–160. [CrossRef]
10. Fargues, J.; Smits, N.; Vidal, C.; Vey, A.; Vega, F.; Mercadier, G.; Quimby, P. Effect of liquid culture media on morphology, growth, propagule production, and pathogenic activity of the Hyphomycete, *Metarhizium flavoviride*. *Mycopathologia* **2002**, *154*, 127–138. [CrossRef]
11. Leopold, J.; Samsinakova, A. Quantitative estimation of chitinase and several other enzymes in fungus *Beauveria-Bassiana*. *J. Invertebr. Pathol.* **1970**, *15*, 34–42. [CrossRef]
12. Charnley, A.K. Fungal pathogens of insects: Cuticle degrading enzymes and toxins. *Adv. Bot. Res.* **2003**, *40*, 241–321. [CrossRef]
13. Wang, C.; Hu, G.; St Leger, R.J. Differential gene expression by Metarhizium anisopliae growing in root exudate and host (*Manduca sexta*) cuticle or hemolymph reveals mechanisms of physiological adaptation. *Fungal Genet. Biol.* **2005**, *42*, 704–718. [CrossRef] [PubMed]
14. Hajek, A.E.; Stleger, R.J. Interactions between fungal pathogens and insect hosts. *Annu. Rev. Entomol.* **1994**, *39*, 293–322. [CrossRef]
15. Gillespie, J.P.; Bailey, A.M.; Cobb, B.; Vilcinskas, A. Fungi as elicitors of insect immune responses. *Arch. Insect Biochem.* **2000**, *44*, 49–68. [CrossRef]
16. Kiguchi, K. Time-Table for the development of the silkworm, *Bombyx-Mori*. *JARQ-Jpn. Agr. Res. Q.* **1983**, *17*, 41–46.
17. Xing, D.X.; Shen, G.W.; Li, Q.R.; Xiao, Y.; Yang, Q.; Xia, Q.Y. Quality formation mechanism of stiff silkworm, *Bombyx* batryticatus using UPLC-Q-TOF-MS-based metabolomics. *Molecules* **2019**, *24*, 3780. [CrossRef]
18. Yan, Q.S.; Ouyang, X.G.; Li, C.B.; Pang, Y. Insect baculovirus-encoded ecdysteroid udp-glucosyltransferases. *Prog. Biochem. Biophys.* **1999**, *26*, 241–242.
19. Chaturvedi, P.; Mishra, M.; Akhtar, N.; Gupta, P.; Mishra, P.; Tuli, R. Sterol glycosyltransferases-identification of members of gene family and their role in stress in *Withania somnifera*. *Mol. Biol. Rep.* **2012**, *39*, 9755–9764. [CrossRef]
20. Oreilly, D.R.; Miller, L.K. A Baculovirus blocks insect molting by producing ecdysteroid udp-glucosyltransferase. *Science* **1989**, *245*, 1110–1112. [CrossRef]
21. Chaturvedi, P.; Misra, P.; Tuli, R. Sterol glycosyltransferases—The enzymes that modify sterols. *Appl. Biochem. Biotech.* **2011**, *165*, 47–68. [CrossRef] [PubMed]
22. Singh, G.; Tiwari, M.; Singh, S.P.; Singh, R.; Singh, S.; Shirke, P.A.; Trivedi, P.K.; Misra, P. Sterol glycosyltransferases required for adaptation of *Withania somnifera* at high temperature. *Physiol. Plantarum* **2017**, *160*, 297–311. [CrossRef] [PubMed]
23. Ramirez-Estrada, K.; Castillo, N.; Lara, J.A.; Arro, M.; Boronat, A.; Ferrer, A.; Altabella, T. Tomato udp-glucose sterol glycosyl-transferases: A family of developmental and stress regulated genes that encode cytosolic and membrane-associated forms of the enzyme. *Front. Plant Sci.* **2017**, *8*, 984. [CrossRef] [PubMed]
24. Pandey, V.; Dhar, Y.V.; Gupta, P.; Bag, S.K.; Atri, N.; Asif, M.H.; Trivedi, P.K.; Misra, P. Comparative interactions of withanolides and sterols with two members of sterol glycosyltransferases from Withania somnifera. *BMC Bioinform.* **2015**, *16*, 120. [CrossRef] [PubMed]
25. Evans, O.P.; O'Reilly, D.R. Purification and kinetic analysis of a baculovirus ecdysteroid UDP-glucosyltransferase. *Biochem. J.* **1998**, *332*, 807–808. [CrossRef]
26. Shikata, M.; Shibata, H.; Sakurai, M.; Sano, Y.; Hashimoto, Y.; Matsumoto, T. The ecdysteroid UDP-glucosyltransferase gene of Autographa californica nucleopolyhedrovirus alters the moulting and metamorphosis of a non-target insect, the silkworm, *Bombyx mori* (Lepidoptera, Bombycidae). *J. Gen. Virol.* **1998**, *79*, 1547–1551. [CrossRef]
27. Shen, G.W.; Wu, J.X.; Wang, Y.; Liu, H.L.; Zhang, H.Y.; Ma, S.Y.; Peng, C.Y.; Lin, Y.; Xia, Q.Y. The expression of ecdysteroid UDP-glucosyltransferase enhances cocoon shell ratio by reducing ecdysteroid titre in last-instar larvae of silkworm, *Bombyx mori*. *Sci. Rep.* **2018**, *8*, 17710. [CrossRef]
28. Reis, B.M.; Henning, V.M.; JoséLima, A.F.; Elíbio, R.; Augusto, S. High frequency gene conversion among benomyl resistant transformants in the entomopathogenic fungus *Metarhizium anisopliae*. *Fems Microbiol. Lett.* **2010**, *142*, 123–127.

29. Mishra, N.C.; Tatum, E.L. Non-mendelian Inheritance of DNA-Induced inositol independence in neurospora. *Proc. Natl. Acad. Sci. USA* **1973**, *70*, 3875–3879. [CrossRef]

30. Hu, G.; St Leger, J. Field studies using a recombinant mycoinsecticide (*Metarhizium anisopliae*) reveal that it is rhizosphere competent. *Appl. Environ. Microbiol.* **2002**, *68*, 6383–6387. [CrossRef]

31. Fang, W.G.; Leng, B.; Xiao, Y.H.; Jin, K.; Ma, J.C.; Fan, Y.H.; Feng, J.; Yang, X.Y.; Zhang, Y.J.; Pei, Y. Cloning of *Beauveria bassiana* chitinase gene Bbchit1 and its application to improve fungal strain virulence. *Appl. Environ. Microbiol.* **2005**, *71*, 363–370. [CrossRef] [PubMed]

32. Fang, W.G.; Feng, J.; Fan, Y.H.; Zhang, Y.J.; Bidochka, M.J.; Leger, R.J.S.; Pei, Y. Expressing a fusion protein with protease and chitinase activities increases the virulence of the insect pathogen *Beauveria bassiana*. *J. Invertebr. Pathol.* **2009**, *102*, 155–159. [CrossRef] [PubMed]

33. Qin, Y.; Ying, S.H.; Chen, Y.; Shen, Z.C.; Feng, M.G. Integration of insecticidal protein Vip3Aa1 into *Beauveria bassiana* enhances fungal virulence to spodoptera litura larvae by cuticle and per Os infection. *Appl. Environ. Microbiol.* **2010**, *76*, 4611–4618. [CrossRef] [PubMed]

34. Peng, G.X.; Jin, K.; Liu, Y.C.; Xia, Y.X. Enhancing the utilization of host trehalose by fungal trehalase improves the virulence of fungal insecticide. *Appl. Microbiol. Biotechnol.* **2015**, *99*, 8611–8618. [CrossRef]

35. Xia, Q.Y.; Wang, J.; Zhou, Z.Y.; Li, R.Q.; Fan, W.; Cheng, D.J.; Cheng, T.C.; Qin, J.J.; Duan, J.; Xu, H.F.; et al. The genome of a lepidopteran model insect, the silkworm *Bombyx mori*. *Insect Biochem. Mol. Biol.* **2008**, *38*, 1036–1045. [CrossRef]

36. Suetsugu, Y.; Futahashi, R.; Kanamori, H.; Kadono-Okuda, K.; Sasanuma, S.; Narukawa, J.; Ajimura, M.; Jouraku, A.; Namiki, N.; Shimomura, M.; et al. Large scale full-length cDNA sequencing reveals a unique genomic landscape in a lepidopteran model insect, *Bombyx mori*. *G3-Genes. Genom. Genet.* **2013**, *3*, 1481–1492. [CrossRef] [PubMed]

37. Matsumoto, Y.; Sekimizu, K. Silkworm as an experimental animal for research on fungal infections. *Microbiol. Immunol.* **2019**, *63*, 41–50. [CrossRef]

38. Qiuxiang, O.; Jones, K.K. What Goes Up Must Come Down: Transcription Factors Have Their Say in Making Ecdysone Pulses. In *Current Topics in Developmental Biology*; Shi, Y.B., Ed.; Elsevier: Morristown, NL, USA, 2013; Volume 103, Chapter 2; pp. 35–71.

Journal of
Fungi

Article

Antimicrobial and Antioxidant Activities of Endophytic Fungi Associated with *Arrabidaea chica* (Bignoniaceae)

Raiana Silveira Gurgel [1,2,†], Dorothy Ívila de Melo Pereira [1,2,†], Ana Vyktória França Garcia [2], Anne Terezinha Fernandes de Souza [2,3], Thaysa Mendes da Silva [2], Cleudiane Pereira de Andrade [1,2], Weison Lima da Silva [4], Cecilia Veronica Nunez [3,4], Cleiton Fantin [3,5], Rudi Emerson de Lima Procópio [2,3] and Patrícia Melchionna Albuquerque [1,2,3,5,*]

[1] Programa Graduate Program in Biodiversity and Biotechnology of the Bionorte Network, School of Health Sciences, Amazonas State University, Manaus 69050-010, Brazil; raianagurgel@hotmail.com (R.S.G.); dorothyivila@gmail.com (D.Í.d.M.P.); cleudiane.andrade@hotmail.com (C.P.d.A.)
[2] Research Group on Chemistry Applied to Technology, School of Technology, Amazonas State University, Manaus 69050-020, Brazil; avfg.geq19@uea.edu.br (A.V.F.G.); anne.fernandes13@gmail.com (A.T.F.d.S.); tms.geq18@uea.edu.br (T.M.d.S.); rprocopio@uea.edu.br (R.E.d.L.P.)
[3] Graduate Program in Biotechnology and Natural Resources of the Amazon, School of Health Sciences, Amazonas State University, Manaus 69050-010, Brazil; cvnunez@gmail.com (C.V.N.); cfantin@uea.edu.br (C.F.)
[4] Bioprospection and Biotechnology Laboratory, National Institute of Amazonian Research, Manaus 69067-375, Brazil; weisilva3@gmail.com
[5] Multicentric Graduate Program in Biochemistry and Molecular Biology, School of Health Sciences, Amazonas State University, Manaus 69050-010, Brazil
[*] Correspondence: palbuquerque@uea.edu.br
[†] These authors contributed equally to this work.

Abstract: The endophytic fungal community of the Amazonian medicinal plant *Arrabidaea chica* (Bignoniaceae) was evaluated based on the hypothesis that microbial communities associated with plant species in the Amazon region may produce metabolites with interesting bioactive properties. Therefore, the antimicrobial and antioxidant activities of the fungal extracts were investigated. A total of 107 endophytic fungi were grown in liquid medium and the metabolites were extracted with ethyl acetate. In the screening of fungal extracts for antimicrobial activity, the fungus identified as *Botryosphaeria mamane* CF2-13 was the most promising, with activity against *E. coli*, *S. epidermidis*, *P. mirabilis*, *B. subtilis*, *S. marcescens*, *K. pneumoniae*, *S. enterica*, *A. brasiliensis*, *C. albicans*, *C. tropicalis* and, especially, against *S. aureus* and *C. parapsilosis* (MIC = 0.312 mg/mL). Screening for antioxidant potential using the DPPH elimination assay showed that the *Colletotrichum* sp. CG1-7 endophyte extract exhibited potential activity with an EC_{50} of 11 μg/mL, which is equivalent to quercetin (8 μg/mL). The FRAP method confirmed the antioxidant potential of the fungal extracts. The presence of phenolic compounds and flavonoids in the active extracts was confirmed using TLC. These results indicate that two of the fungi isolated from *A. chica* exhibit significant antimicrobial and antioxidant potential.

Keywords: endophytes; Amazonian host; phenolic compounds; fungal metabolites; bioprospecting

Citation: Gurgel, R.S.; de Melo Pereira, D.Í.; Garcia, A.V.F.; Fernandes de Souza, A.T.; Mendes da Silva, T.; de Andrade, C.P.; Lima da Silva, W.; Nunez, C.V.; Fantin, C.; de Lima Procópio, R.E.; et al. Antimicrobial and Antioxidant Activities of Endophytic Fungi Associated with *Arrabidaea chica* (Bignoniaceae). *J. Fungi* **2023**, *9*, 864. https://doi.org/10.3390/jof9080864

Academic Editor: Baojun Xu

Received: 8 June 2023
Revised: 4 August 2023
Accepted: 7 August 2023
Published: 21 August 2023

1. Introduction

Popularly known in Brazil as crajiru, pariri and carajuru, among other names, *Arrabidaea chica* (Bonpl.) B. Verl. (1868) is a native species of the Amazon region that belongs to the Bignoniaceae family. It is characterized as a climbing shrub, and can reach 2 m in height [1]. Its astringent, emollient properties, and its red pigment, which is due to the presence of 3-deoxyanthocyanin (carajurine) [2,3], have been widely exploited in the production of cosmetics [4]. In addition, the species *A. chica* has pharmacological properties, such as antimicrobial [5], anti-inflammatory, antiangiogenic and antiproliferative [6], wound healing [7], antiparasitic [8] and antioxidant properties [9–11]. However, although it is widely studied and its biological activities have already been described, there are still no studies on the biotechnological potential of metabolites produced by the endophytic microorganisms associated with *A. chica*.

Endophytic microorganisms are those that live inside plant tissues without causing harm to the host species. They have the ability to interact with the plant at complex levels and, due to this interaction, plants can modulate the metabolic process of these endophytes to produce molecules that have protective functions in relation to the microbe and the host [12].

In recent decades, endophytic fungi have gained prominence as a rich source of natural compounds with interesting pharmacological activities [13–15]. Therefore, exploring endophytic fungi that inhabit medicinal plants provides ample opportunities to discover new metabolites with potential bioactivity [16–22]. In addition, investigations regarding endophytes from tropical plants are still limited, especially when considering the pharmacological potential of these isolates [23].

Endophytic fungi are an unlimited source of new metabolites, and the use of crude extracts from these microorganisms may be a promising alternative since its bioactive compounds can be produced on an industrial scale, thus contributing to both a reduction in cost of the final product and the preservation of plant species [11]. In this sense, endophytes from plants that grow in special ecological niches, such as in the Amazon biome, may have the ability to produce a myriad of secondary metabolites. The bioactive substances resulting from the secondary metabolism of these microorganisms directly contribute to the adaptation of species and their survival, and they are often produced in stress situations [13,24,25]. Flavonoids, alkaloids, steroids, terpenoids, isocoumarins and phenols are among the classes of substances produced by endophytic fungi that present numerous biological activities such as hormonal, antitumor, cytotoxic, antiviral, immunosuppressive, antiparasitic, antimicrobial and antioxidant activities, among others [15,21].

Thus, considering the environmental conditions under which *A. chica* lives in the Amazon rainforest, such as high humidity, constant rainfall and high temperatures, as well as the presence of parasites and natural competitors, it is assumed that the endophytic fungi resident in this plant have the ability to produce bioactive substances to protect its host. Therefore, in this study, we evaluated the production of antimicrobial and antioxidant secondary metabolites produced by the endophytic fungi isolated from leaves and branches of the Amazonian species *A. chica* and identified the most promising fungal species as new sources of bioactive metabolites. This is the first report on the bioactivity of the metabolites of endophytic fungi that inhabit the aerial parts of *A. chica*. Furthermore, this study contributes to the increase in knowledge regarding the biodiversity of the Amazon.

2. Materials and Methods

2.1. Reagents

The materials consisted of culture media, bacterial and fungal strains, commercial antibiotics as well as analytical grade reagents. Potato dextrose agar (PDA) and Mueller–Hinton broth were purchased from Kasvi (Kasvi, São José dos Pinhais, Brazil). Sabouraud broth and yeast extract were supplied by Himedia (Himedia, Thane, India). Microorganisms' strains were acquired from Cefar (Cefar Diagnóstica, Jardim Taquaral, Brazil). Levofloxacin, terbinafine and chloramphenicol were obtained from EMS (EMS pharma, Hortolândia, Brazil). Dextrose, methanol, ethyl acetate, $FeCl_3$, dimethyl sulfoxide (DMSO), 2,3,5-triphenyltetrazoic chloride (TTC) and Folin–Ciocalteu solution were purchased from Dinâmica (Dinâmica, Indaiatuba, Brazil). Resazurin, 2,2-diphenyl-1-picrylhydrazyl (DPPH), quercetin, ascorbic acid, Trolox and gallic acid were obtained from Sigma-Aldrich (Sigma-Aldrich, Saint Louis, MO, USA). 2,4,6-Tripyridyl-s-triazine (TPTZ) was supplied by Merck (Merck, Darmstadt, Germany).

2.2. Endophytic Fungi

The endophytic fungi used in this study were isolated from the aerial parts of three shrubs of *A. chica* (variety II), which were obtained in February 2019 from Embrapa Western Amazon, located on highway AM-010, KM 29 (Manaus–Itacoatiara highway). The plant exsiccate was deposited in the herbarium of the Amazonas Federal Institute (IFAM), under code EAFM2901.

Samples of branches and leaves from the three specimens were washed with detergent under tap water, fragmented into 10×12 cm pieces and subjected to a sequence of submersions in different solutions in the following order and times: (i) for the leaves, 70% alcohol for 1 min; sodium hypochlorite 3% for 2.5 min, 70% alcohol for 30 s and sterile distilled water for 2 min; (ii) for the branches, 70% alcohol for 1 min; sodium hypochlorite 4% for 3 min, 70% alcohol for 30 s and sterile distilled water for 2 min [26].

For the isolation of endophytic fungi, the plant material was cut into pieces of approximately 6 mm^2 and inoculated in Petri dishes (6 fragments in each plate) containing PDA with 50 mg/mL of chloramphenicol added. The plates were incubated at 26 °C for 15 days. According to the cultivable endophytes that had been grown, these were transferred to inclined test tubes containing PDA medium [26]. The isolated fungi were deposited in the Central Microbiological Collection (CCM) of the Amazonas State University (UEA). Access to genetic heritage was registered in the National System for the Management of Genetic Heritage and Associated Traditional Knowledge under code A0B4857.

One hundred and seven fungi, stored using the Castellani [27] method, were reactivated by inoculating a fragment of the stock culture in Petri dishes containing PDA, with subsequent incubation in a microbiological chamber (BOD) at 28 °C for 5–7 days. Table 1 presents information on the 107 endophytic fungi used in the present study for assessing their antioxidant and antimicrobial activity.

Table 1. Endophytic fungi isolated from the leaves and branches of the three specimens of *Arrabidaea chica* used in this study.

Endophytic Fungus Code				Specimen	Plant Part	Number of Isolates
CF1-1	CF1-12	CF1-22	CF1-31			
CF1-2	CF1-13	CF1-23	CF1-32			
CF1-3	CF1-15	CF1-24	CF1-34			
CF1-4	CF1-16	CF1-25	CF1-35			
CF1-5	CF1-17	CF1-26	CF1-36	1	Leaves	33
CF1-6	CF1-18	CF1-27	CF1-37			
CF1-7	CF1-19	CF1-28				
CF1-9	CF1-20	CF1-29				
CF1-11	CF1-21	CF1-30				
CG1-1	CG1-5	CG1-9	CG1-12			
CG1-2	CG1-7	CG1-10	CG1-14	1	Branches	11
CG1-4	CG1-8	CG1-11				
CF2-1	CF2-7	CF2-13	CF2-17			
CF2-2	CF2-9	CF2-14	CF2-18			
CF2-3	CF2-11	CF2-15	CF2-19	2	Leaves	16
CF2-6	CF2-12	CF2-16	CF2-20			
CG2-2	CG2-5	CG2-10	CG2-15			
CG2-3	CG2-7	CG2-11	CG2-16	2	Branches	11
CG2-4	CG2-8	CG2-12				

Table 1. *Cont.*

Endophytic Fungus Code				Specimen	Plant Part	Number of Isolates
CF3-1	CF3-9	CF3-15	CF3-21			
CF3-2	CF3-10	CF3-16	CF3-23			
CF3-4	CF3-11	CF3-17	CF3-24	3	Leaves	22
CF3-5	CF3-13	CF3-18	CF3-26			
CF3-6	CF3-12	CF3-19				
CF3-7	CF3-14	CF3-20				
CG3-1	CG3-8	CG3-13	CG3-18			
CG3-3	CG3-10	CG3-15	CG3-19			
CG3-4	CG3-11	CG3-16		3	Branches	14
CG3-7	CG3-12	CG3-17				
					Total	107

2.3. Production of Fungal Metabolites

Three fungal mycelium fragments (5 × 5 mm in diameter) that were removed from the PDA plates were inoculated into 250 mL Erlenmeyer flasks with 150 mL of liquid medium of the following composition: white potato broth (200 g/L); dextrose (10 g/L); yeast extract (2.0 g/L); NaCl (5.0 g/L); pH 5.0. The cultures were carried out under static conditions at 30 °C for 14 days, according to the methodology of Bose et al. [28], with modifications.

After cultivation, the metabolites were extracted with ethyl acetate in a 1:1 ratio for 4 h at room temperature and shaking at 120 rpm. After this period, the mycelium was removed via filtration in a Büchner funnel, and the fractions were separated in a separation funnel. The solvent was evaporated via the fume hood and the extracts were resuspended at a concentration of 10 mg/mL with a 10% DMSO solution and frozen (−18 °C) for later use in the biological tests [29].

2.4. Antimicrobial Testing

The microdilution technique was used according to the Clinical and Laboratory Standards Institute (CLSI) [30], which involved reducing resazurin for the antibacterial tests and reducing TTC for the antifungal tests. For the preliminary screening, fungal metabolites were tested against commercially acquired strains: *Staphylococcus aureus* CCCD-S009, *Escherichia coli* CCCD-E005 and *Candida albicans* CCCD-CC001. The extracts that showed activity against at least one of the two bacteria tested were evaluated against other strains: *Pseudomonas aeruginosa* CCCD-P004, *Proteus mirabilis* CCCD-P001, *Bacillus subtilis* CCCD-B005, *Staphylococcus epidermidis* CCCD-S010, *Enterococcus faecalis* CCCD-E002, *Serratia marcescens* CCCD-S005, *Klebsiella pneumoniae* CCCD-K003 and *Salmonella enterica* CCCD-S003. For the extracts that showed activity against *C. albicans*, these were also tested against *C. tropicalis* CCCD-CC002, *C. parapsilosis* CCCD-CC004 and *Aspergillus brasiliensis* CCCD-AA001.

For the test, 96-well microplates were used, which contained 100 µL of the extract at different concentrations (10, 5.0, 2.5, 1.25, 0.625 and 0.312 mg/mL) and 100 µL of the microbial inoculum. The microbial inoculum was prepared from colonies grown for 24 h. The microbial suspension was standardized at 0.5 on the McFarland Scale (10^8 CFU/mL) and diluted in the culture medium (Mueller–Hinton broth for the bacteria and Sabouraud broth for the fungi) until reaching 5×10^5 CFU/mL.

The positive control used for the bacteria was levofloxacin at 0.25 mg/mL and, for the fungi, terbinafine was used at 0.40 mg/mL. As a negative control, only the microbial inoculum was inserted and, for sterility control, 100 µL of the sterile culture medium that was used for the preparation of the inoculum was placed in the wells. A blank test was also performed, containing DMSO at different concentrations (1–100%), to ensure that the solvent used to dilute the extracts (DMSO 10%) did not present antimicrobial activity.

Subsequently, the plates were incubated at 37 °C for 24 h (bacteria) and 48 h (fungi) in a BOD chamber. After adding 30 µL resazurin at 0.01% or TTC at 1%, the plates were incubated again at 37 °C for 1–2 h to verify the change in color as a result of the reduction of the dyeing reagents.

For the extracts that showed activity, the minimum inhibitory concentration (MIC) was determined by successive dilutions of the samples. The lowest concentration of the extracts that inhibited microbial growth was considered the MIC.

2.5. Antioxidant Assays

Antioxidant activity was determined using the 2,2-diphenyl-1-picrylhydrazyl radical sequestration (DPPH•) method. The DPPH• solution was prepared at a concentration of 0.06 mmol/L, with methanol P.A., protected from exposure to direct light [31]. The assay was performed in 96-well microplates, with 40 μL of extract and the addition of 250 μL of the DPPH• solution. For the negative control, 40 μL of 10% DMSO and 250 μL of DPPH• solution were added [32]. The microplate was protected from exposure to direct light and, after 10 min, the absorbance readings were performed in a microplate reader (Molecular Devices, Spectramax Plus) at 517 nm. Experiments were performed in triplicate.

Fungal extracts were first tested at a single concentration of 10 mg/mL. Quercetin was used as the standard, at a concentration of 40 μg/mL. The percentage of sequestration of DPPH• radicals was calculated using Equation (1), using the values of the absorbance decay of the sample (Abs_{sample}) and the control ($Abs_{control}$):

$$AA\ (\%) = \frac{Abs_{control} - Abs_{sample}}{Abs_{control}} \times 100 \tag{1}$$

For the samples that showed activity (AA > 70%), the efficient concentration for the sequestration of 50% of the DPPH• radicals (EC_{50}) was determined, which was calculated from the successive dilutions of the samples and the generation of a linear regression graph. Extracts were evaluated in a concentration range from 10,000 to 4.88 μg/mL. Quercetin was tested from 100 to 3.125 μg/mL.

The antioxidant activity was also tested via the ferric reducing antioxidant power (FRAP) method, as described by Benzie and Strain [33] with modifications. The FRAP reagent consists of 100 mL of acetate buffer (0.3 mM), 10 mL of TPTZ (10 mM) and 10 mL of aqueous solution of ferric chloride (20 mM). To test the antioxidant activity, 2.45 mL of the FRAP reagent was incubated with 0.35 mL of the sample for 30 min, at 37 °C, under protection from light. The samples were then analyzed in a spectrophotometer UV–Vis (UV 1800, Shimadzu, Kyoto, Japan) at 595 nm. All experiments were performed in triplicate.

Samples of the extracts were tested at a single concentration of 10 mg/mL. A standard curve was constructed with Trolox in 10% DMSO. The results were expressed in μmol of Trolox equivalent per g of extract (μmol TE/g). Ascorbic acid was used as the standard, at a concentration of 40 μg/mL. DMSO 10% was used as a blank.

2.6. Chemical Profile of Fungal Extracts

The fungal extracts considered most promising in relation to biological activity tests were subjected to analysis via thin layer chromatography (TLC) in order to identify the main chemical classes present in the active metabolites. Then, 20 mg of the extract was solubilized in 2 mL of methanol. With the aid of capillaries, 2 μL of the samples were placed on a silica gel chromatographic plate (TLC aluminum sheets, Macherey-Nagel, 20 × 20 cm, silica gel 60 matrix, fluorescent indicator). Dichloromethane:acetone (9:1) was used as the mobile phase (10 mL volume).

To detect the chemical classes, ultraviolet light at 254 nm and 365 nm and the following chemical developers were used: *p*-anisaldehyde, ferric chloride, aluminum chloride and Dragendorff's reagent. To prepare the *p*-anisaldehyde developer, 0.5 mL of *p*-anisaldehyde was mixed into 10 mL of acetic acid, 85 mL of methanol and 5 mL of concentrated H_2SO_4. The ferric chloride was obtained by diluting 3 g of $FeCl_3$ in 100 mL of ethyl alcohol. Aluminum chloride was prepared with 1 g of $AlCl_3$ in 100 mL of ethyl alcohol. Dragendorff's reagent was prepared with 5 mL of Solution I (0.85 g of basic bismuth nitrate in 10 mL of glacial acetic acid added to 40 mL of distilled water) and 5 mL of Solution II (8 g of potassium iodide in 20 mL of distilled water), with the addition of 20 mL of acetic acid and distilled water to complete the volume to 100 mL [34].

2.7. Dosage of Total Phenolic Content

The samples that were considered most promising had their total phenolics measured using the Folin–Ciocalteu methodology, as described by Singleton and Rossi [35], with modifications. An aliquot of 0.25 mL of the fungal extract (10 mg/mL) and 2.75 mL of the 3% Folin–Ciocalteu solution were used. The samples were vortexed for 10 s, followed by 5 min of rest. After this period, 0.25 mL of the 10% Na_2CO_3 solution was added and the mixtures were incubated at room temperature, while protected from exposure to light for one hour. Subsequently, the absorbance was determined at 765 nm in a UV–Vis spectrometer (UV-1800, Shimadzu, Kyoto, Japan). A gallic acid solution at 200 mg/mL in ethanol was used to prepare the standard curve (0, 25, 50, 75, 100, 150 and 200 mg/mL) and ethanol was used as the blank. The gallic acid calibration curve was used to quantify the total phenolics. The results were expressed in equivalents of gallic acid per 100 mg of the extract (mg GAE/100 mg). All experiments were performed in triplicate.

2.8. Identification of the Most Promising Fungi

The fungi selected as being the most promising were identified by classical taxonomy, as well as using molecular tools. The macromorphological features were analyzed after the fungi were cultured during 7 days at 28 °C in Petri dishes (10 mm × 90 mm) containing PDA. The macroscopic vegetative characteristics, which were color, texture, topography, diffuse pigmentation, color, border and topography of the back of the colony were analyzed [36]. The micromorphology (hyphae and reproductive structures) was assessed using the microculture technique in PDA for 5–7 days. The macroscopic glass slides were stained with lactophenol blue and analyzed in an optical microscope (40×) [37]. The obtained results were compared with taxonomic keys [38,39].

The most promising fungi were also identified by sequencing the three DNA loci-internal transcribed spacer (ITS), β-tubulin (βtub) and calmodulin (CaM). The genomic DNA was extracted using the CTAB method [40], with modifications, following the protocol described by Oetari et al. [41], with modifications: DNA amplification via PCR (polymerase chain reaction) had a final reaction volume of 15 µL: 3 mM $MgCl_2$, 0.2 mM dNTPs, 1× buffer 10×, 0.2 mM forward primer, 0.2 mM reverse primer, 1 U Taq polymerase and 50 ng fungal genomic DNA. The same protocol was used for each of the primers used in this study: ITS1 and ITS4 [42], βtub3 and βtub4r [43], CaM-228F and CaM-737R [44]. The amplification conditions consisted of the following steps: denaturation for 5 min at 95 °C; annealing with 35 cycles at 95 °C, for 30 s, 35 cycles at 54 °C (CaM) or 62 °C (ITS and βtub), depending on the specific hybridization temperature for each primer, for 30 s, followed by 35 cycles at 72 °C, for 1 min and a final extension at 72 °C, for 5 min. The same cycling conditions were used for all primers. The amplified product was subjected to an electrophoretic run on 1.5% agarose gel to verify its efficiency.

The amplified PCR product was purified with PEG 8000 20% and the sequences were read in an automatic sequencer (ABI 3130xl Genetic Analyzer, Applied Biosystems, Thermo Fisher, Waltham, MA, USA). The sequences were manually checked, aligned, edited and analyzed with the help of Bioedit v.7.2.6 [45]. As the reference standards, we used the sequences deposited in GenBank (to obtain the preliminary identification) and the sequences with high similarity with the type specimens for the phylogenetic analysis. The sequences obtained were deposited in GenBank. For the construction of the genetic tree, the sequences of the combined loci (ITS + βtub + CaM) were aligned using the MAFFT program version 7 "https://mafft.cbrc.jp/alignment/software/ (accessed on 10 April 2023)" and phylogenetic analyses were conducted via the MEGA program version X [46], using the maximum-likelihood method, with 1000 bootstrap replicates.

2.9. Statistical Analysis

The experimental data obtained from antioxidant activity and total phenolic content were submitted to analysis of variance (ANOVA) for homogeneous samples; and to the Bonferroni test with a 95% confidence interval to distinguish the differences. The software used was BioEstat v. 5.0.

3. Results

3.1. Antimicrobial Activity of Fungal Extracts

The metabolic extracts of the 107 endophytic fungi isolated from *A. chica* were evaluated for antimicrobial activity. Of the total extracts evaluated, 18 showed activity against at least one of the microbial strains tested (Table 2). The fungus CF2-13, isolated from the *A. chica* leaves, was the most promising regarding the production of metabolites with antimicrobial activity. This fungus was able to inhibit the growth of Gram-negative (*E. coli*; *S. enterica, P. mirabilis, K. pneumoniae* and *S. marcescens*) and Gram-positive (*S. aureus, S. epidermidis* and *B. subtilis*) bacteria; as well as fungi (*C. albicans, C. tropicalis, C. parapsilosis* and *A. brasiliensis*).

Table 2. Minimum inhibitory concentrations (MICs, mg/mL) of the extracts of endophytic fungi from *Arrabidaea chica* that presented antimicrobial activity.

Endophytic Fungi Code	MIC (mg/mL)													
	EC	SA	CA	PA	PM	BS	SEp	EF	SM	KP	Sen	CT	CP	AB
CF1-2	5.00	-	-	-	-	5.00	-	-	-	-	NT	NT	NT	NT
CF1-26	5.00	-	-	-	-	-	-	-	-	-	5.00	NT	NT	NT
CF1-29	5.00	-	-	-	-	-	-	-	-	-	2.50	NT	NT	NT
CF1-37	-	5.00	1.25	-	-	-	-	-	-	-	-	1.25	2.50	-
CG1-10	-	5.00	-	-	-	5.00	-	5.00	-	-	2.50	NT	NT	NT
CG1-1	5.00	1.25	1.25	5.00	5.00	-	-	-	-	-	-	5.00	5.00	-
CG1-2	-	5.00	-	5.00	5.00	-	-	-	2.50	5.00	5.00	NT	NT	NT
CG1-5	5.00	5.00	-	-	-	-	-	5.00	-	-	5.00	NT	NT	NT
CG1-8	5.00	-	-	-	-	5.00	5.00	5.00	-	5.00	5.00	NT	NT	NT
CG1-9	5.00	-	-	-	-	-	5.00	5.00	-	-	-	NT	NT	NT
CF2-11	-	5.00	2.50	-	-	2.50	1.25	-	-	-	-	5.00	5.00	-
CF2-13	**2.50**	**0.312**	**1.25**	**-**	**5.00**	**2.50**	**1.25**	**-**	**5.00**	**5.00**	**2.50**	**1.25**	**0.312**	**2.50**
CF2-16	-	-	1.25	NT	NT	NT	NT	NT	NT	NT	NT	2.50	1.25	-
CG2-5	5.00	5.00	-	-	-	-	5.00	2.50	-	-	5.00	NT	NT	NT
CF3-5	-	-	1.25	NT	NT	NT	NT	NT	NT	NT	NT	2.50	1.25	-
CF3-9	-	5.00	-	NT	NT	NT	NT	NT	NT	NT	NT	NT	NT	NT
CF3-14	5.00	5.00	-	-	-	-	2.50	-	-	-	-	NT	NT	NT
CF3-26	-	-	5.00	NT	NT	NT	NT	NT	NT	NT	NT	5.00	5.00	-

EC = *Escherichia coli*; SA = *Staphylococcus aureus*; CA = *Candida albicans*; PA = *Pseudomonas aeruginosa*; PM = *Proteus mirabilis*; BS = *Bacillus subtilis*; SEp = *Staphylococcus epidermidis*; EF = *Enterococcus faecalis*; SM = *Serratia marcescens*; KP = *Klebsiella pneumoniae*; SEn = *Salmonella enterica*. CT = *Candida tropicalis*; CP = *Candida parapsilosis*; AB = *Aspergillus brasiliensis*; "-" = no antimicrobial activity; NT = Not tested. Levofloxacin was used as positive control for bacterial strains. MICs (µg/mL): EC = 0.05; SA = 0.50; PA = 0.50; PM = 0.25; BS = 0,50; Sep = 0.25; EF = 1.00; SM = 0.25; KP = 4.00; Sen = 0.25. Terbinafine was used as positive control for fungal strains MIC (µg/mL): CA = 6.25; CT = 6.25; CP = 0.125; AB = 25.0. The extract with the most promising antimicrobial activity is in bold.

3.2. Antioxidant Activity of Fungal Extracts

Of the fungal extracts evaluated at 10 mg/mL, 70 showed antioxidant activity (AA) over 70% and were considered active (Table 3). The ANOVA results indicated that there are significant differences ($p < 0.05$) between the mean values of AA. Of the 70 active extracts, 11 showed the highest AA values, between 97.75 and 100%, without statistical difference between them and when compared to the quercetin reference standard at 40 µg/mL ($p < 0.05$).

Table 3. Antioxidant activity of the extracts of endophytic fungi from *Arrabidaea chica*. Antioxidant activity (AA) obtained using the DPPH method, the efficient concentration for the sequestration of 50% of the DPPH• free radicals (EC_{50}), and the ferric reducing antioxidant power (FRAP).

Endophytic Fungi Code	AA *(%)	EC_{50} (µg/mL)	FRAP * (µmol TE/g)	Endophytic Fungi Code	AA * (%)	EC_{50} (µg/mL)	FRAP * (µmol TE/g)
CF1-3	98.61 [a]	5490	110.6 [B]	CF2-11	100.0 [a]	1080	128.5 [B]
CF1-4	100.0 [a]	6870	77.4 [C]	CF2-12	86.32 [b]	6450	93.6 [C]
CF1-7	92.38 [b]	1250	174.4 [A]	**CF2-13**	**92.47 [b]**	**360**	71.9 [C]
CF1-9	90.74 [b]	5530	143.2 [B]	CF2-14	90.22 [b]	1620	167.9 [A]
CF1-12	90.74 [b]	5270	95.2 [C]	**CF2-16**	94.81 [b]	1170	**218.6 [A]**
CF1-13	91.43 [b]	6480	**214.0 [A]**	CF2-17	83.03 [c]	6820	48.0 [C]
CF1-15	94.20 [b]	2710	171.0 [A]	CF2-18	70.74 [d]	7540	46.5 [C]
CF1-16	92.81 [b]	6190	72.4 [C]	CF2-20	82.68 [c]	1660	192.5 [A]
CF1-18	88.14 [b]	2840	178.6 [A]	CG2-2	95.06 [b]	3200	103.3 [B]
CF1-19	82.77 [c]	6590	47.5 [C]	CG2-4	91.95 [b]	1520	151.3 [A]
CF1-20	**100.0 [a]**	**990**	121.9 [B]	CG2-5	89.87 [b]	2720	51.0 [C]
CF1-23	94.20 [b]	5730	154.0 [B]	CG2-7	95.76 [b]	4710	59.7 [C]
CF1-24	89.70 [b]	3420	39.1 [C]	**CG2-10**	**98.01 [a]**	**740**	**191.2 [A]**
CF1-25	93.59 [b]	1060	188.4 [A]	CG2-12	95.75 [b]	4750	169.6 [A]
CF1-26	81.13 [c]	6720	177.3 [A]	CG2-16	89.78 [b]	3250	86.1 [C]
CF1-27	95.93 [b]	1460	125.2 [B]	CF3-1	95.06 [b]	2390	133.4 [B]
CF1-28	93.77 [b]	5020	181.7 [A]	CF3-4	90.04 [b]	6280	197.3 [A]
CF1-29	93.85 [b]	2260	109.1 [B]	**CF3-5**	**86.23 [c]**	**940**	**161.9 [B]**
CF1-30	88.83 [b]	3060	197.5 [A]	CF3-9	100.0 [a]	2480	142.3 [B]
CF1-31	100.0 [a]	1410	55.9 [C]	CF3-11	80.26 [b]	8310	53.2 [C]
CF1-36	88.23 [b]	2550	82.6 [C]	CF3-13	98.35 [a]	65,050	53.5 [C]
CF1-37	**93.68 [b]**	**680**	**171.1 [A]**	CF3-14	95.76 [b]	5400	152.8 [B]
CG1-1	90.56 [b]	2920	131.7 [B]	CF3-16	91.52 [b]	6100	158.2 [B]
CG1-2	90.39 [b]	3560	148.6 [B]	CF3-17	76.97 [c]	7610	56.9 [C]
CG1-4	89.78 [b]	1060	199.0 [A]	CF3-18	100.0 [a]	5650	175.8 [A]
CG1-5	92.38 [b]	5770	166.0 [A]	CF3-20	90.30 [b]	6250	148.5 [B]
CG1-7	**100.0 [a]**	**11**	109.5 [B]	CF3-21	85.89 [c]	6820	139.7 [B]
CG1-10	89.44 [b]	6420	44.7 [C]	CF3-26	88.48 [b]	3080	172.7 [A]
CG1-11	87.97 [b]	6660	128.2 [B]	CG3-3	80.00 [c]	5200	83.1 [C]
CG1-12	88.92 [b]	3700	89.5 [C]	CG3-4	90.04 [b]	4830	114.4 [B]
CG1-14	81.99 [c]	2880	154.8 [B]	**CG3-7**	**79.91 [c]**	**500**	**167.0 [A]**
CF2-2	75.06 [c]	6170	106.7 [B]	CG3-8	89.27 [b]	7300	79.9 [C]
CF2-6	80.43 [c]	5850	82.0 [C]	CG3-13	77.40 [c]	7610	169.3 [A]
CF2-7	78.61 [c]	4100	59.2 [C]	CG3-18	97.75 [a]	5670	158.5 [B]
CF2-9	94.98 [b]	5630	78.9 [C]	CG3-19	91.17 [b]	6210	96.2 [C]
Quercetin	**98.00 [a]**	**8**	NT	**Ascorbic acid**	NT	NT	163.1 [A]

* Assays carried out with the fungal extracts at a concentration of 10 mg/mL. Quercetin and ascorbic acid were tested at 40 µg/mL. NT = not tested. Results are expressed as means of the experiments in triplicate. Means that do not share a letter are significantly different ($p < 0.05$) according to the Bonferroni test. The extracts with the most promising antioxidant action are in bold.

Seven extracts showed an EC_{50} of <1000 µg/mL, which can be considered as promising in the case of crude extracts. The metabolites produced by the fungus CG1-7, isolated from the branches of *A. chica*, presented the lowest EC_{50} value, and were able to sequester 50% of the DPPH• free radicals at a concentration of 11 µg/mL. This result is comparable to the EC_{50} value found for the quercetin reference standard (8 µg/mL). The isolate CF2-13, selected as the best producer of antimicrobial substances, presented an EC_{50} of 360 µg/mL.

The fungal extracts that presented AA of >70% were also evaluated via the FRAP method, which was used as a second indicator of antioxidant activity. The FRAP results were obtained from a calibration curve ($y = 0.0034x - 0.0448$, $R^2 = 0.9927$) of Trolox (0–140 µM) and expressed in Trolox equivalents (TEs) per g of extract. The results of the

ANOVA indicated that there is significant difference between the means of FRAP results ($p < 0.05$).

The results of the FRAP assay corroborate the antioxidant potential of the metabolites produced by the endophytic fungi isolated from *A. chica*. Of the 70 extracts evaluated, 20 presented higher antioxidant power (166.0–218.6 µmol TE/g), without statistic difference between them and when compared to the ascorbic acid reference standard ($p < 0.05$). The isolates CF1-13 and CF2-16, both from *A. chica* leaves, presented the highest FRAP values (>200 µmol TE/g). However, these two endophytic fungi had an $EC_{50} > 1000$ µg/mL, i.e., low DPPH free-radical scavenging potential. The endophytes CF1-37, CG2-10, CF3-5 and CG3-7 presented promising results for both antioxidant methods (DPPH and FRAP) and should be further investigated in order to explore its production of metabolites since they may be new sources of antioxidant molecules. Thus, since the EC_{50} was considered the main parameter in our screening for antioxidant activity, the fungus CG1-7 was selected.

3.3. Chemical Profile of Promising Fungal Extracts

The extracts of the fungi CF2-13 and CG1-7 were analyzed using TLC in order to identify the chemical classes present in the active extracts. After staining the chromatographic plates with *p*-anisaldehyde, purple spots were observed in both extracts, which indicate the presence of terpenes; and a red intense spot in the CG1-7 extract, which indicates the presence of flavonoids (Figure 1a). The visualization of the chromatographic plates under UV light at 254 nm revealed spots that indicate the presence of conjugated double bonds (Figure 1b). The brown spot on the CF2-13 sample, when developed with ferric chloride, indicates the presence of phenolic compounds (Figure 1c) and the fluorescence, when the samples were treated with aluminum chloride and exposed to UV light at 365 nm, indicating the presence of flavonoids in both extracts (Figure 1d). The presence of alkaloids in the bioactive extracts was not detected in the TLC.

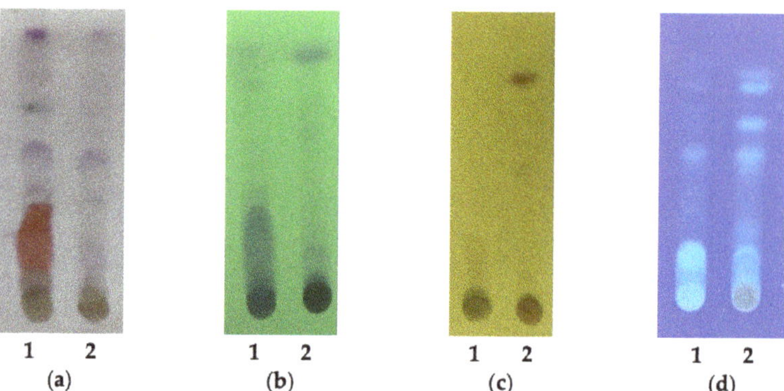

Figure 1. Thin layer chromatography of the metabolic extracts of endophytic fungi from *Arrabidaea chica* CG1-7 (1) and CF2-13 (2). (**a**) Stained with *p*-anisaldehyde, indicating the presence of terpenes (purple spots) and flavonoids (red spot); (**b**) UV light at 254 nm, indicating the presence of conjugated double bonds; (**c**) stained with ferric chloride, indicating the presence of phenolic compounds; (**d**) stained with aluminum chloride and exposure under UV light at 365 nm, indicating the presence of flavonoids.

3.4. Total Phenolic Content of the Most-Active Fungal Extracts

Phenolic compounds are important secondary metabolites with redox properties that are responsible for antioxidant activity. The total phenolic content was measured in the metabolic extracts produced by the endophytic fungi CF2-13 and CG1-7. The total phenolic results were obtained from a calibration curve ($y = 0.0035x - 0.0403$, $R^2 = 0.9901$) of gallic acid (0–100 µg/mL) and expressed in gallic acid equivalents (GAEs) per 100 mg of extract.

The content of phenolic compounds in bioactive extracts of CF2-13 and GC1-7 was found to be 3.70 and 5.28 mg GAE/100 mg of extract, respectively. The extract produced by CG1-7 presented a higher concentration of phenolic compounds than the extract produced by CF2-13 ($p < 0.05$), which is in accordance with the antioxidant activity results.

3.5. Identification of Endophytic Fungi That Produce Bioactive Substances

The fungi that produced the most-promising bioactive extracts were identified using classical taxonomy. The macro and micromorphological characteristics of the most-promising isolates can be seen in Figure 2.

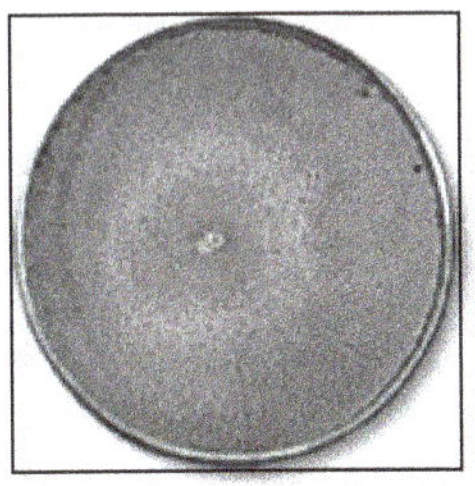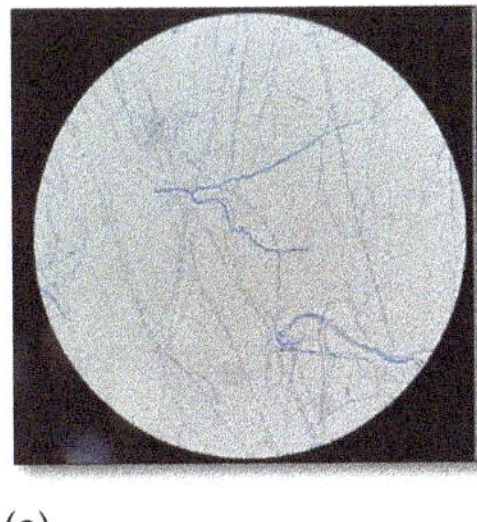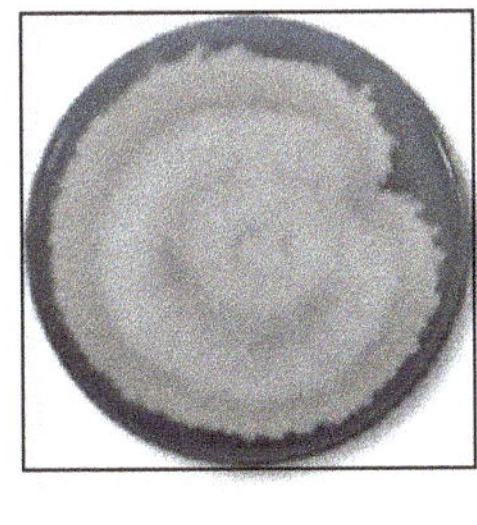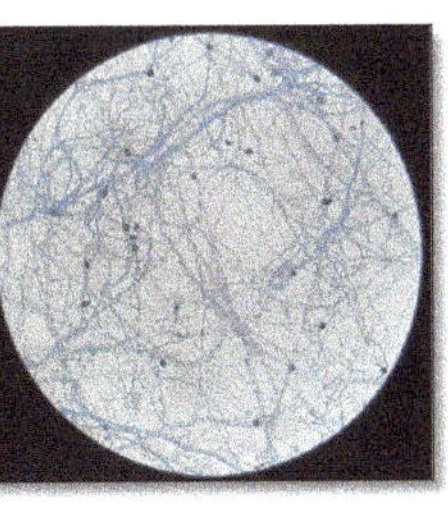

(a)   (b)

Figure 2. Macro- and micromorphological characteristics of endophytic fungi from *Arrabidaea chica* whose metabolic extracts showed the most-promising biological activities: antimicrobial-*Botryosphaeria mamane* CF2-13 (**a**); and antioxidant-*Colletotrichum* sp. CG1-7 (**b**).

The endophytic fungus CF2-13 isolated from the leaves of *A. chica* showed a colony of rapid vegetative growth that was cottony, irregular, flat, initially creamy-white in color, darkening with the aging of the inoculum to lead-gray, with a dark reverse side, moderately dense mycelia culture and slow sporulation. Micromorphological characteristics: septate, hyaline hyphae; ellipsoid, fusiform conidia with both ends straight, smooth-walled, hyaline, aseptate and scarce (Figure 2a).

The endophytic fungus CG1-7 isolated from the branches of *A. chica* presented a colony with fast vegetative growth that was cottony, irregular, flat, whitish to gray towards the center, with a dark reverse side, and dense mycelia culture, but slow sporulation. The micromorphological characteristics include septate, hyaline, thin and dense hyphae; cylindrical, fusiform conidia with both obtuse ends, smooth-walled, hyaline, aseptate and scarce; presence of appressoria (abundant) globose, some clavate, complex with irregular lobes, aseptate and brown in color (Figure 2b).

These fungal isolates were subjected to molecular identification using combined loci (ITS + βtub + CaM) to determine their identity. The isolate CF2-13, obtained from the leaves of *A. chica*, was identified with 99% of maximum likelihood as *Botryosphaeria mamane* (= *Cophinforma mamane*) from the combined analysis (ITS + βtub). However, for this isolate, there was no resolution for the calmodulin locus.

The fungus CG1-7, obtained from *A. chica* branches, was identified with 82% of maximum likelihood as *Colletotrichum siamense* from the combined analysis (ITS + βtub + CaM). Considering that the maximum likelihood was 82%, the fungal species was not confirmed and, therefore, other locus should be used for proper molecular identification.

Figures 3 and 4 show the result of the phylogenetic trees obtained for the isolates *B. mamane* CF2-13 and *Colletotrichum* sp. CG1-7, respectively, revealing the evolutionary history of the sequences analyzed using the maximum-likelihood method and the Tamur–Nei model. The sequences obtained in this study were deposited in GenBank (Table 4).

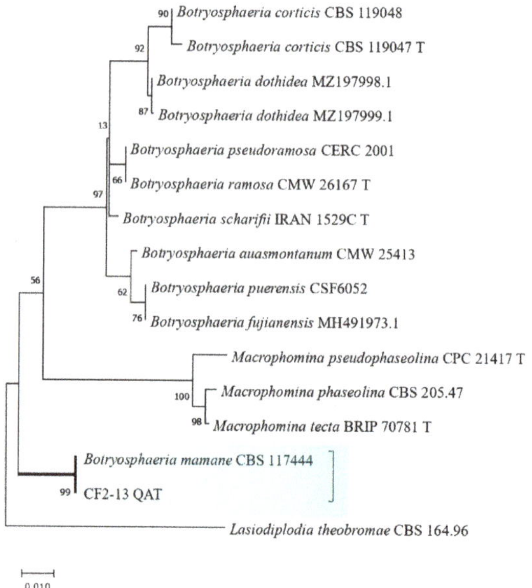

Figure 3. Combined phylogeny of the endophytic fungus CF2-13 isolated from *Arrabidaea chica* leaves using ITS and β-tubulin. The branch with the thicker line indicates the isolate that was sequenced in this study. The scale bar indicates nucleotide substitutions per site, using the neighbor-joining method via maximum-likelihood analysis. The numbers of the nodes indicate the bootstrap values of 1000 replicates. The tree was rooted in *Lasiodiplodia theobromae* CBS 164.96.

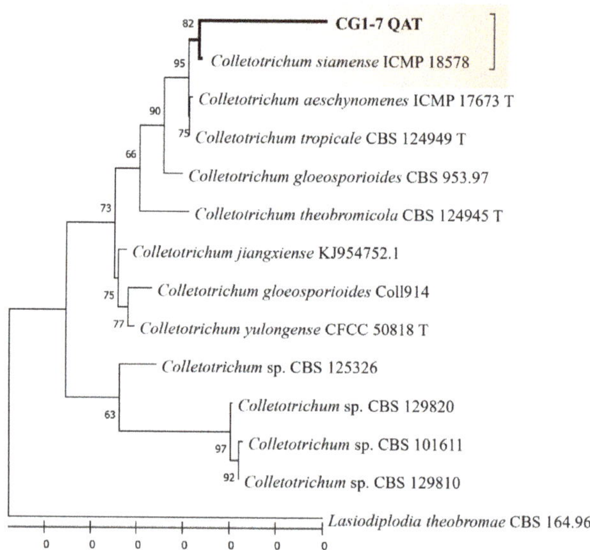

Figure 4. Combined phylogeny of the endophytic fungus CG1-7 isolated from *Arrabidaea chica* branches using ITS, β-tubulin and CaM. The branch with the thicker line indicates the isolate that was sequenced in this study. The scale bar indicates nucleotide substitutions per site, using the neighbor-joining method via maximum-likelihood analysis. The numbers of nodes indicate the bootstrap values of 1000 replicates. The tree was rooted in *Lasiodiplodia theobromae* CBS 164.96.

Table 4. GenBank accession numbers for the fungal isolates from *Arrabidaea chica* producing bioactive metabolites. Newly deposited sequences are shown in bold.

			GenBank Accession Number		
Isolate	**Species**	**Source**	**ITS**	**Btub**	**CaM**
CG1-7	*Colletotrichum* sp.	*A. chica*	**OQ390099**	**OQ412637**	OQ412636
CF2-13	*Botryosphaeria mamane*	*A. chica*	**OQ696843**	**OQ703591**	*-

* No resolution. ITS = internal transcribed spacer region. βtub = β-tubulin. CaM = calmodulin.

4. Discussion

The biological/biochemical role of endophytic fungi in a plant and how they interact with the host and with other endophytes and organisms associated with the plant species is still unclear [13,47]. However, the microbial diversity that different plant species harbor, together with the chemodiversity of the metabolites that endophytic fungi produce, offer the opportunity for the discovery of new bioactive molecules with different biotechnological applications [21]. In addition, several studies have demonstrated the usefulness of endophytic microorganisms in host survival, since endophytes directly influence plant metabolism in order to, for example, resist extreme temperatures and periods of drought, as well as the presence of phytopathogens [48]. Therefore, the traditional use of the plant and the region in which it inhabits are important criteria to be considered for the isolation of endophytes [49,50].

The species *A. chica* is widely cultivated in the Amazon and its leaves are traditionally used as astringents, genital disinfectants, in the treatment of inflammations, skin diseases and wound healing, intestinal colic, dysentery, leukorrhea and anemia [51], and it has proven antimicrobial and antioxidant activity [5,9,11]. Thus, taking into account that the need for new compounds with antimicrobial potential and antioxidant activity is inevitable, we studied the population of endophytic fungi of *A. chica* for its antimicrobial and antioxidant potential.

The antimicrobial activity of extracts of endophytic fungi isolated from *A. chica* was investigated against strains of bacteria and fungi known to be pathogenic to humans. Of the total number of extracts evaluated, 19% inhibited the growth of at least two pathogens. We noticed that the fungi isolated from the leaves of *A. chica* proved to be more promising regarding antimicrobial activity, when compared to endophytes isolated from the branches. Eleven isolates from the leaves produced antimicrobial metabolites, while only seven fungi from the branches showed activity. These results are in accordance with those observed by Santos et al. [50]. The authors found that endophytic fungi from the leaves of *I. suffruticosa* produced bioactive metabolites with antimicrobial activity.

It was observed that the extract of the fungus identified as *Botryosphaeria mamane* CF2-13 exhibited the best antimicrobial activity against the bacteria *S. aureus* and the yeast *C. parapsilosis* (MIC = 0.325 mg/mL). The species *B. mamane* was first described by Gardner [52] as a phytopathogen, associated with witches' brooms on *Sophora chrysophylla*, in Hawai. However, this species has also been isolated as an endophyte and some studies demonstrate the potential of its bioactive metabolites. *B. mamane* was isolated as an endophyte of *Garcinia mangostana* and showed promise as a producer of metabolites with antimicrobial activity against *S. aureus* and methicillin-resistant *S. aureus* [53], thus corroborating the results of the present study. In another study, Oliveira et al. [54] identified volatile organic compounds produced by the endophyte *B. mamane* isolated from plants collected in the Caatinga biome of northeastern Brazil. More recently, Triastuti et al. [55] used *B. mamane* isolated from the medicinal plant *Bixa orellana* L. to observe how histone diacetylase inhibitors alter the production of its secondary metabolites. This fungal species has been renamed *Cophinforma mamane* [56] (*Botryosphaeria mamane* D.E. Gardner = *Cophinforma mamane* (D.E. Gardner) A. J. L. Phillips and A. Alves) and is considered a rich source of new bioactive substances, though it is still poorly studied for biotechnological applications.

Other species of *Botryosphaeria* isolated as endophytes also have antimicrobial potential. Silva et al. [22] evaluated the antimicrobial activity of metabolites from *B. fabicerciana* that were isolated from *Morus nigra* L. and observed an MIC of 64 µg/mL for *S. aureus*, and an MIC of 1000 µg/mL for *E. coli*. Xiao et al. [57] identified 17 of the metabolites produced by *B. dothidea*. The substance pycnophorin, which is produced by the endophyte, inhibited the growth of *B. subtilis* and *S. aureus*, with an MIC of 25 µM; while stemphyperylenol demonstrated high antifungal activity against the phytopathogen *Alternaria solani* (MIC = 1.57 µM). It has been observed that fungi belonging to this genus produce exometabolites such as jasmonic acid and its derivatives, polyketides such as lasiodiplodin and isocofumarin, chaetoglobosins and alternariol analogues, among others, with potential bioactivity [57–60]. These findings justify the investigation of species of this genus as possible sources of antimicrobial-producing fungi and also corroborate the results presented in this work.

In the present study, *B. mamane* CF2-13 produced metabolites with promising antifungal activity, especially for yeasts of the genus *Candida*. *C. albicans* is the most common fungal pathogen in humans; it causes invasive infections and is a serious problem, especially in immunosuppressed patients. However, the epidemiology of fungal infections is evolving rapidly. Other *Candida* species have emerged as the main opportunistic pathogens, are associated with oral mucosa and have been identified as being commensal for a minority of healthy individuals [61]. The rise of the multidrug-resistant fungal pathogen *C. auris*, for instance, poses a global public health menace, and has gained significant attention for its swift and extensive proliferation in the last decade [62]. The anti-*Candida* activity of metabolites of the endophyte *B. mamane* CF2-13, therefore, is worthy of further investigation as a new source of antifungal substances. Additionally, these findings agree with what has been reported for the active metabolites of the host plant [5,63], and for its popular use [50]. According to Matos [64], indigenous tribes in the Amazon use a decoction of *A. chica* leaves for treating fungal infections and for cleaning chronic wounds.

Free radicals are known to induce oxidative damage to the body and, consequently, cause various disorders such as cancer, heart disease, Parkinson's, Alzheimer's, cataracts, diabetes mellitus, arthritis and premature aging [65,66]. Several studies have shown that both medicinal plants and their endophytes can be a potential source of molecules with antioxidant activity [2,14,20]. Some authors even suggest that the medicinal properties of the host plant may be a consequence of the capacity of its endophytic microorganisms to produce biologically active secondary metabolites [50]. The plant *A. chica* is known as a producer of antioxidant substances [9–11]. Our study demonstrates that *A. chica* endophytic fungi are also a promising source of antioxidant compounds.

The screening of extracts produced by *A. chica* endophytic fungi for free-radical scavenging activity via the DPPH• assay showed that, of the extracts evaluated, 65% have potential antioxidant activity, with AA > 70%. In terms of antimicrobial activity, the most promising metabolites for antioxidant activity were produced by isolates from *A. chica* leaves, when compared to the isolates obtained from the branches. These results indicate that there is a more pronounced production of bioactive secondary metabolites in fungi associated with the leaves of *A. chica*, and this could be explained by the metabolites that are produced in *A. chica* leaves. Siraichi et al. [9] identified the phenolic compounds isoscutellarein, 6-hydroxyluteolin, hispidulin, scutellarein, luteolin and apigenin in *A. chica* leaves, and attributed the significant antioxidant activity found in the extract obtained from its leaves to the presence of the mixture of flavonoids, with the main contribution being from scutellarein and apigenin.

The extract of the fungus identified as *Colletotrichum* sp. CG1-7 exhibited the strongest antioxidant activity, with the same potency obtained for the antioxidant reference standard quercetin. The extract of the fungus *B. mamane* GF2-13, selected as the most promising producer of antimicrobial substances, also presented radical scavenging action and showed itself to be a potential new source of bioactive compounds. The genus *Colletotrichum* is one of the most commonly isolated as an endophyte [13]. Despite being known to cause anthracnose in cereals, vegetables and fruit trees, fungi of this genus produce a variety of bioactive secondary metabolites. Molecules containing nitrogen, sterols, terpenes, pyrones, phenolics and fatty acids have been identified among the metabolites of these fungi [67]. Chithra et al. [68] isolated piperine, a substance considered a potent antioxidant, from the metabolites of the endophytic fungus *C. gloeosporioides* isolated from *Piper nigrum*. Tianpanich et al. [69] found that two isocoumarins produced by the endophytic fungus *Colletotrichum* sp. eliminated DPPH• free radicals (EC_{50} of 23.4 and 16.4 µM). On the other hand, contrary to what was observed in the present study with the isolate of *A. chica*, Mahmud et al. [70] observed the low antioxidant activity of the extract obtained from *C. siamense* fungus, which was isolated from *Justicia gendarussa*. The different methods of obtaining the fungal extract, as well as the different host, its location, climatic conditions and the isolation site, may explain the difference in the bioactivity of the metabolites of endophytic fungi.

Several studies have demonstrated the antioxidant potential of endophytic fungi. Budiono et al. [71] observed the antioxidant activity of metabolites from the endophytic fungi of *Syzygium samarangense* L., with the fungus *Lasiodiplodia venezuelensis*, which was isolated from leaves and was shown to be the most promising (EC_{50} = 49.96 µg/mL for the fungal extract). Alves et al. [72] evaluated the activity of endophytic fungi isolated from *Jatropha curcas* L. and the species *Aspergillus nidulans* presented highly promising metabolites in DPPH• sequestration, with an EC_{50} of 5.40 µg/mL. Druzian et al. [73] evaluated the extracts of metabolites produced by the species *Botryosphaeria dothidea* and obtained an EC_{50} of 206 µg/mL. Thus, it can be claimed that the endophytes of *A. chica* are promising sources of antioxidant substances, since seven isolates presented an EC_{50} of < 1000 µg/mL in their metabolic extracts.

The antioxidant potential of extracts from *A. chica* endophytic fungi was also evaluated using the FRAP method. The fungal extracts showed themselves to be able to convert Fe^{3+}-tripyridyltriazine into Fe^{2+}-tripyridyltriazine. The isolates CG2-13 and CF2-16 showed promising FRAP results; however, their metabolites were not effective in the DPPH scavenging activity assay. Pulido et al. [74] suggest that the FRAP test is a useful tool for studying the antioxidant effectiveness of various extracts and pure substances. However, this test may not accurately reflect the process of radical elimination in lipid systems and may not correlate well with other measurements of antioxidant activity. Therefore, it is recommended to combine the FRAP test with other methods in order to better understand the dominant mechanisms of different antioxidants [75].

The DPPH test and FRAP test are often applied in antioxidant investigation, with sets of experiments linked to electrons or radical scavenging. They work based on the reduction process. The DPPH test is used to estimate the antioxidant activity based on the process through which antioxidants limit lipid oxidation, resulting in DPPH free-radical scavenging and therefore determining free-radical scavenging potential. The ferric reducing antioxidant method is performed on electron-transfer processes wherein a ferric salt is applied as an oxidant. The oxidation of ferric 2,4,6-tripyridyl-s-triazine to the colorful ferrous state is the reaction mechanism [76].

According to Aguirre et al. [77], some of the natural antioxidant compounds abundantly produced by endophytic fungi are from the class of phenolic compounds, such as flavonoids and phenolic acids, which corroborates the results found in this study. The TLC results indicate the presence of these classes of molecules in the extracts produced by the two most promising isolates. The studies by Kaur et al. [78], for example, also observed the presence of phenolic compounds and flavonoids in the metabolic extracts of the endophytic fungus *Aspergillus fumigatus* isolated from *Moringa oleifera*, which showed pronounced antioxidant activity against DPPH• radicals (EC_{50} = 40.07 µg/mL).

The indications of the chemical classes observed in the extracts through the TLC analyses were dictated by the culture medium used in the growth of the fungi. The potato dextrose medium is a rich source of glucose, which in turn is a fundamental substrate in the biochemical pathways of shikimate and acetyl-CoA pathways that participate in the formation of terpenes and phenolic substances [79]. Another factor that enables the production of secondary metabolites was the addition of yeast extract to the medium, since this supplement is rich in vitamin B2, a component that participates in the synthesis of different secondary metabolites [80].

Several studies show that some species, including *B. mamane*, which belongs to the genus *Botryosphaeria*, when cultivated in a culture medium with potato, produce phenolic molecules with different structures [81,82]. In the study performed by Oliveira et al. [54], the fungi of the Botryosphaeriaceae family, including *B. mamane*, was evaluated regarding the production of volatile substances such as terpenes. The authors found that, when using potato medium as a substrate, most of the substances produced by the fungi were sesquiterpenes. Similarly, species of the genus *Colletotrichum*, when cultivated in potato as the carbon source, present a secondary metabolism that is mainly conditioned to the production of terpenes and phenolic molecules [67,83]. In our study, potato broth was used to produce the secondary metabolites of *A. chica* endophytic fungi, and therefore, we also found the production of phenolic compounds and terpenes in the TLC analysis.

Microbial sources are rich in phenolic compounds and these metabolites can be obtained via controlled conditions, and at faster speeds than when obtained from plants, which is an advantage in terms of production costs [84]. The optimization of *A. chica* endophytic fungi cultivation, in order to improve the production of bioactive phenolic compounds, could provide a higher concentration of bioactive molecules, and, therefore, should be pursued in future experiments.

5. Conclusions

This study provides information on the endophytic fungi isolated from the Amazonian species *A. chica*. This is the first report on the bioactivity of the metabolites of endophytic fungi that inhabit the aerial parts of this medicinal plant. The secondary metabolites produced by the *A. chica* endophytic fungi may contain novel and unexplored bioactive compounds. The data obtained show that, of the 107 extracts of endophytic fungi evaluated for antimicrobial and antioxidant potential, the extract of *B. mamane* CF2-13 exhibited significant antimicrobial potential against Gram-positive and Gram-negative bacteria, as well as against fungi. The extract of the isolate *Colletotrichum* sp. CG1-7 had pronounced antioxidant activity, which was equivalent to the reference standard. Both active extracts showed the presence of flavonoids. Further studies aiming at the structural elucidation of the molecules present in bioactive extracts should therefore be carried out.

Author Contributions: Conceptualization, P.M.A., R.S.G. and D.Í.d.M.P.; investigation, R.S.G., D.Í.d.M.P., A.V.F.G., T.M.d.S., C.P.d.A., A.T.F.d.S. and W.L.d.S.; methodology: W.L.d.S., C.V.N., C.F. and R.E.d.L.P.; formal analysis, P.M.A., C.F. and R.E.d.L.P.; data curation, R.S.G., D.Í.d.M.P. and C.P.d.A.; validation, R.S.G. and D.Í.d.M.P.; writing—original draft preparation, R.S.G. and D.Í.d.M.P.; writing—review and editing, P.M.A.; project administration, P.M.A. and C.F.; resources: P.M.A., C.F., C.V.N. and R.E.d.L.P.; funding acquisition, P.M.A. and C.F. All authors have read and agreed to the published version of the manuscript.

Funding: This research was funded by Fundação de Amparo à Pesquisa do Estado do Amazonas (FAPEAM) (grants number 01.02.016301.00568/2021-05 and 062.00165/2020), by POSGRAD/FAPEAM 2022 and by Coordenação de Aperfeiçoamento de Pessoal de Nível Superior (CAPES) (finance code 001 and grant number 88881.510151/2020-01—PDPG Amazônia Legal). The APC was funded by Universidade do Estado do Amazonas (UEA).

Institutional Review Board Statement: Not applicable.

Informed Consent Statement: Not applicable.

Data Availability Statement: No new data were created or analyzed in this study. Data sharing is not applicable to this article.

Acknowledgments: The authors gratefully acknowledge Universidade do Estado do Amazonas-UEA, Instituto Nacional de Pesquisas da Amazônia-INPA, Embrapa Amazônia Ocidental, FAPEAM and CAPES for supporting this research.

Conflicts of Interest: The authors declare no conflict of interest. The funders had no role in the design of the study; in the collection, analyses, or interpretation of data; in the writing of the manuscript; or in the decision to publish the results.

References

1. Ferreira, M.G.R. *Crajiru (Arrabidaea chica Verlot)*; Embrapa: Porto Velho, Brazil, 2005; Available online: https://ainfo.cnptia.embrapa.br/digital/bitstream/item/24786/1/folder-crajiru.pdf (accessed on 11 October 2019).
2. Chapman, E.; Perkin, A.G.; Robinson, R. CCCCII—The colouring matters of Carajura. *J. Chem. Soc.* **1927**, *1927*, 3015–3041. [CrossRef]
3. Zorn, B.; Garcia-Pineres, A.; Castro, V.; Murillo, R.; Mora, G.; Merfort, I. 3-Desoxyanthocyanidins from *Arrabidaea chica*. *Phytochemistry* **2001**, *56*, 831–835. [CrossRef] [PubMed]
4. Barros, F.C.F.; Pohlit, A.M.; Chaves, F.C.M. Effect of substrate and cutting diameter on the propagation of *Arrabidaea chica* (Humb. & Bonpl.) B. Verl. (Bignoniaceae). *Rev. Bras. Plantas Med.* **2008**, *10*, 38–42.
5. Mafioleti, L.; da Silva Junior, I.F.; Colodel, E.M.; Flach, A.; Martins, D.T. Evaluation of the toxicity and antimicrobial activity of hydroethanolic extract of *Arrabidaea chica* (Humb. & Bonpl.) B. Verl. *J. Ethnopharmacol.* **2013**, *150*, 576–582. [CrossRef]
6. Michel, A.F.; Melo, M.M.; Campos, P.P.; Oliveira, M.S.; Oliveira, F.A.S.; Cassali, G.D.; Ferraz, V.P.; Cota, B.B.; Andrade, S.P.; Souza-Fagundes, E.M. Evaluation of anti-inflammatory, antiangiogenic and antiproliferative activities of *Arrabidaea chica* crude extracts. *J. Ethnopharmacol.* **2015**, *165*, 29–38. [CrossRef] [PubMed]
7. Aro, A.A.; Simões, G.F.; Esquisatto, M.A.M.; Foglio, M.A.; Carvalho, J.E.; Oliveira, A.L.R.; Gomes, L.; Pimentel, E.R. *Arrabidaea chica* extract improves gait recovery and changes collagen content during healing of the Achilles tendon. *Injury* **2013**, *44*, 884–892. [CrossRef]
8. Miranda, N.; Gerola, A.P.; Novello, C.R.; Ueda-Nakamura, T.; de Oliveira Silva, S.; Dias-Filho, B.P.; Hioka, N.; de Mello, J.C.P.; Nakamura, C.V. Pheophorbide a, a compound isolated from the leaves of *Arrabidaea chica*, induces photodynamic inactivation of *Trypanosoma cruzi*. *Photodiagn. Photodyn. Ther.* **2017**, *19*, 256–265. [CrossRef]
9. Siraichi, J.T.G.; Felipe, D.F.; Brambilla, L.Z.R.; Gatto, M.J.; Terra, V.A.; Cecchini, A.L.; Cortez, L.E.R.; Rodrigues-Filho, E.; Cortez, D.A.G. Antioxidant capacity of the leaf extract obtained from *Arrabidaea chica* cultivated in Southern Brazil. *PLoS ONE* **2013**, *8*, 72733. [CrossRef]
10. Martins, F.J.; Caneschi, C.A.; Vieira, J.L.F.; Barbosa, W.; Raposo, N.R.B. Antioxidant activity and potential photoprotective from amazon native flora extracts. *J. Photochem. Photobiol. B* **2016**, *161*, 34–39. [CrossRef]
11. Ribeiro, F.M.; Volpato, H.; Lazarin-Bidóia, D.; Desoti, V.C.; de Souza, R.O.; Fonseca, M.J.V.; Ueda-Nakamura, T.; Nakamura, C.V.; Silva, S.O. The extended production of UV-induced reactive oxygen species in L929 fibroblasts is attenuated by posttreatment with *Arrabidaea chica* through scavenging mechanisms. *J. Photochem. Photobiol. B* **2018**, *178*, 175–181. [CrossRef]
12. Kusari, S.; Hertweck, C.; Spiteller, M. Chemical ecology of endophytic fungi: Origins of secondary metabolites. *Chem. Biol.* **2012**, *19*, 792–798. [CrossRef] [PubMed]
13. Strobel, G. The emergence of endophytic microbes and their biological promise. *J. Fungi* **2018**, *4*, 57. [CrossRef] [PubMed]
14. Gurgel, R.S.; Rodrigues, J.G.C.; Matias, R.R.; Barbosa, B.N.; Oliveira, R.L.; Albuquerque, P.M. Biological activity and production of metabolites from Amazon endophytic fungi. *Afr. J. Microbiol. Res.* **2020**, *14*, 85–93. [CrossRef]
15. Omomowo, I.O.; Amao, J.A.; Abubakar, A.; Ogundola, A.F.; Ezediuno, L.O.; Bamigboye, C.O. A review on the trends of endophytic fungi bioactivities. *Sci. Afr.* **2023**, *20*, e01594. [CrossRef]
16. Strobel, G.A. Endophytes as sources of bioactive products. *Microbes Infect.* **2003**, *5*, 535–544. [CrossRef]
17. Banhos, E.F.; Souza, A.Q.; Andrade, J.C.; Souza, A.D.L.; Koolen, H.H.F.; Albuquerque, P.M. Endophytic fungi from *Myrcia guianensis* at the Brazilian Amazon: Distribution and bioactivity. *Braz. J. Microbiol.* **2014**, *45*, 153–161. [CrossRef]
18. Gouda, S.; Das, G.; Sen, S.K.; Shin, H.S.; Patra, J.K. Endophytes: A treasure house of bioactive compounds of medicinal importance. *Front. Microbiol.* **2016**, *7*, 1538. [CrossRef]
19. Nagarajan, D. In vitro antioxidant potential of endophytic fungi isolated from *Enicostemma axillare* (Lam.) Raynal. and *Ormocarpum cochinchinense* (Lour.) Merr. *J. Pharmacogn. Phytochem.* **2019**, *8*, 1356–1363.
20. Dhayanithy, G.; Subban, K.; Chelliah, J. Diversity and biological activities of endophytic fungi associated with *Catharanthus roseus*. *BMC Microbiol.* **2019**, *19*, 22. [CrossRef]
21. Fadiji, A.E.; Babalola, O.O. Elucidating mechanisms of endophytes used in plant protection and other bioactives with multifunctional prospects. *Front. Bioeng. Biotechnol.* **2020**, *8*, 467. [CrossRef]
22. Silva, A.A.; Polonio, J.C.; Bulla, A.M.; Polli, A.D.; Castro, J.C.; Soares, L.C.; Oliveira-Junior, V.A.; Vicentini, V.E.P.; Oliveira, A.J.B.; Gonçalves, J.E.; et al. Antimicrobial and antioxidant activities of secondary metabolites from endophytic fungus *Botryosphaeria fabicerciana* (MGN23-3) associated to *Morus nigra* L. *Nat. Prod. Res.* **2021**, *36*, 3158–3162. [CrossRef]

23. Batista, B.N.; Matias, R.R.; Oliveira, R.L.E.; Albuquerque, P.M. Hydrolytic enzyme production from açai palm (*Euterpe precatoria*) endophytic fungi and characterization of the amylolytic and cellulolytic extracts. *World J. Microbiol. Biotechnol.* **2022**, *38*, 30. [CrossRef] [PubMed]

24. Elawady, M.E.; Hamed, A.A.; Alsallami, W.M.; Gabr, E.Z.; Abdel-Monem, M.O.; Hassan, M.G. Bioactive metabolite from endophytic *Aspergillus versicolor* SB5 with anti-acetylcholinesterase, anti-inflammatory and antioxidant activities: In vitro and in silico studies. *Microorganisms* **2023**, *11*, 1062. [CrossRef] [PubMed]

25. Zhang, X.Y.; Li, X.; Han, M.M.; Cai, Z.Y.; Gao, X.; Pang, M.X.; Qi, J.H.; Wang, F. Separating and purifying of endophytic fungi from *Ginkgo biloba* and screening of flavonoid-producing strains. *IOP Conf. Ser. Earth Environ. Sci.* **2019**, *371*, 042052. [CrossRef]

26. Araújo, W.L.; Lima, A.O.S.; Azevedo, J.L.; Marcon, J.; Kuklinsky, S.J.; Lacava, P.T. *Manual: Isolamento de Microrganismos Endofíticos*; Universidade de São Paulo: Piracicaba, Brazil, 2002.

27. Castellani, A. Viability of some pathogenic fungi in distilled water. *J. Trop. Med. Hyg.* **1939**, *42*, 225–226.

28. Bose, P.; Gowrie, S.U.; Chathurdevi, G. Optimization of culture conditions of growth and production of bioactive metabolites by endophytic fungus—*Aspergillus tamarii*. *Int. J. Pharm. Biol. Sci.* **2019**, *9*, 469–478. [CrossRef]

29. Sharma, D.; Pramanik, A.; Agrawal, P.K. Evaluation of bioactive secondary metabolites from endophytic fungus *Pestalotiopsis neglecta* BAB-5510 isolated from leaves of *Cupressus torulosa* D. Don. *3 Biotech* **2016**, *6*, 210. [CrossRef] [PubMed]

30. CLSI. *Reference Method for Broth Dilution Antifungal Susceptibility Testing of Yeasts. CLSI Standard M27*; Clinical and Laboratory Standards Institute: Wayne, PA, USA, 2017.

31. Molyneux, P. The use of the stable free radical diphenylpicrylhydrazyl (DPPH) for estimating antioxidant activity. *Songklanakarin J. Sci. Technol.* **2004**, *26*, 211–219.

32. Duarte-Almeida, J.M.; Dos Santos, R.J.; Genovese, M.I.; Lajolo, F.M. Evaluation of the antioxidant activity using the b-carotene/linoleic acid system and the DPPH scavenging method. *Food Sci. Technol.* **2006**, *26*, 446–452. [CrossRef]

33. Benzie, I.F.F.; Strain, J.J. The Ferric Reducing Ability of Plasma (FRAP) as a Measure of "Antioxidant Power": The FRAP Assay. *Anal. Biochem.* **1996**, *239*, 70–76. [CrossRef]

34. Lopes, J.L.C. Cromatografia em Camada Delgada. In *Fundamentos de Cromatografia*; Collins, C.H., Braga, G.L., Bonato, P.S., Eds.; Editora da Unicamp: Campinas, Brazil, 2006; pp. 67–86.

35. Singleton, V.L.; Rossi, J.A. Colorimetry of total phenolics with phosphomolybdic-phosphotungstic acid reagents. *Am. J. Enol. Vitic.* **1965**, *16*, 144–158. [CrossRef]

36. Watanabe, T. *Pictorial Atlas of Soil and Seed Fungi: Morphologies of Cultured Fungi and Key to Species*; CRC Press: Boca Raton, FL, USA, 2002; 504p.

37. Procop, G.W.; Church, D.L.; Hall, G.S.; Janda, W.S.; Koneman, E.W.; Schreckenberger, P.C.; Woods, G.L. *Koneman's Color Atlas and Textbook of Diagnostic Microbiology*; Jones & Bartlett Learning: Burlington, NJ, USA, 2016; 1830p.

38. Barnett, H.L.; Hunter, B.B. *Illustrated Genera of Imperfect Fungi*; Burgess Publishing Company: Minneapolis, MN, USA, 1972.

39. Hanlin, R.T.; Menezes, M. *Gêneros Ilustrados de Ascomicetos*; Imprensa da Universidade Federal Rural de Pernambuco: Recife, Brazil, 1996; 244p.

40. Doyle, J.; Doyle, J.L. Isolation of plant DNA from fresh tissue. *Focus* **1990**, *12*, 13–15.

41. Oetari, A.; Rahmadewi, M.; Rachmania, M.K.; Sjamsuridzal, W. Molecular identification of fungal species from deteriorated old chinese manuscripts in Central Library Universitas Indonesia. *AIP Conf. Proc.* **2018**, *2023*, 020122. [CrossRef]

42. White, T.J.; Bruns, T.; Lee, S.J.W.T.; Taylor, J. Amplification and direct sequencing of fungal ribosomal RNA genes for phylogenetics. In *PCR Protocols: A Guide to Methods and Applications*; Innis, M.A., Gelfand, D.H., Sninsky, J.J., White, T.J., Eds.; Academic Press: New York, NY, USA, 1990; pp. 315–322.

43. Glass, N.L.; Donaldson, G.C. Development of primer sets designed for use with the PCR to amplify conserved genes from filamentous ascomycetes. *Appl. Environ. Microbiol.* **1995**, *61*, 1323–1330. [CrossRef] [PubMed]

44. Carbone, I.; Kohn, L.M. A method for designing primer sets for speciation studies in filamentous ascomycetes. *Mycologia* **1999**, *91*, 553–556. [CrossRef]

45. Hall, T.A. BioEdit: A user-friendly biological sequence alignment editor and analysis, program for Windows 95/98/NT. *Nucleic Acids Symp. Ser.* **1999**, *41*, 95–98.

46. Kumar, S.; Stecher, G.; Li, M.; Knyaz, C.; Tamura, K. Mega X: Molecular Evolutionary Genetics Analysis across computing platforms. *Mol. Biol. Evol.* **2018**, *35*, 1547–1549. [CrossRef]

47. Suryanarayanan, T.S. Endophyte research: Going beyond isolation and metabolite documentation. *Fungal Ecol.* **2013**, *6*, 561–568. [CrossRef]

48. Xia, Y.; Liu, J.; Chen, C.; Mo, X.; Tan, Q.; He, Y.; Wang, Z.; Yin, J.; Zhou, G. The Multifunctions and future prospects of endophytes and their metabolites in plant disease management. *Microorganisms* **2022**, *10*, 1072. [CrossRef]

49. Verpoorte, R. Exploration of nature's chemodiversity: The role of secondary metabolites as leads in drug development. *Drug Disc. Today* **1998**, *3*, 232–238. [CrossRef]

50. Santos, I.P.; Silva, L.C.N.; Silva, M.V.; Aragão, J.M.; Cavalcanti, M.S.; Lima, V.L. Antibacterial activity of endophytic fungi from leaves of *Indigofera suffruticosa* Miller (Fabaceae). *Front. Microbiol.* **2015**, *6*, 350. [CrossRef]

51. Matias, J.N.; Souza, G.A.; Joshi, R.K.; Marqui, S.V.; Guiguer, E.L.; Araújo, A.C.; Otoboni, A.M.M.B.; Marineli, P.; Barbalho, M. *Arrabidaea chica* (Humb. And Bonpl.): A plant multipurpose medicinal applications. *Int. J. Herb. Med.* **2021**, *9*, 77–87.

52. Gardner, D.E. *Botryosphaeria mamane* sp. nov. associated with witches'-brooms on the endemic forest tree *Sophora chrysophylla* in Hawaii. *Mycologia* **1997**, *89*, 298–303. [CrossRef]
53. Pongcharoen, W.; Rukachaisirikul, V.; Phongpaichit, S.; Sakayaroj, J. A new dihydrobenzofuran derivative from the endophytic fungus *Botryosphaeria mamane* PSU-M76. *Chem. Pharm. Bull.* **2007**, *55*, 1404–1405. [CrossRef]
54. Oliveira, F.C.; Barbosa, F.G.; Mafezoli, J.; Oliveira, M.C.F.; Camelo, A.L.M.; Longhinotti, E.; Lima, A.C.A.; Câmara, M.P.S.; Gonçalves, F.J.T.; Freired, F.C.O. Volatile organic compounds from filamentous fungi: A chemotaxonomic tool of the Botryosphaeriaceae family. *J. Braz. Chem. Soc.* **2015**, *26*, 2189–2194. [CrossRef]
55. Triastuti, A.; Vansteelandt, M.; Barakat, F.; Trinel, M.; Jargeat, P.; Fabre, N.; Amasifuen, C.; Mejia, K.; Valentin, A.; Haddad, M. How histone deacetylase inhibitors alter the secondary metabolites of *Botryosphaeria mamane*, an endophytic fungus isolated from *Bixa orellana*, L. Chem. *Biodivers.* **2019**, *16*, e1800485. [CrossRef]
56. Phillips, A.J.L.; Alves, A.; Abdollahzadeh, J.; Slippers, B.; Wingfield, M.J.; Groenewald, J.Z.; Crous, P.W. The Botryosphaeriaceae: Genera and species known from culture. *Stud. Mycol.* **2013**, *76*, 51–167. [CrossRef] [PubMed]
57. Xiao, J.; Zhang, Q.; Gao, Y.-Q.; Tang, J.-J.; Zhang, A.-L.; Gao, J.-M. Secondary metabolites from the endophytic *Botryosphaeria dothidea* of *Melia azedarach* and their antifungal, antibacterial, antioxidant, and cytotoxic activities. *J. Agric. Food Chem.* **2014**, *62*, 3584–3590. [CrossRef]
58. Rukachaisirikul, V.; Arunpanichlert, J.; Sukpondma, Y.; Phongpaichit, S.; Sakayaroj, J. Metabolites from the endophytic fungi *Botryosphaeria rhodina* PSU-M35 and PSU-M114. *Tetrahedron* **2009**, *65*, 10590–10595. [CrossRef]
59. Linares, A.M.P.; Hernandes, C.; França, S.C.; Lourenço, M.V. Phytoregulatory activity of jasmonates produced by *Botryosphaeria rhodina*. *Hortic. Bras.* **2010**, *28*, 430–434. [CrossRef]
60. Xu, Y.; Lu, C.-H.; Zheng, Z.-H.; Shen, Y.-M. New polyketides isolated from *Botryosphaeria australis* strain ZJ12-1A. *Helv. Chim. Acta* **2011**, *94*, 897–902. [CrossRef]
61. Miceli, M.H.; Díaz, J.A.; Lee, S.A. Emerging opportunistic yeast infections. *Lancet Infect. Dis.* **2011**, *11*, 142–151. [CrossRef] [PubMed]
62. Du, H.; Bing, J.; Hu, T.; Ennis, C.L.; Nobile, C.J.; Huang, G. *Candida auris*: Epidemiology, biology, antifungal resistance, and virulence. *PLoS Pathog.* **2020**, *16*, e1008921. [CrossRef] [PubMed]
63. Torres, C.A.; Zamora, C.M.P.; Nuñez, M.B.; Gonzalez, A.M. In vitro antioxidant, antilipoxygenase and antimicrobial activities of extracts from seven climbing plants belonging to the Bignoniaceae. *J. Integr. Med.* **2018**, *16*, 255–262. [CrossRef]
64. Matos, F.J.A. *Plantas Medicinais no Brasil: Nativas e Exóticas*; Jardim Botânico Plantarum: Nova Odessa, Brazil, 2021.
65. Prior, R.L.; Wu, X. Diet antioxidant capacity: Relationship to oxidative stress and health. *Am. J. Biomed. Sci.* **2013**, *5*, 126–139. [CrossRef]
66. Putri, A.R.; Salni, S.; Widjajanti, H. Antioxidant activity of the secondary metabolites produced by endophytic fungi isolated from Jeruju (*Acanthus ilicifolius* L.) plant. *Biovalentia Biol. Res. J.* **2019**, *5*, 14–19. [CrossRef]
67. Kim, J.W.; Shim, S.H. The fungus *Colletotrichum* as a source for bioactive secondary metabolites. *Arch. Pharm. Res.* **2019**, *42*, 735–753. [CrossRef]
68. *Colletotrichum gloeosporioides* isolated from *Piper nigrum*. *Phytomedicine* **2014**, *21*, 534–540. [CrossRef]
69. Tianpanich, K.; Prachya, S.; Wiyakrutta, S.; Mahidol, C.; Ruchirawat, S.; Kittakoop, P. Radical scavenging and antioxidant activities of isocoumarins and a phthalide from the endophytic fungus *Colletotrichum* sp. *J. Nat. Prod.* **2011**, *74*, 79–81. [CrossRef]
70. Mahmud, S.M.N.; Sohrab, M.H.; Begum, M.N.; Rony, S.R.; Sharmin, S.; Moni, F.; Akhter, S.; Mohiuddin, A.K.M.; Afroz, F. Cytotoxicity, antioxidant, antimicrobial studies and phytochemical screening of endophytic fungi isolated from *Justicia gendarussa*. *Ann. Agric. Sci.* **2020**, *65*, 225–232. [CrossRef]
71. Budiono, B.; Elfita, E.; Muharni, M.; Yohandini, H.; Widjajanti, H. Antioxidant activity of *Syzygium samarangense* L. and their endophytic fungi. *Molekul* **2019**, *14*, 48–55. [CrossRef]
72. Alves, D.R.; Silva, W.M.B.; Santos, D.L.; Freire, F.C.O.; Vasconcelos, F.R.; Morais, S.M. Antioxidant, anticolinesterasic and cytoxic activities of endophytic fungus metabolites. *Braz. J. Dev.* **2020**, *6*, 73684–73691. [CrossRef]
73. Druzian, S.P.; Pinheiro, L.N.; Susin, N.M.B.; Dal Prá, V.; Mazutti, M.A.; Kuhn, R.C.; Terra, L.M. Production of metabolites with antioxidant activity by *Botryosphaeria dothidea* in submerged fermentation. *Bioprocess Biosyst. Eng.* **2019**, *43*, 13–20. [CrossRef] [PubMed]
74. Pulido, R.; Bravo, L.; Saura-Calixto, F. Antioxidant activity of dietary as determined by a modified ferric reducing/antioxidant power assay. *J. Agr. Food Chem.* **2000**, *48*, 3396–3402. [CrossRef] [PubMed]
75. Munteanu, I.G.; Apetrei, C. Analytical methods used in determining antioxidant activity: A review. *Int. J. Mol. Sci.* **2021**, *22*, 3380. [CrossRef]
76. Baliyan, S.; Mukherjee, R.; Priyadarshini, A.; Vibhuti, A.; Gupta, A.; Pandey, R.P.; Chang, C.-M. Determination of antioxidants by DPPH radical scavenging activity and quantitative phytochemical analysis of *Ficus religiosa*. *Molecules* **2022**, *27*, 1326. [CrossRef]
77. Aguirre, J.J.; De La Garza, T.H.; Zugasti, C.A.; Belmares, C.R.; Aguilar, C.N. The optimization of phenolic compounds extraction from cactus pear (*Opuntia ficus-indica*) skin in a reflux system using response surface methodology. *Asian Pac. J. Trop. Biomed.* **2013**, *3*, 436–442. [CrossRef]
78. Kaur, N.; Arora, D.S.; Kalia, N.; Kaur, M. Antibiofilm, antiproliferative, antioxidant and antimutagenic activities of on endophytic fungus *Aspergillus fumigatus* from *Moringa oleifera*. *Mol. Biol. Rep.* **2020**, *47*, 2901–2911. [CrossRef]

79. Santos, R.I. Metabolismo Básico e Origem dos Metabólitos Secundários. In *Farmacognosia—da Planta ao Medicamento*, 5th ed.; Simões, C.M.O., Schenkel, E.P., Gosmann, G., Mello, J.C.P., Mentz, L.A., Petrovick, P.R., Eds.; UFRGS/UFSC: Porto Alegre/Florianópolis, Brazil, 2003; pp. 403–434.
80. Dewick, P.M. *Medicinal Natural Products: A Biosinthetic Approach*; John Wiley & Sons: Hoboken, NJ, USA, 2011.
81. Venkatasubbaiah, P.; Sutton, T.B.; Chilton, W.S. Effect of phytotoxins produced by *Botryosphaeria obtusa*, the cause of black rot of apple fruit and frogeye leaf spot. *Phytopathology* **1991**, *81*, 243–247. [CrossRef]
82. Djoukeng, J.D.; Polli, S.; Larignon, P.; Abou-Mansour, E. Identification of phytotoxins from *Botryosphaeria obtusa*, a pathogen of black dead arm disease of grapevine. *Eur. J. Plant Pathol.* **2009**, *124*, 303–308. [CrossRef]
83. Chen, X.W.; Yang, Z.D.; Sun, J.H.; Song, T.T.; Zhu, B.Y.; Zhao, J.W. Colletotrichine A, a new sesquiterpenoid from *Colletotrichum gloeosporioides* GT-7, a fungal endophyte of *Uncaria rhynchophylla*. *Nat. Prod. Res.* **2018**, *32*, 880–884. [CrossRef] [PubMed]
84. Chandra, P.; Sharma, R.K.; Arora, D.S. Antioxidant compounds from microbial sources: A review. *Food Res. Int.* **2020**, *129*, 108849. [CrossRef] [PubMed]

Journal of
Fungi

Article

Marine-Derived Fungi as a Valuable Resource for Amylases Activity Screening

Di Zhang [1], Lan Liu [1,2] and Bi-Shuang Chen [1,2,*]

[1] School of Marine Sciences, Sun Yat-Sen University, Zhuhai 519080, China
[2] Southern Marine Science and Engineering Guangdong Laboratory (Zhuhai), Zhuhai 519080, China
* Correspondence: chenbsh23@mail.sysu.edu.cn; Tel.: +86-0756-3668775

Abstract: Marine microbial enzymes including amylases are important in different industrial production due to their properties and applications. This study was focused on the screening of marine-derived fungi for amylase activities. First, we isolated a number of fungi from the sediments of the South China Sea. By the method of dish screening (in vitro), we subsequently obtained a series of amylase-producing fungal strains. The cell-lysate activities of amylases produced by marine fungi toward starch hydrolysis were achieved with the dinitrosalyicylic acid (DNS) method. In addition, the effect of pH and temperature on amylase activities, including thermal and pH stability were discussed. Results showed that out of the 57 isolates with amylase-producing activities, fungi *Aspergillus flavus* 9261 was found to produce amylase with the best activity of 10.7482 U/mg (wet mycelia). The amylase of *Aspergillus flavus* 9261 exhibited remarkable thermostability and pH stability with no activity loss after incubation at 50 °C and pH 5.0 for 1 h, respectively. The results provide advances in discovering enzymes from marine-derived fungi and their biotechnology relevance.

Keywords: marine-derived fungi; amylase; starch hydrolysis; enzyme activity; thermal and pH stability

1. Introduction

The world's oceans cover more than 70% of Earth's area, and are considered to be a great reservoir of diverse microbial communities [1]. Marine microbial communities include bacteria, fungi, viruses, algae, plankton and etc. are essential ecological components in marine environments [2]. They usually play an important role in the biogeochemical process, for example, as intermediaries of energy, as decomposers of dead and decaying organic matter in nutrient regeneration cycles [3]. Thus, marine microorganisms have been attracting more and more attention as a resource for novel molecules and enzymes [4], due to their unique metabolic capabilities. Among marine microorganisms, fungi are a large group of eukaryotic and non-vascular organisms, and are widely distributed in marine environments [5]. Indeed, a great diversity of marine fungi have been recovered not only from seawater and sediment, invertebrates, decaying wood and mangrove detritus, but also as symbionts in marine lichens, plants, and algae [6].

Based on the different sources, those microorganisms exclusively from a marine or estuarine habitat are classified as obligate marine fungi, while those from freshwater or terrestrial origin that are able to grow (and possibly sporulate) in marine environments are classified as facultative marine fungi [7]. Nevertheless, the term "marine-derived fungi" is commonly used as a more general classification, due to the fact that most of the fungi isolated from marine samples are not clearly distinguished as obligate or facultative marine microorganisms [3,8]. Studies have demonstrated that marine-derived fungi are ecologically relevant as decomposers of organic matter [9]. In this sense, marine-derived fungi can be considered as a source of hydrolytic and oxidative enzymes with novel physiological characteristics (such as high salt tolerance, thermostability, barophilicity), which can be used in industrial and/or environmental processes.

Citation: Zhang, D.; Liu, L.; Chen, B.-S. Marine-Derived Fungi as a Valuable Resource for Amylases Activity Screening. *J. Fungi* **2023**, *9*, 736. https://doi.org/10.3390/jof9070736

Academic Editor: Baojun Xu

Received: 7 June 2023
Revised: 6 July 2023
Accepted: 7 July 2023
Published: 9 July 2023

Microbial enzymes received great attention from industrial production, and amylase catalyzing the hydrolysis of starch-rich materials is one of the most prominent enzymes [10]. Currently, the amylases mainly used in industrial production (such as sugar, baking, brewing and preparation of digestive aids) are those derived from *Bacillus* strains [11,12]. The optimum activity of those *Bacillus* amylases was reported at a remarkably high temperature and pH optima ranging from 55 to 70 °C, and 3.0 to 11.0, respectively. However, their activity decreases significantly at room or lower temperatures, which limits the development of amylases related industries. Amylases from halophilic producing microorganisms (such as marine microorganisms) may harbor more stability regarding their notable structural features and distinctive functional characteristics [13–15]. Among different microorganisms' sources, marine-derived fungi are the major branch contributing to expanding the number of amylases, especially those with optimum activity at low temperatures [16]. Indeed, versatile amylases have been reported from marine fungi. For example, strain *Aureobasidium pullulans* isolated from the Pacific Ocean sediments was capable of producing exocellular amylases [17]; A novel amylase was isolated from marine *Mucor* sp. [18]; The production of the enzyme α-amylase was first reported in marine *Pseudoalteromonas undina* NKMB 0074 with a wide range of pH (7–11) [19]. Those microbial sources fulfilled the industrial demands and substituted other hydrolysis approaches. Nevertheless, amylases-producing microorganisms are still the major and desirable source for the discovery of novel enzymes [20,21].

Continuing our longstanding interest in the mining of marine microbial enzymes [22–24], and in conjunction with our recent interest in the discovery of novel enzymes [25,26], we engaged in the exploration of highly efficient amylase-producing fungal strains from marine habitats. In the present study, 199 fungi were isolated from the sediments collected from the South China Sea, and 57 amylolytic fungi were successfully obtained. Based on the amylase activity measurement, one representative strain was chosen to characterize the amylase properties and stability. It needs to emphasize that in this study, fungal extracellular amylases were assessed qualitatively (by a visual screening for amylolytic fungi in a plate assay). Additionally, intracellular amylase activities were quantified spectrophotometrically within fungal cell lysates (fungal cell-disrupted mycelia).

2. Material and Methods

2.1. Marine Sediment Samples

Sediment samples at about 200–2492 m depth was collected on the South China Sea Open Cruise in January 2022. Latitudes and longitudes of the collected sediment samples 1–6 are 112.90 °E and 19.48 °N, 112.69 °E and 18.56 °N, 112.63 °E and 18.31 °N, 112.56 °E and 18.00 °N, 112.48 °E and 17.67 °N, 112.36 °E and 17.19 °N, respectively (Figure 1). Sediment sampling was performed using the combination of the Remote Operated Vehicle, a box corer, and an alcohol-sterilized PVC cylinder of 5 cm inner diameter, to keep the collected samples undisturbed and compact. The intubation depth of sediment cores obtained from these locations was about 28–30 cm. After collection, the sediment samples were directedly introduced into sterile plastic bags filled with N_2 to avoid any aerial contamination [27]. The bags were sealed and then placed in a box filled with ice and quickly transferred to the microbiology laboratory on board and stored at −4 °C for future analyses.

2.2. Fungal Isolation

About 1 g of sediment from each location was transferred into a sterile vial for fungi isolation using a flame-sterilized spatula. For the isolation of fungi, a modified particle plating technique reported by Zhang [28] was used. The components of media for fungi isolation contained malt extract agar (MEA), Czapek Dox agar (CDA), glucose peptone starch agar (GPSA), and potato glucose agar (PDA), with a strength of 1/5 to simulate the low nutrient conditions in the seawater [29,30]. To each basic media, 0.5 g/L benzylpenicillin and 0.03 g/L rose bengal were added, avoiding the growth of bacteria. The isolation was performed on the inoculated plates in the dark at a temperature of 30 °C for different times until the fungi could be distinguished. Based on the morphological observations (such as growth characteristics, mycelia, and diffusible pigments), fungal isolates were identified and transferred into new agar plates for pure culture, and for amylase-producing fungal strain screening.

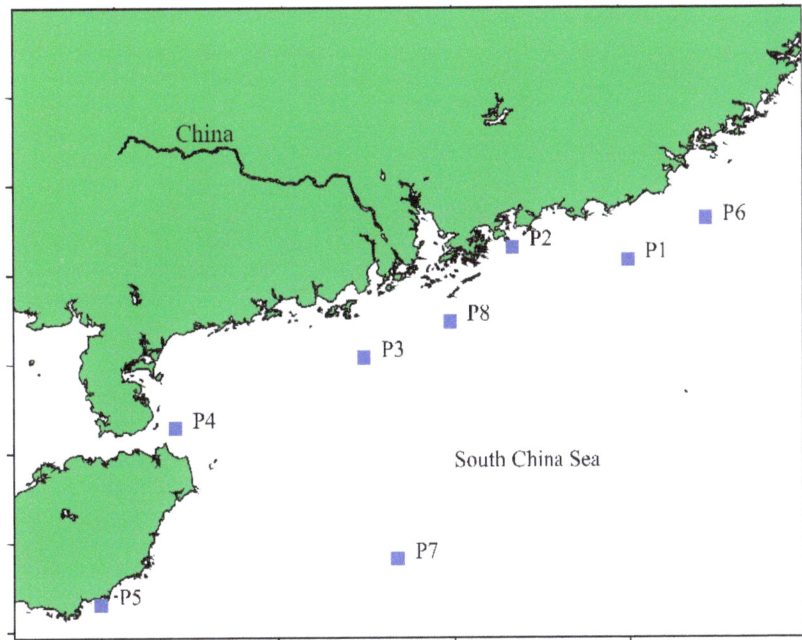

Figure 1. Map of the South China Sea, location of the sampling sites.

2.3. Visual Screening for Amylotyc Fungi

Marine fungi strain isolated above were sub-cultured in Dox-medium for amylolytic fungi isolation on agar plates. The Dox-medium contained (per liter in seawater): $NaNO_3$ (2 g), K_2HPO_4 (1 g), $MgSO_4 \cdot 7H_2O$ (0.5 g), KCl (0.5 g), $FeSO_4 \cdot 7H_2O$ (traces), and agar (20 g). The final pH was adjusted to 5, supplemented with soluble starch (1% w/v) and incubated for 72 h at 30 °C. Thereafter, 1% (w/v) iodine solution was dropped into plates waiting for the amylase to provide clear zones. Experiments were performed in triplicate, and the average values were shown as the final results.

2.4. Fungi Identification and Phylogenetic Analysis

The combination of morphological differences and internal transcribed spacersequences (ITS) gave the identification of amylolytic fungi strains. The steps for Fungal ITS gene sequencing and identification are listed below: Total genomic DNA was extracted from all selected fungal strains by using DNA Isolation Kit; From the genomic DNA, nearly full-length ITS sequences were amplified by polymerase chain reaction (PCR) using primers ITS1 (5′-TCCGTAGGTGAACCTGCGG-3′) and ITS4 (5′-TCCTCCGCTTATTGATATGC-3′); The PCR products were sequenced by BGI Technology CO., Ltd. (Shenzhen, China); The sequence results were compared in GenBank by BLAST (Basic Local Alignment Search Tool).

Nearly full-length ITS sequences of amylolytic fungi strains were aligned using Clustal X (1.83) in MEGA 5.0 software, applying the default parameters. Phylogenetic trees of ITS sequences were created using the neighbor-joining (NJ) method with bootstrap analysis using 1000 replicates. This study showed that incorporating fungal sequences with the highest homology of sequences existed in National Center for Biotechnology Information (NCBI).

2.5. Amylase Production and Protein Estimation

The obtained fungal spores from a 7-day-old slant culture were sporulated (0.5 cm $\times$ 0.5 cm) in flasks with 50 mL aliquots of the sterilized Dox-medium (attuned to pH 5.0). The sporulated mixtures were incubated continuously at 30 °C for 7 days. After filtration through Whatman No. 1 filter paper, the fermented broth was separated, giving the filtrate and the mycelia. The mycelia were collected and washed twice with 100 mM of sodium citrate buffer (pH 5.5) and resuspended in the same buffer. The mycelia were disrupted using a French press (2.05 kBar, 2 shots). Cell-free extract and cell debris were separated by centrifugation for 40 min at 10,000$\times$ g at 4 °C. The protein content of clear supernatant was determined by Lowry et al. [31], and the dinitrosalyicylic acid (DNS) method [32] was applied for measuring amylase activity. One unit (1 U) of enzyme activity is defined as the number of mycelia (wet weight) catalyzing the hydrolysis of 1 mg starch within 1 min. Experiments were performed in triplicate and the average values were shown as final results. It needs to mention that enzyme activity means the cell-lysate enzyme activities of the cell-disrupted mycelia.

2.6. Characterization of Enzyme Stability

The optimum pH was determined by standard activity assay [33] at different pH in the range of 5.0–11.0, with sodium citrate buffer (50 mM) for a pH range from 5.0–6.0, potassium phosphate buffer (50 mM) for a pH range from 6.0–9.0, glycine-NaOH buffer (50 mM) for pH 9.0–11.0. For the determination of the temperature optima, a standard activity assay was performed at different temperatures in the range of 25–75 °C. The reaction mixtures were kept at each temperature for 5 min before starch and crude enzyme solution were added to initiate the reaction. And activity measure at standard conditions (pH 5.5, 25 °C) was taken as 100%. Experiments were performed in triplicate, and the average values were shown as the final results.

In order to determine its thermostability, the crude enzyme solution was incubated at different temperatures (37, 45, 50, 55, 60 and 65 °C) for 1 h, and the residual activity was measured by standard activity assay. The activity of the enzyme without incubation at the given temperature was defined as 100%. In addition, pH stability was recognized according to testing variant gradient buffer values ranging from pH = 2 to pH =13 (for 60 min) at room temperature. Residual activities were examined following the standard conditions. Experiments were performed in triplicate, and the average values were shown as the final results.

3. Results and Discussion

3.1. Fungal Isolation

We started our investigation with fungi isolation using conventional culture-dependent methods. A total of six marine sediment samples of the South China Sea were processed for fungi isolation on extract agar (MEA), Czapek Dox agar (CDA), glucose peptone starch agar (GPSA), and potato glucose agar (PDA), respectively. Although there is always difficult to recover culturable marine-derived fungi from sediment samples, a total of 199 fungi isolates were successfully obtained in this study (Table 1). The discovery of culturable marine-derived fungi might be due to the use of effective isolation media (MEA, CDA, GPSA and PDA), which have been successfully used for isolating deep-sea fungi from the Central Indian Basin [34] and the South China Sea [28]. To identify the obtained fungi isolates, ITS sequence analysis and observation of morphological characteristics were combined. Although, several isolates had relatively lower similarities to their closest matches, most of the isolates had 95–100% sequence similarities to the existing fungal ITS sequences in the NCBI database. The identified fungi bellowed to 11 different genera, including *Aspergillus, Chaetomium, Cladosporium, Cyphellophora, Fusarium, Meyerozyma, Neodeightonia, Penicillium, Saccharomyces, Ochroconis* and *Trichobotrys*. Unfortunately, none of the discovered fungi were new from sediment of deep-sea environments, probably due to the fact that a significantly increasing number of fungal communities had been identified with metabarcoding studies. As a result, it is challenging to isolate novel strains from sediment of deep-sea environment based on the method of isolation and cultivation. Nevertheless, it needs to mention that the fungal communities were different from different sampling sites, for example, 41 and 42 isolated were obtained from sites P2 and P4, respectively, while only 10 isolates were obtained from site P7. To achieve a greater diversity of fungal signatures in marine sediments, a combination of culture-dependent and culture-independent approaches might be beneficial, which is under way in our laboratory.

Table 1. Diversity and distribution of fungi isolated from six marine sediment samples of the South China Sea.

Fungal Species	Fungal Families	Number of Fungal Isolated							
		P1	P2	P3	P4	P5	P6	P7	P8
Aspergillus niger	*Trichocomaceae*		2		2	3	1		1
A. sydowii	*Trichocomaceae*	1	2	1	1		2		
A. terreus	*Trichocomaceae*		3	2	4	1	1		
A. oryzae	*Trichocomaceae*	1	1		2		1	1	
A. tubingensis	*Trichocomaceae*		2	1	2			1	1
A. nomiae	*Trichocomaceae*	2	1	2		2	1	1	2
A. flavus	*Trichocomaceae*		2	1	3	1	2		
A. versicolor	*Trichocomaceae*	1			2	1	1		1
A. phoenicis	*Trichocomaceae*		1		1				
A. westerdijkiae	*Trichocomaceae*		2		2				1
Chaetomium globosum	*Chaetomiaceae*		1		1	1			
Cladosporium cladosporioides	*Davidiellaceae*	1	1						
C. colombiae	*Davidiellaceae*		2		3	3	2		2
C. oxysporum	*Davidiellaceae*		2	1	2	1		1	1
C. sphaerospermun	*Davidiellaceae*	1	1		3				
C. uredinicola	*Davidiellaceae*			3					2
Cyphellophora fusarioides	*Cyphellophoraceae*		1		2	1			
Fusarium proliferatum	*Nectriaceae*	2		3					
F. oxysporum	*Nectriaceae*	1	2	3					1
F. nirenbergiae	*Nectriaceae*		2		2	3			
F. solani	*Nectriaceae*	2	1		1		1	1	
Meyerozyma caribbica	*Debaryomycetaceae*		2	1	1				1
Neodeightonia subglobosa	*Botryosphaeriaceae*	1	2		3	2			
Penicillium aurantiogriseum	*Trichocomaceae*		1	2	2			1	1
P. chrysogenum	*Trichocomaceae*	2	3	3	1	1	1	1	
P. album	*Trichocomaceae*		2	1		4		3	2
Saccharomyces cerevisiae	*Saccharomycetaceae*	1	1	2	1		2		1
Ochroconis mirabilis	*Sympoventuriaceae*		1		1	1	1		
Trichobotrys effusa	*Pleosporales*				2	4			1
Total number of fungal isolates		16	41	27	42	29	16	10	18

3.2. Screening for Amylotyc Fungi (Qualitative Determination of Extracellular Enzyme Activity by Visual Plate Assay)

The ocean is considered to be a great reservoir of biodiversity, leading to bioactive metabolites from marine-derived fungi are considered an attractive point in the field of drug discovery and production. Indeed, there is a number of research highlighting the diversity of fungi derived from different marine environments as well as their potential as producers of bioactive marine natural products. Unfortunately, the use of marine microorganisms as sources of novel enzymes is still rare.

The growing commercial demand for new and more efficient enzymes to implement and optimize industrial processes, which highly depends on new renewable and environmentally sustainable enzyme sources. The adaptability of marine-derived fungi to oceanic conditions makes them an attractive source of unique enzymes. Therefore, we turned our attention to exploring marine-derived fungi from the South China Sea for amylase activity screening. Thus, all the 199 fungi strains isolated above were sub-cultured in Dox-medium for amylolytic fungi isolation on agar plates, as described in Section 2.3.

The amylase activity was visible as clear zones on the agar plates created by the fungal isolates (Figure 2). The larger the transparent circle produced by marine fungi on the plate containing starch, the stronger the ability of fungi to produce amylase. In this study, only clear zones of a diameter of more than 2 cm were considered as the positive starch zone hydrolysis by tested marine-derived fungi. Gratifyingly, the results of amylase activity screening showed that 57 isolates (Figure 3 and Table S1) showed amylase activity, of which 30 belonged to the genera *Aspergillus*, suggesting that *Aspergillus* sp. are the main sources of amylase-producing fungi. This is supported by the findings of Barakat, who proposed that the genera *Aspergillus* are essential sources of amylases-producing fungi [33]. Indeed, *Aspergillus* species was characterized as the most pervasive and easy to culture microorganisms, Thus, such species are the major and desirable source for the discovery of novel amylases.

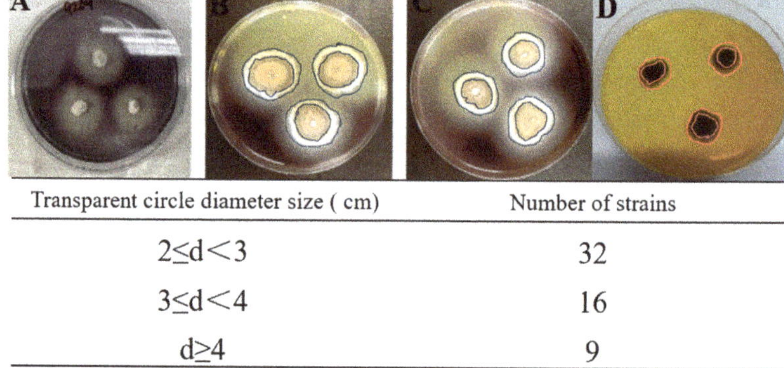

Transparent circle diameter size (cm)	Number of strains
$2 \leq d < 3$	32
$3 \leq d < 4$	16
$d \geq 4$	9

Figure 2. Examples of transparent circles produced by marine fungi with amylase activity, (**A**): fungi with no amylase activity; (**B**,**C**): fungi with amylase activity; (**D**): the use of *Aspergillus fumigatus* as a positive control.

3.3. Characteristics of Cell-Lysate Enzyme Activity (Quantitative Determination of Intracellular Enzyme Activities Spectrophotometrically)

Encouraged by the discovery of 57 amylase-producing fungi strains, we continuously evaluated the cell-lysate extract enzyme activity toward starch hydrolysis. In order to have a clear idea about the amylase activities of the 57 marine fungi obtained above, the dinitrosalyicylic acid (DNS) method [32] was applied for measuring the amylase activity. The mycelia of the 57 amylase-producing fungi strains were collected and washed twice with 100 mM of sodium citrate buffer (pH 5.5) and resuspended in the same buffer. The mycelia were disrupted using a French press (2.05 kBar, 2 shots). Cell-free extract and cell debris were separated by centrifugation for 40 min at 10,000 rpm at 4 °C. The clear supernatant was collected and used for the measurement of the protein content and the amylase activity. It needs to mention that one unit (1 U) of enzyme activity is defined as the number of mycelia (wet weight) catalyzing the hydrolysis of 1 mg starch within 1 min. It needs to mention that enzyme activity means the cell-lysate enzyme activity of the cell-disrupted mycelia.

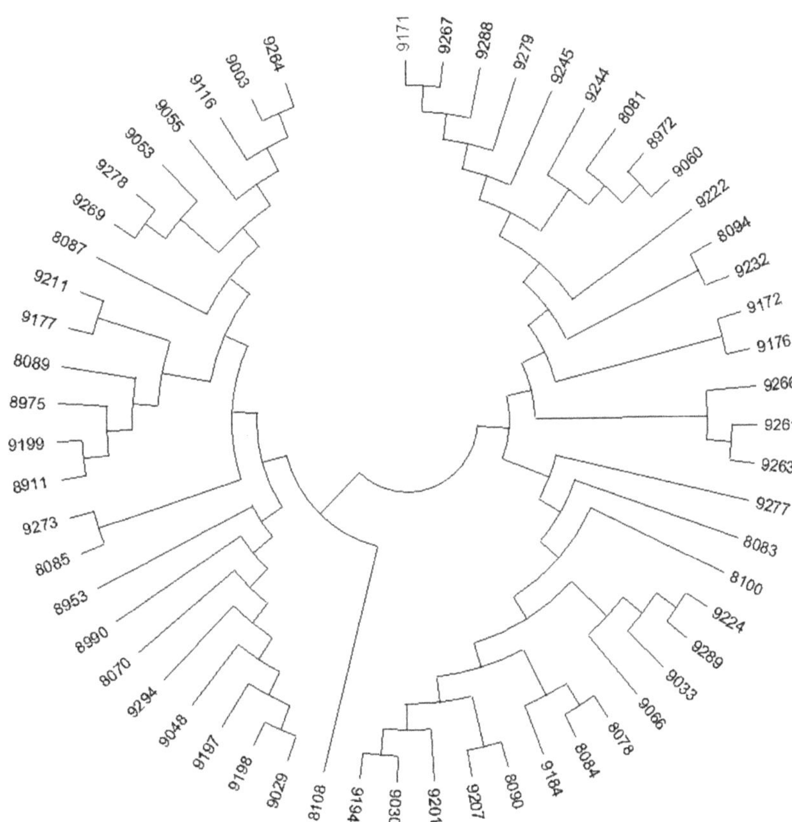

Figure 3. The phylogenetic tree of 57 marine fungi with significant amylase activity. The names of all the fungi are available in Table S1 of the Supporting Information.

The results are summarized in Table 2. Notably, the amylase enzyme produced by marine fungi showed good to excellent activity. The strain *Fusarium nirenbergiae* 9245 produced amylases with a poor activity of 0.062 U/mg towards starch hydrolysis, while the amylase produced by the strain *Aspergillus flavus* 9261 showed the best catalytic activity of 10.7482 U/mg. The observation highlights that *Aspergillus* sp. has been considered the main source of amylases-producing fungi. Notably, *Pezizomycotina* sp. 8081 is a potential source for amylase production of significant starch hydrolysis activity, which was found for the first time. Taking a closer look at the starch-degrading activity of *Aspergillus flavus* 9261, we found that the cell-lysate enzyme activity of the marine *Aspergillus terreus* SS isolated from the Red Sea was shown to be 5.326 U of per mg wet mycelia (intracellular enzyme activities) [35]. As a result, we can make the conclusion that the marine fungi *Aspergillus flavus* 9261 exhibited an activity of 10.7482 U of per mg mycelia, which is superior to *Aspergillus terreus* SS for the production of amylase enzymes. Taking the activity, cultural conditions and available genome sequence into account, we decided to continue to use the marine fungi *Aspergillus flavus* 9261 for all further studies.

Table 2. The amylase activity produced by cell-lysate extracts from marine fungi [a,b].

Fungi	Activity (U/mg)	Fungi	Activity (U/mg)	Fungi	Activity (U/mg)
8070	0.0921 ± 0.0001	9030	0.2790 ± 0.0002	9211	0.1140 ± 0.0001
8078	0.1571 ± 0.0001	9033	0.0964 ± 0.0003	9222	0.0811 ± 0.0002
8081	1.3895 ± 0.0002	9048	0.0880 ± 0.0001	9224	0.0533 ± 0.0004
8083	0.1997 ± 0.0001	9053	0.1793 ± 0.0002	9232	0.3304 ± 0.0001
8084	0.1777 ± 0.0001	9055	0.0499 ± 0.0001	9244	0.1198 ± 0.0001
8085	0.1640 ± 0.0003	9060	0.1140 ± 0.0003	9245	0.0262 ± 0.0003
8087	0.1592 ± 0.0002	9066	0.1146 ± 0.0002	9261	10.7482 ± 0.0002
8089	0.0447 ± 0.0001	9116	0.0293 ± 0.0001	9263	0.5694 ± 0.0001
8090	0.3540 ± 0.0001	9171	0.6806 ± 0.0003	9264	0.2164 ± 0.0001
8094	0.2288 ± 0.0002	9172	0.1606 ± 0.0004	9266	0.3230 ± 0.0001
8100	0.1365 ± 0.0001	9176	0.0723 ± 0.0002	9267	0.4128 ± 0.0002
8108	0.0625 ± 0.0003	9177	0.2946 ± 0.0005	9269	0.1125 ± 0.0001
8911	0.0315 ± 0.00002	9184	0.3517 ± 0.0002	9273	0.0996 ± 0.0002
8953	0.0874 ± 0.0002	9194	0.1654 ± 0.0002	9277	0.1752 ± 0.0001
8972	0.1265 ± 0.0001	9197	0.1629 ± 0.0003	9278	0.1430 ± 0.0001
8975	0.1165 ± 0.0001	9198	0.2528 ± 0.0002	9279	0.2446 ± 0.0002
8990	0.0423 ± 0.0002	9199	0.0641 ± 0.0001	9288	0.2030 ± 0.0003
9003	0.0604 ± 0.0001	9201	0.1991 ± 0.0001	9289	0.4982 ± 0.0002
9029	0.1571 ± 0.0002	9207	0.0668 ± 0.0002	9294	0.1679 ± 0.0002

[a]: One unit (1 U) of enzyme activity is defined as the number of mycelia (wet-weight) catalyzing the hydrolysis of 1 mg starch within 1 min. [b]: Mean values ± standard deviation of cell-lysate enzyme activity. Experiments were performed in triplicate.

3.4. Effect of pH and Temperature on Amylase Activities of Aspergillus flavus 9261

To fully assess the potential of the marine fungi *Aspergillus flavus* 9261 as amylase-producing strain, the temperature and pH optima was determined. The influence of temperature and pH on the cell-lysate enzyme activity were investigated by monitoring the change in cell-lysate enzyme activity toward starch hydrolysis (Figure 4). For the determination of the temperature optima, a standard activity assay was performed at different temperatures in the range of 25–75 °C. The reaction mixtures were kept at each temperature for 5 min before starch and crude enzyme solution were added to initiate the reaction. Activity measure at standard conditions (pH 5.5, 25 °C) was taken as 100%. Testing a broad range of temperature scales (25–65 °C) revealed that the enzyme exhibited an optimal activity at 55 °C, while it steeply decreased over 60 °C (Figure 4A). On the other hand, testing a broad range of pH scales (2–13) revealed the optimum pH of 5 for the amylolytic activity (Figure 4C). Thus, the amylase produced by *Aspergillus flavus* 9261 was found to have the best activity at pH = 5 and 55 °C, which was supported by the findings (*Aspergillus flavus* AUMC10636 has the optimum activity toward starch hydrolysis at pH = 5.0 and 60 °C) of Ali and workers [36].

In order to determine its thermostability, the crude enzyme solution was incubated at different temperatures (37, 45, 50, 55, 60 and 65 °C) for given times before the substrate was added to initiate the reaction, and the residual activity was measured by standard activity assay. The activity of the enzyme without incubation at the given temperature was defined as 100%. The results showed that the amylase produced by *Aspergillus flavus* 9261 was relatively stable under 50 °C for 1 h, but lost its activity almost completely after incubation at 60 °C for 30 min. In addition, pH stability was recognized according to testing variant gradient buffer values ranging from pH = 2 to pH =13 (for 30 min) at room temperature. Residual activities were examined following the standard conditions. It can be clearly seen that the amylase produced by *Aspergillus flavus* 9261 maintained most of its activity after incubation at pH of 5.0 for 1 h (Figure 4D). Therefore, the ideal condition and incubation period were 50 °C, pH = 5.0 and 1 h, respectively.

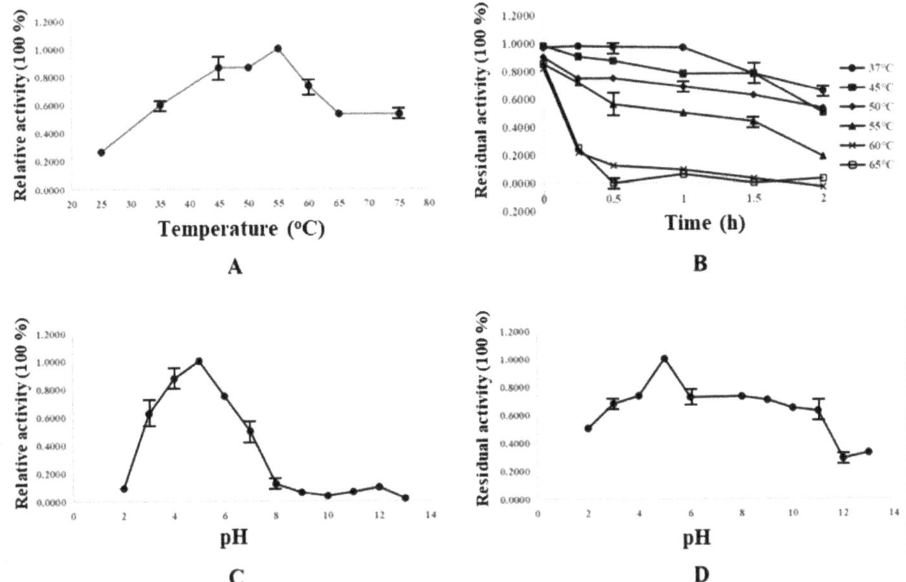

Figure 4. Temperature optima (**A**), thermostability (**B**), pH optima (**C**) and pH stability (**D**) of the amylases produced by *Aspergillus flavus* 9261. (**A**): standard activity assay was performed at different temperatures in the range of 25–75 °C. The reaction mixtures were kept at each temperature for 5 min before starch and crude enzyme solution were added to initiate the reaction. And activity measure at standard condition (pH 5.5, 25 °C) was taken as 100%. (**B**): the crude enzyme solution was incubated at different temperatures (37, 45, 50, 55, 60 and 65 °C) for given times, and the residual activity was measured by standard activity assay. The activity of the enzyme without incubation at the given temperature was defined as 100%. (**C**): standard activity assay was performed at different pH values of 2–11. The reaction mixtures were kept at each pH for 5 min before starch and crude enzyme solution were added to initiate the reaction. And activity measure at standard condition (pH 5.5, 25 °C) was taken as 100%. (**D**): pH stability was recognized according to testing variant gradient buffer values ranging from pH = 2 to pH =13 (for 30 min) at room temperature. Residual activities were examined following the standard conditions. Experiments were performed in triplicate, and the average values were shown as final results.

4. Conclusions

Microbial enzymes received great attention from industrial production, and amylase catalysing the hydrolysis of starch-rich materials is one of the most prominent enzymes. Currently, the amylases mainly used in industrial production are those derived from Bacillus strains. The optimum activity of those Bacillus amylases was reported at a remarkably high temperature ranging from 55 to 70 °C. However, their activity decreases significantly at room or lower temperatures, which limits the development of amylases related industries. Thus, amylases-producing microorganisms (especially fungi) are still the major and desirable source for the discovery of novel enzymes.

The adaptability of marine-derived fungi to oceanic conditions makes them an attractive source of unique enzymes. In this manuscript, we focus on the screening of marine-derived fungi for amylase activity.

In conclusion, this research successfully obtained 199 marine-derived fungi from the South China Sea, of which 57 isolates were found to have amylase activity toward starch hydrolysis by providing clear zones on agar plates. The application of the dinitrosalyicylic acid (DNS) method for measuring the amylase activity revealed that all the tested 57 isolates exhibited an amylase activity of 0.0262–10.7482 U per mg of mycelia (wet weight), which is good to excellent compared to those of reported fungi. The marine fungi *Aspergillus flavus* 9261 was found to produce amylase with the best activity of 10.7482 U/mg (wet mycelia), and is moderately tolerant towards temperature as well as thermal and pH stability. The amylase of *Aspergillus flavus* 9261 exhibited remarkable thermostability and pH stability with no activity loss after incubation at 50 °C and pH 5.0 for 1 h, respectively. In a sense, the potential of amylases produced from marine fungi was considered highly appreciated and of economic value. This opens up an entirely new approach to the discovery of marine-derived enzymes and investigations into their possible applications.

Supplementary Materials: The following supporting information can be downloaded at: https://www.mdpi.com/article/10.3390/jof9070736/s1, Table S1: Information of the amylase-producing fungus strains.

Author Contributions: Conceptualization, B.-S.C. and D.Z.; methodology, D.Z. and B.-S.C.; formal analysis, B.-S.C. and D.Z.; writing—original draft preparation, B.-S.C. and L.L.; writing—review and editing, D.Z.; L.L. and B.-S.C.; supervision, L.L. and B.-S.C.; project administration, L.L. and B.-S.C.; funding acquisition, L.L. and B.-S.C. All authors have read and agreed to the published version of the manuscript.

Funding: This research was funded by Guangdong Basic and Applied Basic Research Foundation (Grant No. 2023A1515030226).

Institutional Review Board Statement: Not relevant to the study, not ethical approval required.

Informed Consent Statement: Study NOT involving humans.

Data Availability Statement: No data beyond the data presented in the manuscript.

Conflicts of Interest: The authors declare no conflict of interest.

References

1. Beygmoradi, A.; Homaei, A. Marine microbes as a valuable resource for brand new industrial biocatalysts. *Biocatal. Agric. Biotechnol.* **2017**, *11*, 131–152. [CrossRef]
2. Sowell, S.M.; Norbeck, A.D.; Lipton, M.S.; Nicora, C.D.; Callister, S.J.; Smith, R.D. Proteomic analysis of stationary phase in the marine bacterium "Candidatus Pelagibacter ubique". *Appl. Environ. Microbiol.* **2017**, *74*, 4091–4100. [CrossRef] [PubMed]
3. Bonugli-Santos, R.C.; dos Santos Vasconcelos, M.R.; Passarini, M.R.Z.; Vieira, G.A.L.; Lopes, V.C.P.; Mainardi, P.H.; dos Santos, J.A.; de Azevedo Duarte, L.; Otero, I.V.R.; da Silva Yoshida, A.M.; et al. Marine-derived fungi: Diversity of enzymes and biotechnological applications. *Front. Microbiol.* **2015**, *6*, 269. [CrossRef] [PubMed]
4. Zeng, J.; Guo, J.; Tu, Y.; Yuan, L. Functional study of C-terminal domain of the thermoacidophilic raw starch-hydrolyzing α-amylase Gt-amy. *Food Sci. Biotechnol.* **2020**, *29*, 409–418. [CrossRef] [PubMed]
5. Farias, T.C.; Kawaguti, H.Y.; Koblitz, M.G.B. Microbial amylolytic enzymes in foods: Technological importance of the *Bacillus* genus. *Biocatal. Agric. Biotechnol.* **2021**, *35*, 102054. [CrossRef]
6. de Eugenio, L.I.; Murguiondo, C.; Galea, S.; Prieto, A.; Barriuso, J. Fungal–lactobacteria consortia and enzymatic catalysis for polylactic acid production. *J. Fungi* **2023**, *9*, 342. [CrossRef]
7. Li, Q.; Wang, G. Diversity of fungal isolates from three Hawaiian marine sponges. *Microbiol. Res.* **2009**, *164*, 233–241. [CrossRef]
8. Osterhage, C. Isolation, Structure Determination and Biologicalactivity Assessment of Secondary Metabolites Frommarine-Derived Fungi. Technischen Universität Carolo-Wilhelmina: Braunschweig, Germany, 2001.
9. Arifeen, M.Z.U.; Liu, C.-H. Novel enzymes isolated from marine-derived fungi and its potential applications. *J. Biotechnol. Bioeng.* **2018**, *2*, 1–12.
10. Bai, X.F.; Guo, Y.; Kaustubh, B.; Li, C.; Gu, Z.B.; Hu, K. The Global amylase research trend in food science technology: A data-driven analysis. *Food Rev. Int.* **2021**. [CrossRef]
11. Kikani, B.A.; Singh, S.P. Amylases from thermophilic bacteria: Structure and function relationship. *Crit. Rev. Biotechnol.* **2021**, *3*, 325–341. [CrossRef]
12. El-Kady, E.M.; Asker, M.S.; Hassanein, S.M. Optimization, production, and partial purification of thermostable α-amylase produced by marine *Bacterium bacillus* sp. NRC12017. *Int. J. Pharmaceut. Clin. Res.* **2017**, *9*, 558. [CrossRef]
13. Mafakher, L.; Ahmadi, Y.; Fard, J.K.; Yazdansetad, S.; Gomari, S.R.; Far, B.E. Alpha-Amylase Immobilization: Methods and Challenges. *Pharm. Sci.* **2022**, *2*, 144–155.
14. Wang, Y.C.; Hu, H.F.; Ma, J.W. A novel high maltose-forming a-amylase from *Rhizomucor miehei* and its application in the food industry. *Food Chem.* **2020**, *305*, 125447. [CrossRef] [PubMed]

15. Paul, J.S.; Dikshit, B.; Santi, L. Aspects and recent trends in microbial α-amylase: A review. *Appl. Biochem. Biotechnol.* **2021**, *193*, 2649. [CrossRef]
16. Farooq, M.A.; Ali, S.; Hassan, A.; Tahir, H.M.; Mumtaz, S.; Mumtaz, S. Biosynthesis and industrial applications of alpha-amylase: A review. *Arch. Microbiol.* **2021**, *4*, 1281–1292. [CrossRef] [PubMed]
17. Abdel-Fattah, Y.R.; Soliman, N.A.; El-Toukhy, N.M. Production, purification, and characterization of thermostable α-Amylase produced by *Bacillus licheniformis* isolate AI20. *J. Chem.* **2013**, *1*, 4615.
18. Nandy, S.K. Bioprocess technology governs enzyme use and production in industrial biotechnology: An overview. *Enz. Engineer.* **2016**, *5*, 1000144. [CrossRef]
19. Ullah, I.; Khan, M.S.; Khan, S.S. Identification and characterization of thermophilic amylase producing bacterial isolates from the brick kiln soil. *Saudi J. Biol. Sci.* **2021**, *28*, 970. [CrossRef]
20. Zafar, A.; Aftab, M.N.; Iqbal, I. Pilot-scale production of a highly thermostable α-amylase enzyme from *Thermotoga petrophila* cloned into *E. coli* and its application as a desizer in textile industry. *RSC Adv.* **2019**, *9*, 984. [CrossRef]
21. Viader, R.P.; Yde, M.; Hartvig, J.W. Optimization of beer brewing by monitoring α-amylase and β-amylase activities during mashing. *Baverages* **2021**, *7*, 12.
22. Liu, H.; Duan, W.-D.; de Souza, F.Z.R.; Liu, L.; Chen, B.-S. Asymmetric ketone reduction by immobilized *Rhodotorula mucilaginosa*. *Catalysts* **2018**, *8*, 165. [CrossRef]
23. Liu, H.; Chen, B.-S.; de Souza, F.Z.R.; Liu, L. A Comparative study on asymmetric reduction of ketones using the growing and resting cells of marine-derived fungi. *Mar. Drugs* **2018**, *16*, 62. [CrossRef] [PubMed]
24. Liu, H.; de Souza, F.Z.R.; Liu, L.; Chen, B.-S. The use of marine-derived fungi for preparation of enantiomerically pureal cohols. *Appl. Microbiol. Biotechnol.* **2018**, *102*, 1317–1330. [CrossRef] [PubMed]
25. Zeng, Y.-Y.; Yin, X.; Liu, L.; Zhang, W.; Chen, B. Comparative characterization and physiological function of putative fatty acid photodecarboxylases. *Mol. Catal.* **2022**, *532*, 112717. [CrossRef]
26. Yin, X.; Gong, W.; Zhan, Z.; Wei, W.; Li, M.; Jiao, J.; Chen, B.; Liu, L.; Li, W.; Gao, Z. Mining and engineering of valine dehydro genases from a hot spring sediment metagenome for the synthesis of chiral non-natural L-amino acids. *Mol. Catal.* **2022**, *533*, 112767. [CrossRef]
27. Zhang, X.-Y.; Wang, G.-H.; Xu, X.-Y.; Nong, X.-H.; Wang, J.; Amin, M.; Qi, S.-H. Exploring fungal diversity in deep-sea sediments from Okinawa trough using high-throughput Illumina sequencing. *Deep-Sea Res.* **2016**, *116*, 99–105. [CrossRef]
28. Zhang, X.; Zhang, Y.; Xu, X.; Qi, S. Diverse deep-sea fungi from the South China Sea and their antimicrobial activity. *Curr. Microbiol.* **2013**, *67*, 525–530. [CrossRef]
29. Cathrine, S.J.; Raghukumar, C. Anaerobic denitrification in fungi from the coastal marine sediments off Goa, India. *Mycol. Res.* **2009**, *113*, 100–109. [CrossRef]
30. Damare, S.; Raghukumar, C.; Raghukumar, S. Fungi in deep sea sediments of the Central Indian Basin. *Deep Sea Res.* **2006**, *53*, 14–27. [CrossRef]
31. Lowry, O.H.; Rosebrough, N.J.; Farr, A.L.; Randall, R.J. Protein measurement with the Folin phenol reagent. *J. Biological Chem.* **1951**, *193*, 265–275. [CrossRef]
32. Miller, G.L. Use of dinitrosalyicylic acid reagent for determination of reducing sugar. *Anal. Chem.* **1959**, *31*, 426–428. [CrossRef]
33. Beltagy, E.A.; Abouelwafa, A.; Barakat, K.M. Bioethanol production from immobilized amylase produced by marine *Aspergillus flavus* AUMC 10636. *Egypt. J. Aquat. Res.* **2022**, *48*, 325–331. [CrossRef]
34. Nagano, Y.; Nagahama, T.; Hatada, Y.; Nunoura, T.; Takami, H.; Miyazaki, J.; Takai, K.; Horikoshi, K. Fungal diversity in deep-sea sediments – the presence of novel fungal groups. *Fungal Ecol.* **2010**, *3*, 316–325. [CrossRef]
35. Ali, I.; Akbar, A.; Yanwisetpakdee, B.; Prasongsuk, S.; Lotrakul, P.; Punnapayak, H. Purification, characterization, and potential of saline waste watervremediation of a polyextremophilic α-amylase from an obligate halophilicv *Aspergillus gracilis*. *BioMed Res. Int.* **2014**, *2014*, 106937. [CrossRef] [PubMed]
36. Ali, E.H.; El-Nagdy, M.A.; Al-Garni, S.M.; Ahmed, M.S.; Rawaa, A.M. Enhancement of alpha amylase production by *Aspergillus flavus* AUMC 11685 onmandarin (*Citrus reticulata*) peel using submerged fermentation. *Eur. J. Biol. Res.* **2017**, *7*, 154–164.

Journal of
Fungi

Article

Investigation on the Influence of Production and Incubation Temperature on the Growth, Virulence, Germination, and Conidial Size of *Metarhizium brunneum* for Granule Development

Tanja Seib [1], Katharina Fischer [1], Anna Maria Sturm [2] and Dietrich Stephan [1,*]

[1] Julius Kühn-Institute, Federal Research Centre for Cultivated Plants, Institute for Biological Control, Schwabenheimerstraße 101, 69221 Dossenheim, Germany; tanja.seib@julius-kuehn.de (T.S.)

[2] Technical University Darmstadt, Department Biologie, Schnittspahnstraße 4, 64287 Darmstadt, Germany

* Correspondence: dietrich.stephan@julius-kuehn.de

Abstract: Important for the infection of an insect with an entomopathogenic fungus and its use as a plant protection agent are its growth, conidiation, germination, and virulence, which all depend on the environmental temperature. We investigated not only the effect of environmental temperature but also that of production temperature of the fungus. For this purpose, *Metarhizium brunneum* JKI-BI-1450 was produced and incubated at different temperatures, and the factors mentioned as well as conidial size were determined. The temperature at which the fungus was produced affects its subsequent growth and conidiation on granule formulation, the speed of germination, and the conidial width, but not its final germination or virulence. The growth and conidiation was at its highest when the fungus was produced at 25 °C, whereas when the germination was faster, the warmer the fungus was produced. The incubation temperature optimum of JKI-BI-1450 in relation to growth, speed of germination, and survival time was 25–30 °C and for conidiation 20–25 °C. Conidial length decreased with increasing incubation temperature. Although the fungus could not be adapted to unfavorable conditions by the production temperature, it was found that the quality of a biological control agent based on entomopathogenic fungi can be positively influenced by its production temperature.

Keywords: conidial size; germination; granule; growth; *Metarhizium brunneum*; temperature; virulence

Citation: Seib, T.; Fischer, K.; Sturm, A.M.; Stephan, D. Investigation on the Influence of Production and Incubation Temperature on the Growth, Virulence, Germination, and Conidial Size of *Metarhizium brunneum* for Granule Development. *J. Fungi* **2023**, *9*, 668. https://doi.org/10.3390/jof9060668

Academic Editor: Baojun Xu

Received: 24 April 2023
Revised: 5 June 2023
Accepted: 8 June 2023
Published: 14 June 2023

1. Introduction

The aim of this study was to investigate the effects of different temperatures on the growth, conidiation, conidial size, germination and virulence of a *Metarhizium*-based soil granule, which was developed to control pest insects in the soil, especially in potato cultivation. Therefore, we developed a formulation technology for granules based on entomopathogenic fungi [1], and examined whether this technology can be used for *Metarhizium brunneum* strain JKI-BI-1450. To produce our granule, the fungus was initially produced in liquid medium. Subsequently, a thin layer of biomass was coated on autoclaved millet by fluid bed drying.

The advantage of applying a soil granule produced this way is that only after its application in the soil and the associated contact with moisture does the fungus overgrow the granule (Figure 1) and conidiate on the surface, which in turn leads to the pest insect being infected once contact has been established with the conidia [2]. The direct application of conidia into the soil has the advantage that the infectious unit of the fungus is directly present, however the product is dusting, and thus poses a risk for both the user and bystander.

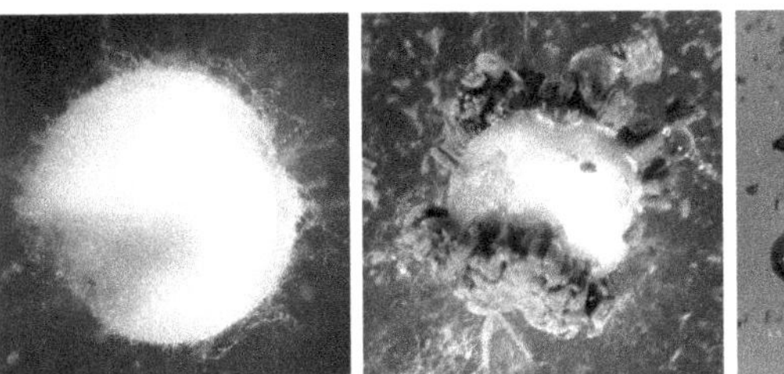

Figure 1. Growth and conidiation of *Metarhizium brunneum* JKI-BI-1450 on a granule grain.

Therefore, the growth and conidiation of the fungus on the granule in the soil are essential for its effectiveness, and for both of these processes, the temperature of the environment is important. Generally, temperatures between 20 to 30 °C are optimal for the growth and sporulation of entomopathogenic fungi, which is also valid for *Metarhizium* [3–6]. However, the temperature optimum varies between different *Metarhizium* species. Kryukov et al. [7] demonstrated that *M. brunneum* exhibited the best growth at 25 °C, whereas the optimum growth temperature of *Metarhizium robertsii* was 30 °C. Only *M. robertsii* grew at 35 or 37.5 °C, whereas *M. brunneum* was not able to grow at these temperatures. Keyser et al. [8] also showed that the examined *M. brunneum* species exhibited their maximum mycelial growth at 24 °C. Strains of *Metarhizium acridum*, *Metarhizium giuzhouense*, and *M. robertsii* had their maximum at 28 °C or even at 32 °C.

One constraint in applying entomopathogenic fungi for controlling soil dwelling pests is the difference in soil temperatures and the temperature optimum of the fungi. The granule investigated in this study should later be used in cultivation at the same time when potato planting, where the soil temperature is lower than the optimum of the fungus [9]. An application at lower temperatures can lead to a reduced growth and conidiation of the fungus [6,10,11]. As conidia are the infection units of the fungus [12], lower levels of conidiation may lead to a reduction in efficacy [13]. We intended to take advantage of the fact that many microorganisms can cope better with stressful conditions when they have already been exposed to them [14–17]. Therefore, we investigated whether the fungus may be adapted to unfavorable temperatures by culturing it several times at these temperatures. Following growth and conidiation of the fungus on the granule, the next step to successful infection is the contact of the conidia with the insect cuticle. After adhesion on the cuticle, the conidia form a germ tube, which penetrates the cuticle through enzymatic and mechanical pressure. Moreover, germination is also heavily influenced by temperature [6,18]. Walstad et al. [19] and Dimbi et al. [3] demonstrated the highest germination for *Metarhizium* after 24 h at 25 °C. Similar results were determined by Hywel-Jones and Gillespie [20], with the fastest and highest germination having being observed after an incubation at 25 to 30 °C. Rath et al. [21] and Skalický et al. [22] showed that germination rates up to 100% were even possible at low temperatures, although such values are obtained slower compared to rates at higher temperatures. After adhesion, germination, and penetration, the fungus can multiply in the hemolymph and kill the insect either by nutrient deprivation or toxins, which some fungi are able to produce [23,24]. Apart from growth and germination, the virulence of entomopathogenic fungi is also temperature-dependent [6,25]. For *Metarhizium*, many studies showed the highest or fastest mortality at higher temperatures of 30 or 35 °C [3,26], or 25 to 30 °C [27,28], respectively.

Investigations on the influence of temperature on the growth, germination, and virulence of entomopathogenic fungi have often been described. The novelty of our study is that we examined both the influence of the production and incubation temperature. Initially, we investigated whether the production temperature of the biomass used for our granule-based formulation affects the growth of the fungus on this granule at different incubation temperatures. The purpose of these experiments was to verify whether the quality of the granule, and its ability to grow at sub-optimal temperatures could be improved by the previous production parameters of the biomass. In our application strategy, the infectious conidia were only formed after application in soil under moist conditions and at sufficient temperatures. Therefore, in the second step, we assessed whether the temperature for conidiation had an influence on their size, germination, and virulence to better understand the potential field conditions.

2. Materials and Methods

2.1. Fungal Strain

Metarhizium brunneum strain JKI-BI-1450 was isolated in 2013 at the Institute for Biological Control, Julius Kühn-Institute, Darmstadt (Germany) from an infected adult *Agriotes lineatus*, which was collected by Jörn Lehmhus in Braunschweig, Germany.

2.2. Maintenance of JKI-BI-1450

The fungal strain JKI-BI-1450 was stored at $-80\ °C$ in cryo-tubes (Microbank, Pro-Lab Diagnostics, Richmond Hill, Canada) and routinely cultured on malt peptone agar (MPA) containing 3% (w/v) malt extract (Merck, Darmstadt, Germany), 0.5% (w/v) peptone from soybean (Merck), and 1.8% (w/v) agar-agar (Roth, Karlsruhe, Germany).

2.3. Colony Size and Conidiation at Different Temperatures

JKI-BI-1450 was spread on MPA and incubated for 2 weeks at 25 °C in the dark. A total of 2 µL of a conidial suspension (1×10^8 conidia/mL in sterile 0.5% (v/v) Tween 80® (Merck)) of JKI-BI-1450 was pipetted in the center of the petri dishes filled with MPA. The dishes were then incubated for 14 days at 15, 20, 25, 30, and 37 °C in the dark, respectively. The diameter of the fungal colony was measured twice, perpendicular to one another. To determine the conidiation, 2 mL of sterile 0.5% (v/v) Tween 80® was added to each dish and the conidia were scraped off with a sterile spatula. The conidia concentration was determined using a hemocytometer, and the conidia per mm^2 was calculated based on the final colony size. The experiment was repeated six times independently with five replicates each.

2.4. Temperature Adaptation

JKI-BI-1450 was cultivated on MPA for 14 days at 15, 20, 25, and 30 °C in the dark, respectively. New MPA plates were then inoculated with the fungi produced at these temperatures and incubated for 14 days at the same temperatures as before in the dark. This subsequent cultivation was repeated three times. After that, the final adapted fungus cultures of each growth temperature were stored at $-80\ °C$.

2.5. Effects of Production and Incubation Temperature on the Fungus Growth

The granules were produced as described by Stephan et al. [1]. For liquid culture inoculation, the temperature-adapted fungus cultures were taken from the $-80\ °C$ deep freezer and cultivated for 21 days on MPA at the corresponding temperature in the dark. For the liquid production of fungal biomass, 50 mL sterile medium containing 2.5% (w/v) glucose (Merck), 2% (w/v) corn steep (Sigma Aldrich, Buchs, Switzerland), and 0.5% (w/v) sodium chloride (Merck) in 100 mL flasks was used. Each flask was inoculated with 5×10^6 conidia suspension of the previously described 21-day-old cultures in 1 mL of sterile 0.5% (v/v) Tween 80® and were then incubated at 130 rpm on a horizontal shaker (Novotron, 50 mm deflection, Infors, Bottmingen, Switzerland). In liquid culture, the conidia of JKI-BI-1450 germinate and form mycelium and submerged spores. The duration of the production process is influenced by the production temperature. To determine the end of the exponential growth phase at each production temperature, the formation of the submerged spores was observed microscopically every 24 h. Based on these results, the incubation time at 15 °C was set to 120 h, and at 30 °C to 144 h, respectively. At 20 and 25 °C the fungus was cultivated for 72 h. The biomass suspension of each cultivation was centrifuged for 10 min at 25 °C and $15,344\times g$. The supernatant was then discarded, and the pellet was resuspended in 0.9% (w/v) sodium chloride solution and centrifuged again. This process was repeated twice. The dry weight of the biomass was determined using a moisture determination balance (Ma30, Sartorius, Göttingen, Germany), and a suspension of 3% dry biomass was prepared by diluting with a 0.9% (w/v) sodium chloride solution. From each biomass suspension a granule was prepared with a laboratory fluid-bed dryer (Strea-1, Aeromatic-Fielder AG, Bubendorf, Switzerland, nozzle diameter: 1 mm, Container volume: 16.5 L). In contrast to the process described by Stephan at al. [1], 100 g of autoclaved millet seeds were placed in the container, and 15 g of the 3% biomass suspension was sprayed by a top spray variant on the millet with a nozzle pressure of 2 bar. The granules produced in this way will be henceforth referred to as 15 °C-granule, 20 °C-granule, 25 °C-granule, and 30 °C-granule throughout this study. For each batch, ten granule grains (millet seeds covered with fungus) were placed on five petri dishes filled with agar (1% (w/v) agar-agar) and were then incubated at 15, 20, 25, and 30 °C for 14 and 28 days in the dark, respectively. After this incubation period, the number of colonized granule grains was determined. Granule grains covered with the mycelium or conidia of JKI-BI-1450 were counted as colonized. To determine the conidia concentration, 1 mL of sterile 0.5% (v/v) Tween 80® was added to each petri dish. The granule grains and the suspension was transferred to a tube. The tubes were placed on a vortexer for 10 s and afterwards in an ultrasonic bath (Sonorex RK 52, 35 kHz, Bandelin electronic GmbH & Co. KG, Berlin, Germany) for 15 min. The conidia concentration was determined using a hemocytometer. Based on the colonization and the determined amount of conidia per plate, the conidia per granule grain were determined by dividing the amount of conidia by the number of colonized granule grains. This allowed to determine the conidiation independently of colonization.

Ten conidia of each petri dish were photographed, and the length and width was measured using the program cellSens Standard, Olympus (Figure 2). The conidial index was calculated by dividing the length by the width. The experiment was repeated independently three times.

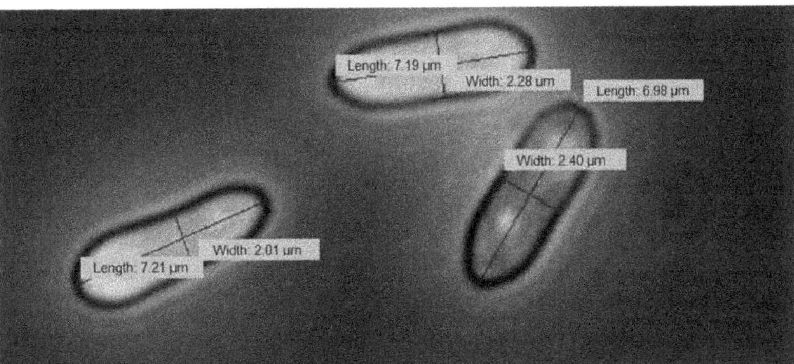

Figure 2. Measurement of conidia.

2.6. Effect of More Unfavorable Incubation Temperature

A 25 °C-granule was produced and formulated as described. Ten granule grains were placed on five separate petri dishes filled with agar (1% (w/v) agar-agar) and were incubated at 5, 10, 25, and 35 °C for 14, 28, and 42 days in the dark, respectively. The granule colonization, the conidia per granule grain, and the conidial size were measured as described above. The experiment was repeated independently three times.

2.7. Effect of Simulated Soil Temperature

To simulate the soil temperature in the field during potato planting as realistically as possible, soil temperature data from a field trial in Uelzen Niedersachsen, Germany were used for adjusting the temperature profile of the incubator. The data were based on temperatures measured in 2018 and 2019, which were obtained from three points inside the potato mound. From these six measurements, averages were calculated every 4 h for 6 weeks, starting from the day of potato planting (7 May 2018 and 8 May 2019) [9]. A 25 °C-granule was produced as described above. Ten granule grains were placed on petri dishes filled with agar (1% (w/v) agar-agar). In total, 60 petri dishes were incubated at the simulated soil temperature, and an additional 60 petri dishes were incubated at a constant temperature of 25 °C in the dark. Ten petri dishes of each treatment were evaluated weekly over 6 weeks. The granule colonization, the conidia per granule grain, and the conidial size were measured as described above. The experiment was repeated independently three times.

2.8. Effect of Temperature on the Germination and the Virulence

2.8.1. Preparation of Fungus Suspension

To determine the germination rate, the temperature-adapted fungus cultures were incubated for 21 days on MPA at the corresponding temperatures in the dark. An undefined number of conidia of each fungus was suspended in a 1.5 mL tube filled with 0.5% (v/v) Tween 80[®]. The tubes were placed on a vortexer for 10 s and afterwards in an ultrasonic bath (Sonorex RK 52, 35 kHz, Bandelin electronic GmbH & Co. KG) for 15 min. The conidia concentration was determined using a hemocytometer, and a conidial suspension of 1×10^6 conidia/mL was made for each adapted fungus by diluting with 0.5% (v/v) Tween 80[®]. The length and width were determined for 100 conidia as previously described.

2.8.2. Speed of Germination

Twenty four droplets of each conidial suspension with a volume of 10 µL were placed on petri dishes filled with MPA, and were incubated at 15, 20, 25, and 30 °C in the dark, respectively. After 3, 6, 9, 12, 15, 18, 21, and 24 h, respectively, three droplets each were cut out of the agar and placed on a slide. On each agar piece, one hundred conidia were observed under the light microscope ($\times 400$), and the percentage of germinated conidia was subsequently determined. Conidia were rated as germinated when the germ tube was longer than the width of the conidia. The experiment was repeated independently three times.

2.8.3. Final Germination Rate

To determine the final germination rate, the experiment was repeated as described above, except that 25 mg/L Benomyl (Sigma Aldrich) was added to the MPA. The germination rate was determined after incubating for 96 h at the corresponding temperatures in the dark. The experiment was repeated independently three times.

2.8.4. Effect of Temperature on the Virulence

The temperature-adapted fungal cultures were incubated for 21 days on MPA at the corresponding temperatures in the dark. Five mL of 0.5% (v/v) Tween 80[®] were added to each plate, following which the suspensions were filtered through four layers of gauze, the supernatant was sonicated (Sonorex RK 52, Bandelin electronic GmbH & Co. KG 35 kHz) for 15 min, and for each adapted fungal culture 5 mL of a conidial suspension with 1×10^7 conidia/mL with 0.5% (v/v) Tween 80[®] was prepared. Afterwards, *Galleria mellonella* larvae (larval stage 5–6) were dipped for 2–3 s into the four fungal suspensions, 0.5% (v/v) Tween 80[®], or sterile deionized water. Each larvae was placed individually into a plastic box (7 cm diameter and 2.5 cm height). Ten larvae per treatment were incubated at 15, 20, 25, and 30 °C in the dark, respectively. The number of dead larvae was determined daily for a period of 14 days. The experiment was independently repeated four times.

2.9. Statistical Analysis

The data was statistically analyzed with the software SAS Studio 3.8. Normality of the data was tested using the Shapiro–Wilk test and the homogeneity of variance was checked by the Levene's test. To compare the colony size and the conidiation at different temperatures, and the effect of extreme incubation temperatures on the fungal growth, the Wilcoxon test ($p < 0.05$) was used. For analyzing the effects of production and incubation temperature, or incubation period on the fungal growth and the conidial size, a generalized linear model with Wald statistics for type 3 analysis and multiple comparison according to the Tukey test (GLMM, $p < 0.05$) was employed. For the analysis of the effect of one parameter (production temperature, incubation temperature, and incubation period), the data of the other two parameters were pooled. The effect of simulated soil temperature on the fungal growth was analyzed for each week by the non-parametric Mann–Whitney U-test ($p < 0.05$). Furthermore, the data of each incubation temperature was compared using a generalized linear model ($p < 0.05$). The final germination rate was determined to be above 99% in all treatments. Due to the method that was used, a more exact determination of the values was not possible, and therefore a statistical evaluation was not conducted. To compare the germination progress, the results of the germination speed and final germination rates were assembled, and τ was calculated for each treatment and every replicate. For the calculation of τ, a non-linear regression according to Dantigny et al. [29] (Formula (1)) was used.

Formula (1)—Calculation of τ:

$$P = P_{max}\left[1 - \frac{1}{1} + \left(\frac{t}{\tau}\right)^{\mathrm{d}}\right] \tag{1}$$

where P is the percentage of germinated conidia in dependence of the maximum percentage of germination, (P_{max}), t designates the germination time, τ is the point of time when 50% of the maximal germinated conidia have been germinated, and d is the design parameter. The design parameter was selected according to Dantigny et al. [29].

Furthermore, the slopes at the inflection point for each treatment and every replicate were calculated with Formula (2).

Formula (2)—Calculation of the slope at the inflection point:

$$Slope = \frac{(\mathrm{d} \times P_{max} \times q)}{\tau \times q^{\frac{1}{\mathrm{d}}} \times (q+1)^2} \tag{2}$$

With $q = \left(\frac{\mathrm{d}-1}{\mathrm{d}+1}\right)$

For analyzing the effects of the production and incubation temperature on τ and the slope at the infection point a generalized linear model with Wald statistics for type 3 analysis and multiple comparison according to the Tukey test (GLMM, $p < 0.05$) was used. For the analysis of the effect of one parameter, the data of the other parameters were pooled.

To analyze the virulence, the mean survival time (ST_{50}) for each replicate was calculated using a survival analysis (Kaplan–Meier) and compared using a generalized linear model ($p < 0.05$).

For analyzing the effect of the production temperature and both controls, data of all incubation temperatures, whereas for the effect of the incubation temperature, data of all production temperatures were pooled.

3. Results

3.1. Colony Size and Conidiation at Different Temperatures

The incubation temperature had a significant effect on the colony size ($x^2 = 137.3667$; df = 4; $p < 0.0001$) and conidiation ($x^2 = 125.8045$; df = 4; $p < 0.0001$). Moreover, the results demonstrate a significantly larger colony size of *M. brunneum* strain JKI-BI-1450 after incubation at 25 or 30 °C than after incubation at 15 or 20 °C (Figure 3). In contrast, the fungus formed significantly more conidia at 20 or 25 °C than at 15 or 30 °C, respectively.

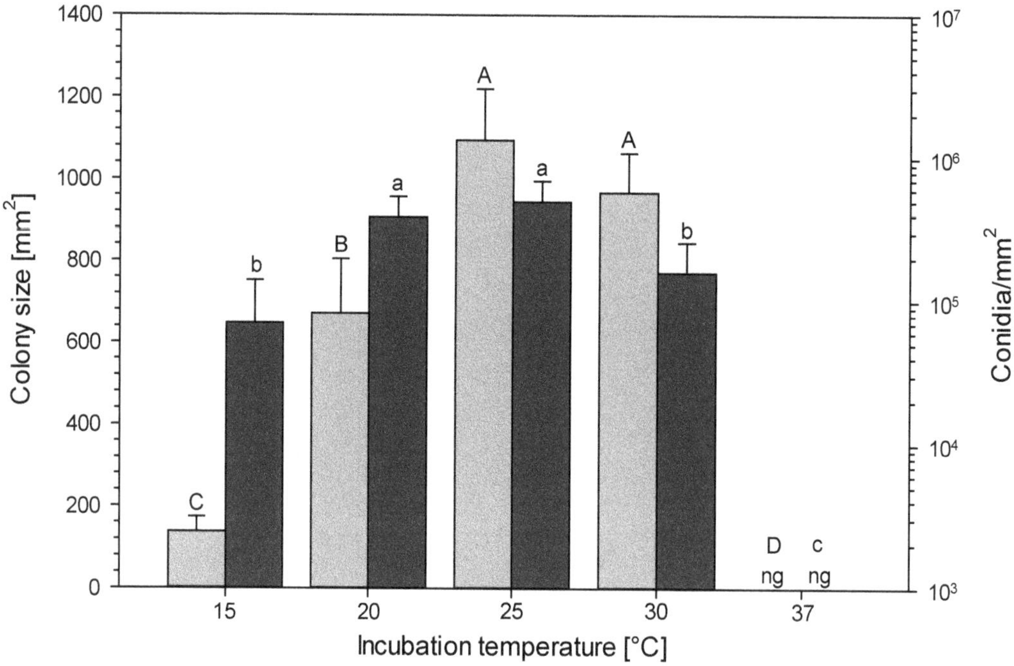

Figure 3. Influence of temperature on colony size (light grey) and conidia per mm^2 (dark grey) of JKI-BI-1450 after 14 days (means and standard deviation). Different capital letters and corresponding lowercase letters indicate significant differences (Wilcoxon test, $p < 0.05$, $n = 30$), ng = no growth.

3.2. Effects of Production and Incubation Temperature on the Fungal Growth on the Soil Granule

The granule colonization by JKI-BI-1450 was found to have been significantly influenced by the production and incubation temperature of the biomass (production temperature: ($x^2 = 2298.94$; df = 3; $p < 0.001$); and incubation temperature ($x^2 = 118.70$; df = 3; $p < 0.001$)) (Table 1). In brief, over all incubation temperatures, granule colonization was 87% for the granules prepared with the biomass produced at 25 °C. At a production temperature of 20 and 30 °C, respectively, granule colonization was significantly reduced but still higher than 70%. When the biomass was produced at 15 °C, only 1% of granule grains were colonized. In contrast, when incubated at 15 °C across all production temperatures combined, maximum colonization was achieved at an average of 72%. With increasing incubation temperatures, the percentage of colonized granule grains decreased continuously, but did not reach less than 50%. The incubation period was found to exhibit no effects on the granule colonization ($x^2 = 0.27$; df = 1; $p = 0.6037$).

Table 1. Effects of the production and incubation conditions on the fungal growth of granules coated with the *Metarhizium brunneum* strain JKI-BI-1450.

Conditions		Growth Parameter				*n*
		Granule Colonization [%]		Conidia per Granule Grain		
Production temperature [°C]	15	1.66 (±4.34) *	D **	$3.85\ (\pm 25.2) \times 10^5$	C	
	20	81.25 (±19.98)	B	$2.47\ (\pm 1.70) \times 10^7$	AB	120
	25	87.17 (±17.40)	A	$2.77\ (\pm 1.66) \times 10^7$	A	
	30	72.25 (±23.24)	C	$2.31\ (\pm 2.00) \times 10^7$	B	
Incubation temperature [°C]	15	72.41 (±42.21)	A	$1.09\ (\pm 1.18) \times 10^7$	C	
	20	61.58 (±37.28)	B	$2.26\ (\pm 1.74) \times 10^7$	B	120
	25	56.92 (±36.53)	B	$3.17\ (\pm 2.45) \times 10^7$	A	
	30	50.92 (±36.69)	C	$1.06\ (\pm 0.97) \times 10^7$	C	
Incubation period [weeks]	2	60.83 (±39.15)	A	$1.32\ (\pm 1.32) \times 10^7$	B	240
	4	60.08 (±38.80)	A	$2.47\ (\pm 2.19) \times 10^7$	A	

* Mean ± standard deviation. ** Means of one condition at one growth parameter with the same letters are not significantly different. (GLMM, $p < 0.05$).

Besides the granule colonization, the conidiation of JKI-BI-1450 on the granule grains was significantly influenced by the production temperature of the biomass ($x^2 = 432.21$; df = 3; $p < 0.0001$). The granule-containing biomass produced at 25 °C formed the most conidia with 2.75×10^7 conidia per granule grain, followed by the 20 °C-granule and 30 °C-granule over all incubation temperatures. The conidiation of the 15 °C-granule was a hundred times lower than the other granules and differed significantly from granules with the other production temperatures. In addition to the production temperature, the incubation temperature of the granules was found to significantly influence the conidiation ($x^2 = 285.84$; df = 3; $p < 0.0001$). The conidia concentration on the granules incubated at 25 °C was significantly the highest with 3.17×10^7 conidia per granule grain, followed by incubation at 20 °C. The conidiation was observed to be the lowest at 15 and 30 °C incubation. They differed significantly from the other incubation temperatures, but not from each other. The incubation period also significantly influenced the conidiation ($x^2 = 122.85$; df = 1; $p < 0.0001$). The conidia per granule grain doubled from the second to the fourth week of incubation, and differed significantly from each other, summarized over all production and incubation temperatures.

In addition to colonization and conidiation, the conidial size was also examined. Since the fungal growth on the 15 °C-granule was extremely low, the conidial size could not be determined for this granule (Table 2). The production temperature of the biomass did not affect the length ($x^2 = 4.24$; df = 2; $p = 0.1199$) but did significantly impact the width ($x^2 = 11.20$; df = 2; $p = 0.0037$) and the conidial index ($x^2 = 13.21$; df = 2; $p = 0.0014$). Significantly wider and more roundish conidia were formed on the granule-containing biomass produced at 20 °C compared to the higher temperatures. The incubation temperature of the granules also influenced the size of the conidia (length: ($x^2 = 542.28$; df = 3; $p < 0.0001$); width: ($x^2 = 166.61$; df = 3; $p < 0.0001$); and index; ($x^2 = 169.62$; df = 3; $p < 0.0001$)). Conidia were significantly longer and wider when the granules were incubated at 15 °C compared to incubation at higher temperatures. The conidial index also revealed that the conidia were significantly more rounded when the granules were incubated at higher rather than lower temperatures. Furthermore, the conidia were significantly longer after an incubation period of 4 weeks than after 2 weeks ($x^2 = 11.24$; df = 1; $p = 0.0008$). The incubation period was found to have no influence on the conidial width ($x^2 = 1.44$; df = 1; $p = 0.2306$) or the conidial index ($x^2 = 1.86$; df = 1; $p = 0.1727$).

Table 2. Effects of the production and incubation conditions on the conidial size of the *Metarhizium brunneum* strain JKI-BI-1450 formed on the granules.

Conditions		Conidial Size					
		Conidial Length [µm]		Conidial Width [µm]		Conidial Index *		*n*

Conditions		Conidial Length [µm]		Conidial Width [µm]		Conidial Index *		*n*
Production temperature [°C]	20	6.70 (±0.58) **	A ***	2.21 (±0.22)	A	3.05 (±0.39)	B	1200
	25	6.75 (±0.58)	A	2.19 (±0.20)	B	3.10 (±0.38)	A	1200
	30	6.73 (±0.67)	A	2.19 (±0.22)	B	3.11 (±0.43)	A	1166
Incubation temperature [°C]	15	7.06 (±0.55)	A	2.26 (±0.22)	A	3.16 (±0.42)	A	900
	20	6.78 (±0.50)	B	2.15 (±0.19)	C	3.18 (+0.39)	A	900
	25	6.51 (±0.52)	C	2.16 (±0.19)	C	3.05 (±0.38)	B	900
	30	6.54 (±0.70)	C	2.22 (±0.23)	B	2.97 (±0.37)	C	866
Incubation period [weeks]	2	6.69 (±0.59)	B	2.19 (±0.21)	A	3.08 (±0.39)	A	1780
	4	6.76 (±0.63)	A	2.20 (±0.22)	A	3.10 (±0.41)	A	1786

* The index is calculated by dividing the length by the width. ** Mean ± standard deviation. *** Means of one condition at one conidial size parameter with the same letters are not significantly different. (GLMM, $p < 0.05$).

3.3. Effect of More Unfavorable Incubation Temperature on the Fungal Growth

Within the investigation period, the fungus was not able to grow or to form conidia on the granule incubated at 5 or 35 °C (Table 3). The four incubation temperatures and three incubation periods resulted in significant differences in colonization ($x^2 = 173.0394$; df = 11; $p < 0.0001$) and conidiation ($x^2 = 160.1709$; df = 11; $p < 0.0001$). Incubating for 2 weeks at 10 °C revealed a significantly lower granule colonization, only reaching 80%, whereas 25 °C achieved nearly 100% granule colonization in the same timeframe. No significant differences were observed between 10 or 25 °C for incubation periods of 4 and 6 weeks, respectively. The conidiation was significantly lower after incubation at 10 °C rather than at 25 °C for each evaluation time point. After 2 weeks, JKI-BI-1450 did not form any conidia on the granule. After 4 weeks, approximately 5×10^3 conidia per granule grain were formed. The conidia concentration increased a hundred-fold after incubating for an additional 2 weeks. When incubated at 25 °C, the conidia concentration was above 10^7 at all evaluation time points.

Table 3. Effects of the incubation temperature and period on the fungal growth of a granule based on biomass produced at 25 °C.

Incubation Temperature [°C]	Incubation Period [weeks]	Growth Parameter				*n*
		Granule Colonization [%] *		Conidia per Granule Grain		
5	2	0 (±0)	C **	0 (±0)	C	15
	4	0 (±0)	C	0 (±0)	C	15
	6	0 (±0)	C	0 (±0)	C	15
10	2	80 (±15.49)	B	0 (±0)	C	15
	4	98.67 (±3.49)	A	$4.67 (±6.94) \times 10^3$	BC	15
	6	99.33 (±2.49)	A	$5.99 (±9.29) \times 10^5$	B	15
25	2	99.33 (±2.49)	A	$2.41 (±1.59) \times 10^7$	A	15
	4	100 (±0)	A	$3.74 (±1.39) \times 10^7$	A	15
	6	99.33 (±2.49)	A	$3.86 (±1.03) \times 10^7$	A	15
35	2	0 (±0)	C	0 (±0)	C	15
	4	0 (±0)	C	0 (±0)	C	15
	6	0 (±0)	C	0 (±0)	C	15

* Mean ± standard deviation. ** Means of one growth parameter with the same letters are not significantly different. (Wilcoxon test, $p < 0.05$).

In addition, significantly longer ($x^2 = 318.15$; df = 1; $p < 0.0001$) and wider ($x^2 = 420.42$; df = 1; $p < 0.0001$) conidia were formed on granules incubated at 10 °C compared to 25 °C (Table 4). The conidial index did not differ between these two incubation temperatures ($x^2 = 0.50$; df = 1; $p = 0.4796$). The incubation period had no effect on the conidial size length:

$(x^2 = 2.57; \mathrm{df} = 1; p = 0.1089)$; width $(x^2 = 0.08; \mathrm{df} = 1; p = 0.7716)$; and index $(x^2 = 2.54; \mathrm{df} = 1; p = 0.1107))$.

Table 4. Effects of the incubation conditions on the conidial size of the *Metarhizium brunneum* strain JKI-BI-1450 on the granules.

Conditions		Conidial Size					
		Conidial Length [μm]		Conidial Width [μm]		Conidial Index *		n
Incubation temperature [°C]	10	7.96 (±1.18) **	A ***	2.75 (±0.38)	A	2.93 (±0.44)	A	210
	25	6.62 (±0.50)	B	2.24 (±0.15)	B	2.97 (±0.30)	A	300
Incubation period [weeks]	4	7.18 (±1.12)	A	2.43 (±0.35)	A	2.97 (±0.37)	A	240
	6	7.16 (±1.03)	A	2.47 (±0.37)	A	2.92 (±0.36)	A	270

* The index is calculated by dividing length by width. ** Mean ± standard deviation. *** Means of one condition with one conidial size parameter and the same letters are not significantly different. (GLMM, $p < 0.05$).

3.4. Effect of Simulated Soil Temperature on the Fungal Growth

First, the colonization at 25 °C was compared with the colonization at simulated soil temperature for each week independently. In this comparison of the incubation temperatures, the granule colonization was found to not differ significantly, except for the incubation periods of two and six weeks (Figure 4). Here, the colonization was significantly higher for the granules that were incubated at the simulated soil temperature compared to 25 °C (week 1: $(x^2 = 2.3677; \mathrm{df} = 1; p = 0.1239)$; week 2: $(x^2 = 8.3617; \mathrm{df} = 1; p = 0.0038)$; week 3: $(x^2 = 0.4229; \mathrm{df} = 1; p = 0.5155)$; week 4: $(x^2 = 3.6004; \mathrm{df} = 1; p = 0.0578)$; week 5: $(x^2 = 1.6223; \mathrm{df} = 1; p = 0.2028)$; and week 6: $(x^2 = 50.7817; \mathrm{df} = 1; p < 0.0001))$. Secondly, the colonization of the fixed and simulated incubation temperature was compared separately over the experimental period. Both the incubation at 25 °C and at the simulated soil temperature showed no significant increases in colonization from the 2nd week onwards and was between 90 and 100% (25 °C: $(x^2 = 12.54; \mathrm{df} = 5; p = 0.0281)$; and simulated soil temperature $(x^2 = 25.22; \mathrm{df} = 5; p < 0.0001))$.

Conidiation at 25 °C was also compared with the conidiation at simulated soil temperature for each week of the experiment. During week 4, the conidia concentration was significantly lower for the granules incubated at the simulated soil temperature rather than at 25 °C (week 1: $(x^2 = 17.1323; \mathrm{df} = 1; p < 0.0001)$; week 2: $(x^2 = 15.5149; \mathrm{df} = 1; p < 0.0001)$; week 3: $(x^2 = 13.6696; \mathrm{df} = 1; p = 0.0002)$; week 4: $(x^2 = 1.3300; \mathrm{df} = 1; p = 0.2488)$; week 5: $(x^2 = 0.2310; \mathrm{df} = 1; p = 0.6308)$; and week 6: $(x^2 = 3.1751; \mathrm{df} = 1; p = 0.0748))$. Thereafter, the conidia per granule grain no longer differed. The conidiation at one incubation temperature was also compared separately over the duration of the experiment. During incubation at 25 °C the maximum conidia concentration was achieved after 2 weeks with $1.86\text{--}2.17 \times 10^7$ conidia per granule grain $(x^2 = 88.47; \mathrm{df} = 5; p < 0.0001)$, while during incubation at simulated soil temperature this was not achieved until after 4 weeks with $1.44\text{--}1.79 \times 10^7$ conidia per granule grain $(x^2 = 197.58; \mathrm{df} = 5; p < 0.0001)$.

Since no conidia were formed after 1 week of incubation at the simulating soil temperature, this time point was not considered when determining the conidial size. The granule formed significantly longer $(x^2 = 862.81; \mathrm{df} = 1; p < 0.0001)$ and wider $(x^2 = 7.45; \mathrm{df} = 1; p = 0.0063)$ conidia during incubation at the simulated soil temperature rather than at 25 °C, where the granule was significantly rounder $(x^2 = 285.16; \mathrm{df} = 1; p < 0.0001)$ (Table 5). The incubation period also influenced the conidial size. After 3 weeks of incubation, the conidia were significantly the longest $(x^2 = 14.83; \mathrm{df} = 4; p = 0.0051)$, whereas the conidia were the widest $(x^2 = 19.60; \mathrm{df} = 4; p = 0.0006)$ after 4 weeks of incubation. In addition, the conidia were significantly most elongated after 2 and 3 weeks of incubation and became more round with increasing incubation time $(x^2 = 23.93; \mathrm{df} = 4; p < 0.0001)$.

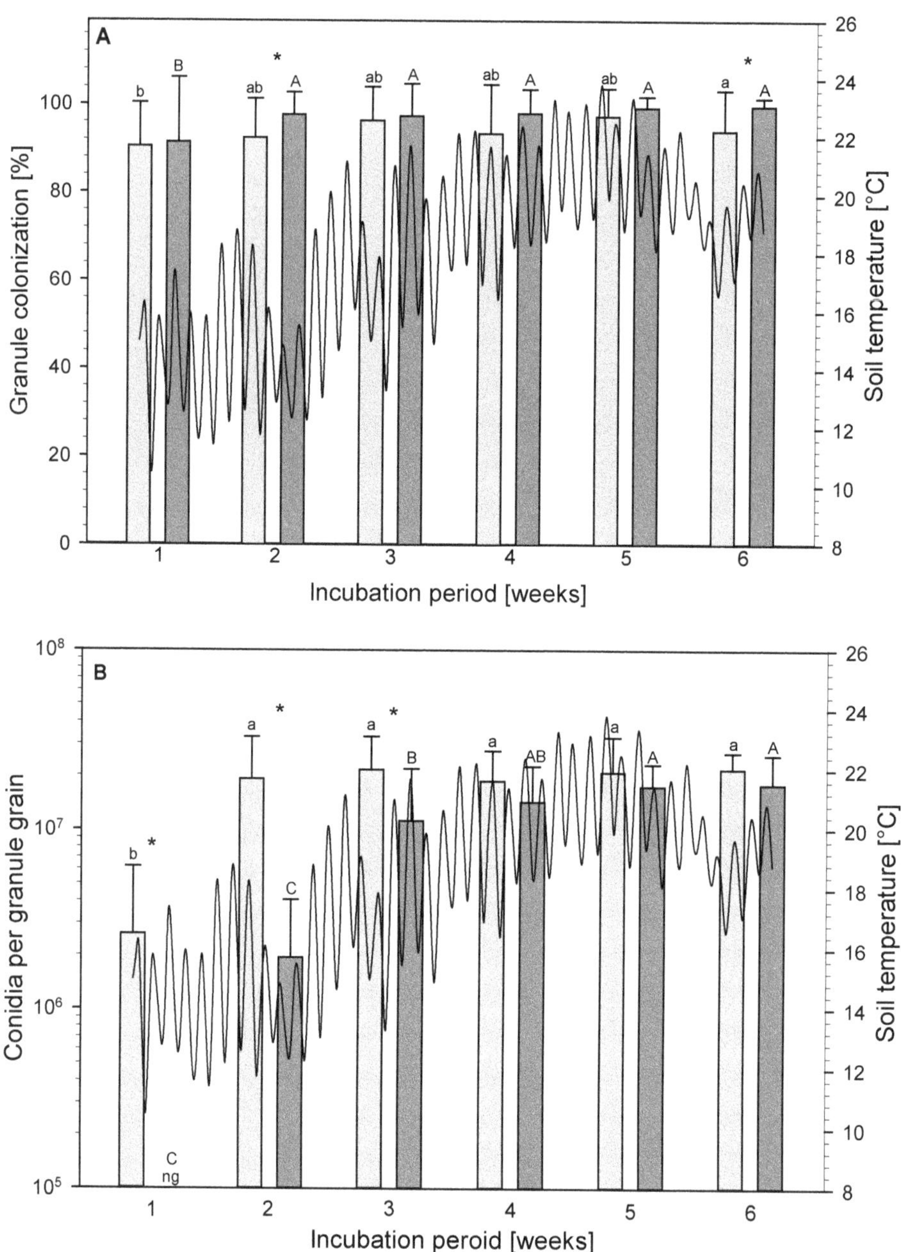

Figure 4. Effect of incubation at 25 °C (light grey) and at simulated soil temperature (dark grey) on the granule colonization (**A**), and on the number of conidia per granule grain (**B**) (means and SD). Means with * at one evaluation time are significantly different (Mann–Whitney U-Test, $p < 0.5$, $n = 15$). The differences between the evaluation times of the incubation at 25 °C are represented by different lower case letters and for the simulated soil temperature by different upper case letters (GLMM, $p < 0.05$, $n = 30$). The black line indicates the simulated soil temperatures from the day the potatoes were planted until 6 weeks thereafter ($n = 3$), ng = no growth.

Table 5. Effects of the incubation conditions on the conidial size of the *Metarhizium brunneum* strain JKI-BI-1450 on the granules.

Condition		Conidial Size						
		Conidial Length [µm]		Conidial Width [µm]		Conidial Index *		*n*
Incubation temperature [°C]	Soil temp.	6.96 (±0.54) **	A ***	2.23 (±0.22)	A	3.15 (±0.40)	A	1500
	25	6.38 (±0.55)	B	2.21 (±0.19)	B	2.91 (±0.38)	B	1500
Incubation period [weeks]	2	6.70 (±0.63)	AB	2.19 (±0.20)	B	3.07 (±0.39)	A	600
	3	6.73 (±0.64)	A	2.22 (±0.21)	AB	3.07 (±0.43)	A	600
	4	6.63 (±0.59)	B	2.23 (±0.22)	A	3.00 (±0.43)	B	600
	5	6.66 (±0.63)	AB	2.24 (±0.21)	A	3.00 (±0.42)	B	600
	6	6.63 (±0.59)	B	2.23 (±0.19)	AB	3.00 (±0.38)	B	600

* The index is calculated by dividing the length by the width. ** Mean ± standard deviation. *** Means of one condition at one conidial size parameter with the same letters are not significantly different. (GLMM, $p < 0.05$).

3.5. Effect of Temperature on Conidial Size and Germination

Incubation at 15 and 20 °C resulted in significantly longer ($x^2 = 234.64$; df = 3; $p < 0.0001$) and rounder ($x^2 = 134.10$; df = 3; $p < 0.0001$) conidia compared to higher temperatures (Table 6). However, the incubation temperature had no significant influence on the conidia width ($x^2 = 6.53$; df = 3; $p = 0.0887$).

Table 6. Effects of the incubation temperature on the conidial size of the *Metarhizium brunneum* strain JKI-BI-1450 on MPA dishes after 21 days.

Condition		Conidial Size						
		Conidial Length [µm]		Conidial Width [µm]		Conidial Index *		*n*
Incubation temperature [°C]	15	7.04 (±0.36) **	A ***	2.26 (±0.15)	A	3.13 (±0.26)	A	300
	20	6.97 (±0.37)	A	2.27 (±0.19)	A	3.09 (±0.30)	AB	300
	25	6.47 (±0.51)	C	2.26 (±0.15)	A	2.87 (±0.29)	C	300
	30	6.82 (±0.68)	B	2.24 (±0.17)	A	3.06 (±0.33)	B	300

* The index is calculated by dividing the length by the width. ** Mean ± standard deviation. *** Means of one condition with one conidial size parameter and the same letters are not significantly different. (GLMM, $p < 0.05$).

The germination was found to be faster when the fungus was produced at higher temperatures (Figure 5). This effect was observed across all incubation temperatures.

The statistical evaluation of this experiment is presented in Table 7. τ was reached significantly fastest with a production temperature of 30 °C (after 12.3 h), closely followed by 25 °C (after 12.8 h) ($x^2 = 1376.52$; df = 3; $p < 0.0001$). When the fungus was produced at 15 and 20 °C, τ was reached significantly slower (3 h later). The incubation temperature also exhibited a significant influence on τ ($x^2 = 2822.81$; df = 3; $p < 0.0001$). The higher the temperatures at which the fungus was incubated, the faster τ was reached. T achieved values between 11 h at 30 °C and 17 h at 15 °C, with a significant difference observed among all incubation temperatures.

The slope at the inflection point was also significantly influenced by the production temperature ($x^2 = 11.99$; df = 3; $p = 0.0074$). When the fungus was produced at 20 °C, the slope was at its steepest, while at 25 °C, the slope was found to be at its lowest (Table 7). The slopes of the other two production temperatures were deemed to be in between. When incubated at the higher temperatures (25 or 30 °C, respectively), the slope of the germination rate was significantly steeper than when incubated at the lower temperatures (15 or 20 °C, respectively) ($x^2 = 92.95$; df = 3; $p < 0.0001$).

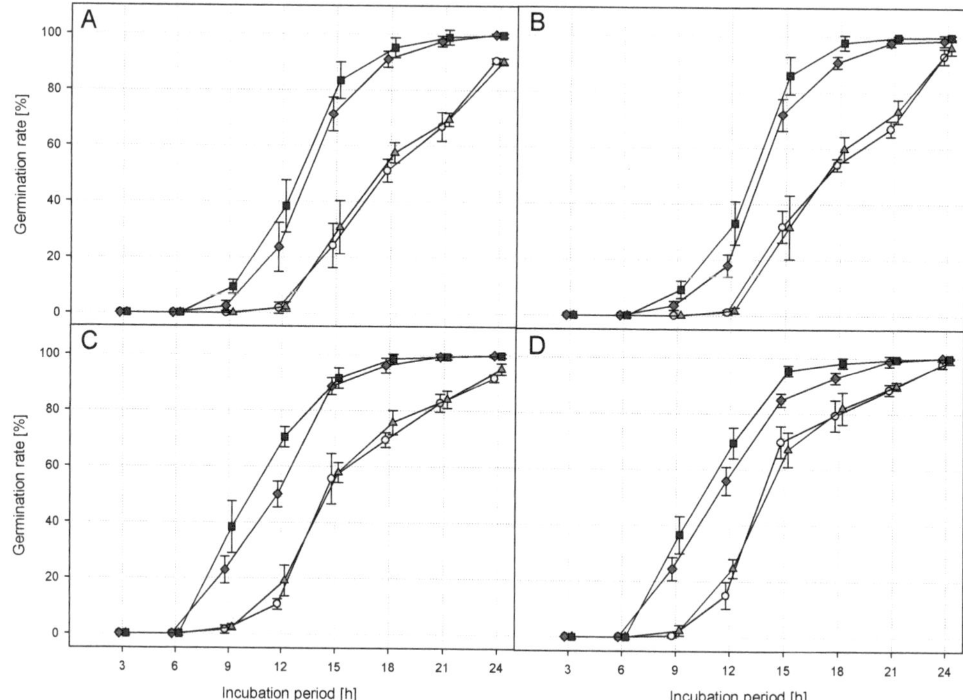

Figure 5. Effects of the production and incubation temperature on the germination rate of *Metarhizium brunneum* JKI-BI-1450 are shown (means and SD). The symbols show the different production temperatures (○ = 15 °C, △ = 20 °C, ◆ = 25 °C, ■ = 30 °C) and the letters correspond to the incubation temperature ((**A**) = 15 °C, (**B**) = 20 °C, (**C**) = 25 °C, and (**D**) = 30 °C). The symbols are slightly offset for better visibility. $n = 9$.

Table 7. Effects of the production and incubation temperature of the germination of JKI-BI-1450 after 96 h, and of the germination process over 24 h. τ is the time point where 50% of the maximal germinated conidia are germinated.

Conditions		Germination after 96 h [%]	τ		Slope Inflection		n
Production temperature [°C]	15	99.53 (±0.70) *	15.42 (±2.44)	C **	13.69 (±4.65)	AB **	9
	20	99.00 (±1.12)	15.32 (±2.15)	C	15.33 (±5.95)	B	9
	25	99.67 (±0.59)	12.81 (±2.14)	B	13.00 (±2.27)	A	9
	30	99.17 (±1.08)	12.31 (±1.70)	A	14.99 (±3.51)	AB	9
Incubation temperature [°C]	15	99.47 (±0.70)	16.18 (±1.83)	D	12.07 (±4.48)	A	9
	20	99.19 (±1.04)	15.71 (±1.64)	C	11.32 (±2.25)	A	9
	25	99.39 (±0.96)	12.57 (±1.18)	B	15.99 (±3.25)	B	9
	30	99.31 (±1.01)	11.40 (±1.39)	A	17.63 (±3.74)	B	9

* Mean ± standard deviation. ** Means of τ and slope inflection at one condition with the same letters are not significantly different. (GLMM, $p < 0.05$).

3.6. Effect of Temperature on the Virulence

The production temperature of the fungus was found to have no influence on the ST$_{50}$ but did differ significantly from the controls ($x^2 = 129.33$; df = 5; $p < 0.0001$) (Table 8). The ST$_{50}$ of the water control was significantly the highest with 10.83 days followed by the Tween 80® control (7.63 days), and the fungi treatments with an average of 4 days. However, it was found that the survival time significantly decreased with increasing incubation temperature (x^2 = 102.65; df = 3; p < 0.0001). ST_{50} was highest when the larvae were incubated at 15 °C, with an average of 6.03 days, and lowest at 30 °C, with an average of 3.16 days, respectively. The final mortality of the fungal treatments in relation to the production temperature did not differ significantly from each other (95–100%) but from the control treatments (x^2 = 560.82; df = 5; p < 0.0001). The final mortality at 15 to 25 °C incubation was approximately 100% and differed significantly from the one observed at 30 °C (x^2 = 17.17; df = 3; p = 0.0007).

Table 8. Effects of the production and incubation temperature on the virulence of JKI-BI-1450 against *Galleria mellonella*.

Conditions		ST_{50} [d]		Final Mortality [%]		n
Control	Water	10.83 (±2.85) *	C **	18.75 (±19.28)	C	16
	0.5% Tween 80®	7.63 (±5.16)	B	55.00 (±21.6)	B	16
Production temperature [°C]	15	3.97 (±1.39)	A	98.13 (±5.44)	A	16
	20	4.03 (±1.35)	A	100 (±0)	A	16
	25	4.54 (±1.84)	A	95.63 (±10.31)	A	16
	30	4.04 (±1.74)	A	98.13 (±7.5)	A	16
Incubation temperature [°C]	15	6.03 (±1.31)	C	99.38(±2.5)	A	16
	20	4.06 (±1.32)	B	100 (±0)	A	16
	25	3.33 (±0.76)	AB	100 (±0)	A	16
	30	3.16 (±0.94)	A	92.5 (±12.38)	B	16

* Mean ± standard deviation. ** Means of ST_{50} and final mortality at one condition with the same letters are not significantly different. (GLMM, p < 0.05).

4. Discussion

The *M. brunneum* strain JKI-BI-1450 can form mycelium and conidia across a wide temperature range. This fungus generated the largest colony size and the most conidia/mm² at 25 °C. Similar results for the radial growth of *M. brunneum* were reported by Keyser et al. [8] and Kryukov et al. [7], respectively. More specifically, the optimum for mycelial growth of JKI-BI-1450 was 25–30 °C, and for conidiation 20–25 °C, respectively. Thomas and Jenkins [30], and Onsongo et al. [27], determined a higher temperature optimum for mycelia growth than for conidia formation for *Metarhizium anisopliae* and *Metarhizium flavoviride* as well. This effect has also been observed for other fungi such as *Mycosphaerella* var. *difformis fijiensis* [31,32], and *Pyrenophora semeniperda* [32]. These results suggest that the optima for mycelial growth and conidiation are slightly different. This could indicate that the fungus does not form mycelium and conidia uniformly in the field at fluctuating temperatures (day and night rhythm). As the soil temperature at the time of potato planting in Germany with 9.6–14.7 °C [9] is far below the optimum temperature of JKI-BI-1450 (25 °C), we tried to improve its growth under sub-optimal conditions. However, the assumption that the growth of the fungus at unfavorable temperatures can be improved by prior exposure to this stress could not be confirmed. We found no indications that the strain JKI-BI-1450 is able to adapt to unfavorable temperatures. Andrade-Linares et al. [17] also showed that not all of the fungi they assessed were able to increase their growth through priming. Most studies of stress adaptation have been conducted with conidia exposed to the stressor immediately prior to incubation. Therefore, further research on the quality of conidia of JKI-BI-1450 directly exposed to lower temperatures would be essential to clarify whether this fungus is able to adapt to stressful conditions. The results of this experiment show that the growth of the fungus and conidiation on the granules, which is crucial for its effectiveness, are dependent on and can be positively influenced by its previous production temperature. When the biomass, which was produced at the optimal growing temperature of 25 °C, was used for the formulation of the granule, both growth parameters were at their highest regardless of the incubation temperature. Therefore, the fungus is better able to grow at unfavorable temperatures on granules produced at this temperature than on ones produced at higher or lower temperatures.

The temperature during liquid production has an influence on the production of secondary metabolites of *Nigrospora* sp., which was previously shown by Arumugam et al. [33]. Furthermore, Kim et al. [34] demonstrated that the incubation temperature at liquid production causes the synthesis of phytases from *Aspergillus* sp. Therefore, it would be interesting to examine in further experiments whether the production temperature can affect the metabolism of JKI-BI-1450, which might have an impact on the growth of the granules. The production temperature can also influence further formulation steps. Jin et al. [35] showed that submerged spores of *Trichoderma harzianum* were able to survive vacuum drying better when produced at 32 °C, rather than at higher or lower temperatures. A varying survival of the biomass during fluid bed-drying could be a possible explanation for our diverse results of the different production temperatures. Another possible explanation could be that the different production temperatures resulted in a varying proportion of mycelia and submerged spores in the suspensions used for the production of the granules. The granules were prepared with a defined biomass concentration, not with a defined spore concentration. Stephan et al. [1] demonstrated that the mycelium, submerged spores, or conidia of *M. brunneum* JKI-BI-1339 were suitable for the production of a granule by fluid bed-drying. When we produced granules only based on mycelium of JKI-BI-1450, the fungus was not able to grow on this granule (Seib, unpublished data). Therefore, further experiments have to be conducted to determine the importance of the mycelium/spore proportion on the granule quality. In our experiments, the submerged spore concentration of the biomass suspensions were determined but the difference in the growth of the fungus on the different granules cannot be due to the amount of submerged spores, but maybe to a difference in the characteristics of the spores themselves. The spore concentration of the biomass suspensions averaged 3.7×10^6 spores/mL.

Our results show the highest granule colonization after an incubation at 15 °C. However, this result is in contrast to both our and other results from different studies where a temperature optimum of *Metarhizium* was observed between 25 to 30 °C [3,5,6,19,20,36–38], respectively. We attribute this to an increased contamination at higher incubation temperatures. At elevated temperatures, the granules were more often colonized by bacteria than after incubation at lower temperatures. This prevented the growth of the fungus as a consequence. During the process of fluid bed-drying, non-sterile air is blown into the cylinder, which could lead to these contaminations. In subsequent experiments, the effects of higher colonization by incubation at lower temperatures was no longer observed. Therefore, and as the germination and virulence tests were performed exclusively with conidia and not with the granules, the results of all the following experiments were not influenced. The highest conidiation was found when incubated at 25 °C, which was also supported by the other experiments we conducted in this study. Thomas and Jenkins [30], and Tefera and Pringle [39], also showed a temperature optimum for conidiation at 24–25 °C for *Metarhizium*. On a 25 °C-granule, JKI-BI-1450 did not grow between 10 and 30 °C. The comparison of growth at 10 and 25 °C showed that after 4 weeks of incubation, there were no significant differences observed concerning granule colonization, whereas the conidia concentration at all evaluation times was lower at 10 °C than at 25 °C, respectively. The mycelium of *Metarhizium* strains can be formed at 10 °C with no or reduced sporulation, as previously confirmed by Skalický et al. [22]. To better understand the growth of the fungus under field conditions, a 25 °C-granule was incubated at a simulated soil temperature as well as at 25 °C, and the growth between the two was compared. Colonization did not differ, but conidiation was slower from the simulated soil temperature. The fungus formed conidia on the granule only after two weeks at simulated soil temperature. Consecutively, the pest insects in the soil can only become infected two weeks after an application of the granule. Since the main aim was to protect the progeny tubers and not the seed potato, a later conidiation would not necessarily present a problem. In fact, it would allow other micro- and macro-organisms in the soil time to colonize or feed on the grains.

An influence of fluctuating temperatures on the colony growth and virulence of *Beauveria bassiana* was investigated by Ghazanfar et al. [40]. The more the temperatures fluctuate, the more the growth was reduced compared to the growth at a constant 25 °C. Furthermore, the effect against *Heliothis virescens* was reduced by incubation at fluctuating temperatures. This effect was not seen for *Spodoptera littoralis*, which was also investigated.

The influence of the incubation temperature on the size of the conidia has been proven across all experiments. For strain JKI-BI-1450, incubation at lower temperatures always led to the formation of longer and partly wider conidia. Similar results were also shown by Glare et al. [41] for 16 *Metarhizium* strains from seven species. For other fungi, this effect were reported by Phillips [42,43] for *Monilinia fructicola*, by Tian and Bertolini [44] for *Monilinia laxa,* and by Tian and Bertolini [45] for *Botryfis allii* and *Penicillium hirsutum,* respectively. In contrast, Maitlo et al. [46] described that *Fusarium oxysporum* f. sp. *ciceris* formed the longest and broadest conidia when incubated at 30 °C, while at lower or higher temperatures the conidia were smaller. This strongly suggests that the relationship between the temperature and conidial size differs among the fungi of different genera and must be examined individually for each fungus. Further research must be conducted to understand the meaning of conidial size for biological activity. Since on the granule, infectious conidia were firstly formed in the soil, we investigated the effects of conidia production temperature, and thus their size, as well as the subsequent incubation temperature on germination and virulence.

We determined in our study that the higher the temperature at which the fungus was produced and incubated, the faster it reached the time at which 50% of the finally germinated conidia were germinated (τ). Our results are in contrast to Phillips [42], where conidia produced at lower temperatures germinated faster. This suggests that the relationship between the temperature and germination is strongly dependent on the fungus, as well as the conidial size. However, our experiments revealed a germination rate of approximately 90% after 24 h, and over 99% after 96 h at all production and incubation temperatures. At lower temperatures, the conidia germinated more slowly, but the amount of germinated conidia after 96 h was not lower than at higher temperatures. A comparable time delay of germination was also shown by Skalický et al. [22] for nine *M. anisopliae* strains comparing the germination rates at 15 and 20 °C after 24 and 48 h, respectively. Dimbi et al. [3] determined a much greater influence of the incubation temperature on the germination rate after 24 h for several *M. anisopliae* strains than within the results of our experiments. All strains showed a maximum germination of about 90% at 25 °C, whereas the germination rate at 15 °C was below 10%, respectively. Results on the influence of the germination speed on the virulence are inconsistent within the literature. Some studies show that rapidly germinating spores are more virulent [47–51], but there are also studies that conclude the opposite [52].

To investigate this further, the influence of the temperature on the virulence was examined. The production temperature, and thus the conidial size and the speed of germination had no effect on the ST_{50} or the final mortality of *G. mellonella*. This contradicts the results from Altre et al. [48] on the virulence of different strains of *Paecilomyces fumosoroseus* against *Plutella xylostella*, where larger conidia were more virulent. Furthermore, since the production temperature of the conidia affects their germination speed but not the final germination and virulence, it appears that a successful infection in this context depends on more than just fast germination. Our experiments showed that the higher the incubation temperature, the lower the ST_{50}. In case of the fungus JKI-BI-1450, a high incubation temperature ranging from 25–30 °C was found to be correlated with a great mycelia growth in regard to colony size, fast germination, and fast virulence. Besides a fast germination, a faster growth of the fungus on the cuticle promotes a successful and faster penetration through the fungus by reaching the susceptible areas faster.

Dimi et al. [3], Bugeme et al. [26], Fargues et al. [28], and Onsongo et al. [27] demonstrated the highest or among the highest mortality of *Metarhizium* when incubated at 30 °C. In contrast, our results indicated a significantly lower mortality at 30 °C compared to the other incubation temperatures, but with a low ST_{50}. The influence of higher incubation temperatures on the speed of insect development could be a possible explanation for this, as earlier molting potentially strips off conidia. To verify this, one would need to repeat the experiment with younger larvae.

In summary, we were able to show that the growth of an entomopathogenic fungus on a soil granule can be influenced and improved by its production conditions. Colonization and conidiation of a granule were at their highest when the fungus was produced at its optimum temperature. Furthermore, it was observed that the fungus was also able to grow on the granules and form conidia at lower temperatures and at simulated soil temperatures at the time of potato planting. This was particularly important, as our fungal strain and formulation process were selected for the control of wireworms in potatoes. However, the growth of the fungus, along with the germination and duration until lethal effect on the insects were found to be prolonged by lower temperatures. If this effect endangers the potential of the granule to control wireworms, an application at higher temperatures, for example in the year before planting the potatoes, would therefore give the fungus time to grow, sporulate, and establish in the soil, and reduce the population of the pest insect before potato cultivation.

Author Contributions: T.S.: Coordinated all experiments, evaluated the data statistically, collected the data, wrote the paper; Co-authors: K.F. and A.M.S.: Collected the data, D.S.: Supervised the study, involved in planning the experiments, involved in writing the paper. All authors have read and agreed to the published version of the manuscript.

Funding: The research was funded by the German Federal Ministry of Food and Agriculture (BMEL), by resolution of the German Federal Parliament under the funding code 2814906415.

Institutional Review Board Statement: Not applicable.

Informed Consent Statement: Not applicable.

Data Availability Statement: The data that support the findings of this study are available from the corresponding author upon reasonable request.

Conflicts of Interest: The authors declare no conflict of interest.

References

1. Stephan, D.; Bernhardt, T.; Buranjadze, M.; Seib, C.; Schäfer, J.; Maguire, N.; Pelz, J. Development of a fluid-bed coating process for soil-granule-based formulations of Metarhizium brunneum, Cordyceps fumosorosea or Beauveria bassiana. *J. Appl. Microbiol.* **2020**, *131*, 307–320. [CrossRef]
2. Ortiz-Urquiza, A.; Keyhani, N.O.; Quesada-Moraga, E. Culture conditions affect virulence and production of insect toxic proteins in the entomopathogenic fungus *Metarhizium anisopliae*. *Biocontrol Sci. Technol.* **2013**, *23*, 1199–1212. [CrossRef]
3. Dimbi, S.; Maniania, N.K.; Lux, S.A.; Mueke, J.M. Effect of constant temperatures on germination, radial growth and virulence of *Metarhizium anisopliae* to three species of African tephritid fruit flies. *Biocontrol* **2004**, *49*, 83–94. [CrossRef]
4. Ouedraogo, A.; Fargues, J.; Goettel, M.S.; Lomer, C.J. Effect of temperature on vegetative growth among isolates of *Metarhizium anisopliae* and *M. flavoviride*. *Mycopathologia* **1997**, *137*, 37–43. [CrossRef]
5. Fargues, J.; Maniania, N.K.; Delmas, J.C.; Smits, N. Influence de la température sur la croissance in vitro d'hyphomycètes entomopathogènes. *Agronomie* **1992**, *12*, 557–564. [CrossRef]
6. Ekesi, S.; Maniania, N.K.; Ampong-Nyarko, K. Effect of Temperature on Germination, Radial Growth and Virulence of *Metarhizium anisopliae* and Beauveria bassiana on *Megalurothrips sjostedti*. *Biocontrol Sci. Technol.* **1999**, *9*, 177–185. [CrossRef]
7. Kryukov, V.; Yaroslavtseva, O.; Tyurin, M.; Akhanaev, Y.; Elisaphenko, E.; Wen, T.-C.; Tomilova, O.; Tokarev, Y.; Glupov, V. Ecological preferences of *Metarhizium* spp. from Russia and neighboring territories and their activity against Colorado potato beetle larvae. *J. Invertebr. Pathol.* **2017**, *149*, 1–7. [CrossRef]
8. Keyser, C.A.; Fernandes, É.K.K.; Rangel, D.E.N.; Roberts, D.W. Heat-induced post-stress growth delay: A biological trait of many Metarhizium isolates reducing biocontrol efficacy? *J. Invertebr. Pathol.* **2014**, *120*, 67–73. [CrossRef] [PubMed]
9. Paluch, M. Biological Control of Wireworms (*Agriotes* spp.) in Potato Cultivation Using the Entomopathogenic Fungus *Metarhizium brunneum*: Factors that Influence the Effectiveness of Mycoinsecticide Formulations. Ph.D. Thesis, Department of Biology, Darmstadt University of Technology, Darmstadt, Germany, 2021.
10. Hallsworth, J.E.; Magan, N. Water and Temperature Relations of Growth of the Entomogenous Fungi Beauveria bassiana, *Metarhizium anisopliae*, and Paecilomyces farinosus. *J. Invertebr. Pathol.* **1999**, *74*, 261–266. [CrossRef] [PubMed]
11. Hallsworth, J.E.; Magan, N. Culture Age, Temperature, and pH Affect the Polyol and Trehalose Contents of Fungal Propagules. *Appl. Environ. Microbiol.* **1996**, *62*, 2435–2442. [CrossRef]

12. Hajek, A.E.; St. Leger, R.J. Interactions between Fungal Pathogens and Insect Hosts. *Annu. Rev. Entomol.* **1994**, *39*, 293–322. [CrossRef]
13. Bharadwaj, A.; Stafford, K.C. Potential of Tenebrio molitor (Coleoptera: Tenebrionidae) as a bioassay probe for *Metarhizium brunneum* (Hypocreales: Clavicipitaceae) activity against *Ixodes scapularis* (Acari: Ixodidae). *J. Econ. Entomol.* **2011**, *104*, 2095–2098. [CrossRef]
14. Kapoor, M.; Sreenivasan, G.M.; Goel, N.; Lewis, J. Development of thermotolerance in Neurospora crassa by heat shock and other stresses eliciting peroxidase induction. *J. Bacteriol.* **1990**, *172*, 2798–2801. [CrossRef] [PubMed]
15. Guan, Q.; Haroon, S.; Bravo, D.G.; Will, J.L.; Gasch, A.P. Cellular memory of acquired stress resistance in Saccharomyces cerevisiae. *Genetics* **2012**, *192*, 495–505. [CrossRef] [PubMed]
16. Andrade-Linares, D.R.; Lehmann, A.; Rillig, M.C. Microbial stress priming: A meta-analysis. *Environ. Microbiol.* **2016**, *18*, 1277–1288. [CrossRef]
17. Andrade-Linares, D.R.; Veresoglou, S.D.; Rillig, M.C. Temperature priming and memory in soil filamentous fungi. *Fungal Ecol.* **2016**, *21*, 10–15. [CrossRef]
18. Carruthers, R.I.; Haynes, D.L. Temperature, Moisture, and Habitat Effects on *Entomophthora muscae* (Entomophthorales: Entomophthoraceae) Conidial Germination and Survival in the Onion Agroecosystem. *Environ. Entomol.* **1986**, *15*, 1154–1160. [CrossRef]
19. Walstad, J.D.; Anderson, R.F.; Stambaugh, W.J. Effects of environmental conditions on two species of muscardine fungi (*Beauveria bassiana* and *Metarrhizium anisopliae*). *J. Invertebr. Pathol.* **1970**, *16*, 221–226. [CrossRef]
20. Hywel-Jones, N.L.; Gillespie, A.T. Effect of temperature on spore germination in *Metarhizium anisopliae* and *Beauveria bassiana*. *Mycol. Res.* **1990**, *94*, 389–392. [CrossRef]
21. Rath, A.C.; Anderson, G.C.; Worledge, D.; Keon, T.B.; Koen, T.B. The Effect of Low Temperatures on the Virulence of *Metarhizium anisopliae* (DAT F-001) to the Subterranean Scarab, *Adoryphorus couloni*. *J. Invertebr. Pathol.* **1995**, *65*, 186–192. [CrossRef]
22. Skalický, A.; Bohatá, A.; Šimková, J.; Osborne, L.S.; Landa, Z. Selection of indigenous isolates of entomopathogenic soil fungus *Metarhizium anisopliae* under laboratory conditions. *Folia Microbiol.* **2014**, *59*, 269–276. [CrossRef] [PubMed]
23. Kershaw, M.J.; Moorhouse, E.R.; Bateman, R.; Reynolds, S.E.; Charnley, A.K. The Role of Destruxins in the Pathogenicity of *Metarhizium anisopliae* for Three Species of Insect. *J. Invertebr. Pathol.* **1999**, *74*, 213–223. [CrossRef]
24. Golo, P.S.; Gardner, D.R.; Grilley, M.M.; Takemoto, J.Y.; Krasnoff, S.B.; Pires, M.S.; Fernandes, É.K.K.; Bittencourt, V.R.E.P.; Roberts, D.W. Production of destruxins from *Metarhizium* spp. fungi in artificial medium and in endophytically colonized cowpea plants. *PLoS ONE* **2014**, *9*, e104946. [CrossRef] [PubMed]
25. Stimmann, M.W. Effect of Temperature on Infection of the Garden Symphylan by *Entomophthora coronata*. *J. Econ. Entomol.* **1968**, *61*, 1558–1560. [CrossRef]
26. Bugeme, D.M.; Knapp, M.; Boga, H.I.; Wanjoya, A.K.; Maniania, N.K. Influence of temperature on virulence of fungal isolates of *Metarhizium anisopliae* and *Beauveria bassiana* to the two-spotted spider mite *Tetranychus urticae*. *Mycopathologia* **2009**, *167*, 221–227. [CrossRef]
27. Onsongo, S.K.; Gichimu, B.M.; Akutse, K.S.; Dubois, T.; Mohamed, S.A. Performance of Three Isolates of *Metarhizium anisopliae* and Their Virulence against *Zeugodacus cucurbitae* under Different Temperature Regimes, with Global Extrapolation of Their Efficiency. *Insects* **2019**, *10*, 270. [CrossRef]
28. Fargues, J.; Ouedraogo, A.; Goettel, M.S.; Lomer, C.J. Effects of Temperature, Humidity and Inoculation Method on Susceptibility of Schistocerca gregaria to Metarhizium flavoviride. *Biocontrol Sci. Technol.* **1997**, *7*, 345–356. [CrossRef]
29. Dantigny, P.; Nanguy, S.P.-M.; Judet-Correia, D.; Bensoussan, M. A new model for germination of fungi. *Int. J. Food Microbiol.* **2011**, *146*, 176–181. [CrossRef]
30. Thomas, M.B.; Jenkins, N.E. Effects of temperature on growth of *Metarhizium flavoviride* and virulence to the variegated grasshopper, *Zonocerus variegatus*. *Mycol. Res.* **1997**, *101*, 1469–1474. [CrossRef]
31. Jacome, L.H.; Schuh, W. Effect of temperature on growth and conidial production in vitro, and comparison of infection and aggressiveness in vivo among isolates of *Mycosphaerella fijiensis* var. *difformis*. *Trop. Agric.* **1993**, *70*, 51–59.
32. Campbell, M.A.; Medd, R.W.; Brown, J.F. Growth and sporulation of *Pyrenophora semeniperda* in vitro: Effects of culture media, temperature and pH. *Mycol. Res.* **1996**, *100*, 311–317. [CrossRef]
33. Arumugam, G.K.; Srinivasan, S.K.; Joshi, G.; Gopal, D.; Ramalingam, K. Production and characterization of bioactive metabolites from piezotolerant deep sea fungus *Nigrospora* sp. in submerged fermentation. *J. Appl. Microbiol.* **2015**, *118*, 99–111. [CrossRef] [PubMed]
34. Kim, D.-S.; Godber, S.J.; Kim, H.-R. Culture conditions for a new phytase-producing fungus. *Biotechnol. Lett.* **1999**, *21*, 1077–1081. [CrossRef]
35. Jin, X.; Taylor, A.G.; Harman, G.E. Development of Media and Automated Liquid Fermentation Methods to Produce Desiccation-Tolerant Propagules of *Trichoderma harzianum*. *Biol. Control* **1996**, *7*, 267–274. [CrossRef]
36. Roberts, D.W.; Campbell, A.S. Stability of entomopathogenic fungi. *Misc. Publ. Entomol. Soc. Am.* **1977**, *10*, 19–76.
37. Alves, S.B.; Risco, S.H.; Almeida, L.C. Influence of photoperiod and temperature on the development and sporulation of *Metarhizium anisopliae* (Metsch.) Sorok. *Z. Angew. Entomol.* **1984**, *97*, 127–129. [CrossRef]
38. Milner, R.J.; Lozano, L.B.; Driver, F.; Hunter, d. 2003-BioControl-483-335-348. *Biocontrol* **2003**, *48*, 335–348. [CrossRef]

39. Tefera, T.; Pringle, K. Germination, Radial Growth, and Sporulation of *Beauveria bassiana* and *Metarhizium anisopliae* Isolates and Their Virulence to *Chilo partellus* (Lepidoptera: Pyralidae) at Different Temperatures. *Biocontrol Sci. Technol.* **2003**, *13*, 699–704. [CrossRef]

40. Ghazanfar, M.U.; Hagenbucher, S.; Romeis, J.; Grabenweger, G.; Meissle, M. Fluctuating temperatures influence the susceptibility of pest insects to biological control agents. *J. Pest Sci.* **2020**, *93*, 1007–1018. [CrossRef]

41. Glare, T.R.; Scholte Op Reimer, Y.; Cummings, N.; Rivas-Franco, F.; Nelson, T.L.; Zimmermann, G. Diversity of the insect pathogenic fungi in the genus Metarhizium in New Zealand. *N. Z. J. Bot.* **2021**, *59*, 440–456. [CrossRef]

42. Phillips, D.J. Change in Conidia of *Monilinia fructicola* in Response to Incubation Temperature. *Phytopathology* **1982**, *72*, 1281–1283. [CrossRef]

43. Phillips, D.J. Effect of Temperature on *Monilinia fructicola* Conidia Produced on Fresh Stone Fruits. *Plant Dis.* **1984**, *68*, 610–612. [CrossRef]

44. Tian, S.P.; Bertolini, P. Effect of Temperature during Conidial Formation of *Monilinia laxa* on Conidial Size, Germination and Infection of Stored Nectarines. *J. Phytopathol.* **1999**, *147*, 635–641. [CrossRef]

45. Tian, S.P.; Bertolini, P. Changes in conidial morphology and germinability of *Botrytis allii* and *Penicillium hirsutum* in response to low-temperature incubation. *Mycol. Res.* **1996**, *100*, 591–596. [CrossRef]

46. Maitlo, S.; Rajput, A.Q.; Syed, R.N.; Khanzada, M.A.; Rajput, N.A.; Lodhi, A.M. Influence of physiological factors on vegetative growth and sporulation of *Fusarium oxysporum* f. sp. ciceris. *Pak. Bot. Soc.* **2017**, *49*, 311–316.

47. Samuels, K.; Heale, J.B.; Llewellyn, M. Characteristics relating to the pathogenicity of *Metarhizium anisopliae* toward *Nilaparvata lugens*. *J. Invertebr. Pathol.* **1989**, *53*, 25–31. [CrossRef]

48. Altre, J.A.; Vandenberg, J.D.; Cantone, F.A. Pathogenicity of *Paecilomyces fumosoroseus* isolates to diamondback moth, *Plutella xylostella*: Correlation with spore size, germination speed, and attachment to cuticle. *J. Invertebr. Pathol.* **1999**, *73*, 332–338. [CrossRef] [PubMed]

49. Al-Aidroos, K.; Seifert, A.M. Polysaccharide and protein degradation, germination, and virulence against mosquitoes in the entomopathogenic fungus *Metarhizium anisopliae*. *J. Invertebr. Pathol.* **1980**, *36*, 29–34. [CrossRef]

50. Dillon, R.J.; Charnley, A.K. A technique for accelerating and synchronising germination of conidia of the entomopathogenic fungus *Metarhizium anisopliae*. *Arch. Microbiol.* **1985**, *142*, 204–206. [CrossRef]

51. Dillon, R.J.; Charnley, A.K. Initiation of germination in conidia of the entomopathogenic fungus, *Metarhizium anisopliae*. *Mycol. Res.* **1990**, *94*, 299–304. [CrossRef]

52. Boucias, D.G.; Pendland, J.C. Nutritional requirements for conidial germination of several host range pathotypes of the entomopathogenic fungus *Nomuraea rileyi*. *J. Invertebr. Pathol.* **1984**, *43*, 288–292. [CrossRef]

Journal of *Fungi*

Article

Filamentous Fungi Are Potential Bioremediation Agents of Semi-Synthetic Textile Waste

Rachel Harper and Suzy Clare Moody *

School of Life Sciences, Pharmacy, and Chemistry, Kingston University, Kingston upon Thames KT1 2EE, UK; r.harper@kingston.ac.uk
* Correspondence: s.moody@kingston.ac.uk

Abstract: Textile waste contributes to the pollution of both terrestrial and aquatic ecosystems. While natural textile fibres are known to be biodegraded by microbes, the vast majority of textiles now contain a mixture of processed plant-derived polymers and synthetic materials generated from petroleum and are commonly dyed with azo dyes. This presents a complex recycling problem as the separation of threads and removal of dye are challenging and costly. As a result, the majority of textile waste is sent to landfill or incinerated. This project sought to assess the potential of fungal bioremediation of textile-based dye as a step towards sustainable and environmentally-friendly means of disposal of textile waste. Successful development of an agar-independent microcosm enabled the assessment of the ability of two fungal species to grow on a range of textiles containing an increasing percentage of elastane. The white rot fungus *Hypholoma fasciculare* was shown to grow well on semi-synthetic textiles, and for the first time, bioremediation of dye from textiles was demonstrated. Volatile analysis enabled preliminary assessment of the safety profile of this process and showed that industrial scale-up may require consideration of volatile capture in the design process. This study is the first to address the potential of fungi as bioremediation agents for solid textile waste, and the results suggest this is an avenue worthy of further exploration.

Keywords: mycoremediation; bioremediation; textile; dye; white rot; brown rot; *Hypholoma*

Citation: Harper, R.; Moody, S.C. Filamentous Fungi Are Potential Bioremediation Agents of Semi-Synthetic Textile Waste. *J. Fungi* **2023**, *9*, 661. https://doi.org/10.3390/jof9060661

Academic Editor: Baojun Xu

Received: 3 May 2023
Revised: 7 June 2023
Accepted: 10 June 2023
Published: 13 June 2023

1. Introduction

The textile industry is a major source of pollution throughout the lifecycle of a garment, from the use of fossil fuels for the production of common synthetic textiles [1] to water contamination from dyes [2], from microplastic generation through washing [3] to air, terrestrial and aquatic pollution from landfill and incineration [4]. Textile waste presents a complex disposal problem due to the heterogeneity of composition, the volume of waste [5], and the disproportionate concentration of both production and disposal in low- and middle-income countries (LMICs) [6]. Most garments are now at least in part, synthetic, whether this is from the thread for seams, elastane to give stretchiness, plastic buttons or zips, or synthetic fibres such as nylon and polyester. While natural fibres such as cotton and wool are known to biodegrade over time [7,8], this can be negatively impacted by dyes and other processing and the end result is the prolonged presence of processed natural microfibres and synthetic microfibres in the environment [9,10]. As both landfill and incineration are unsustainable, a better way of processing the large volume of textile waste currently generated must be found.

The use of biological agents to eliminate polluting substances [11] is a well-established field, with both fungi and fungal enzymes already being investigated for use in the bioremediation of radioactivity, chemical spills, and microplastics [12–15]. Filamentous fungi are so named for their hyphal growth which enables exploration of the ecological niche and capture of nutrient resources. The hyphal morphology suggests this group of fungi as candidates for bioremediation of solid textile waste as they could potentially grow on and through the textiles. Further to this a specific group of filamentous fungi, the wood decay fungi, potentially have the enzymatic capacity to utilize the various components of semi-synthetic textiles. The wood decay fungi are ecological specialists adapted to perform lignocellulose degradation through the action of a comprehensive range of extracellular enzymes. Lignin is a polyphenolic compound of irregular structure that is highly recalcitrant to breakdown and, combined with the polymeric chains of cellulose and hemi-cellulose, forms lignocellulose in plants. Commonly used textiles include cotton (predominantly cellulose), viscose (a highly processed form of cellulose), and petroleum-derived synthetics such as polyester and elastane (aromatic polymers with some structural similarity to lignin) (see Figure 1).

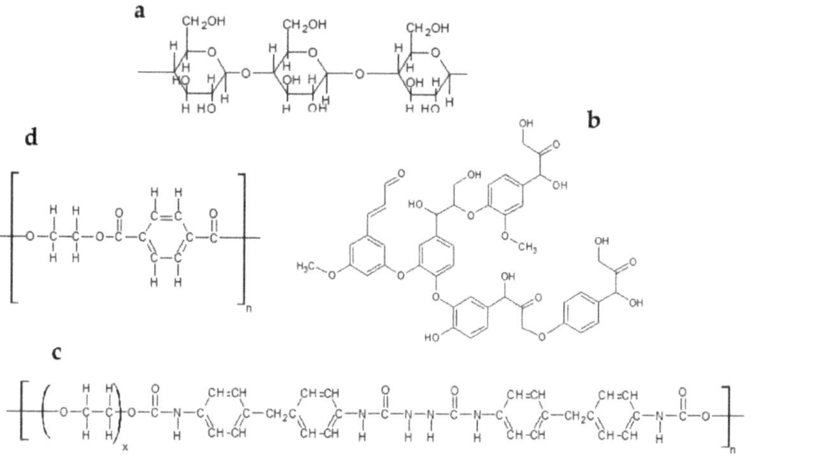

Figure 1. Indicative structures of (clockwise from the top): (**a**) cellulose, (**b**) lignin, (**c**) elastane, (**d**) polyester. Note the commonality of aromatic rings and carbon to oxygen bonding patterns in both natural and petroleum-derived materials.

Two broad mechanisms of wood decay, brown and white rot, have been described as opposite ends of a spectrum of activity. White rot fungi are capable of complete mineralization of lignocellulose through the action of low-specificity peroxidases, laccases, cellulases, and other carbohydrate-active enzymes (cazymes), reducing complex heteropolymers in the environment into smaller compounds which can be assimilated to support growth [16]. Brown rot fungi alter lignin without utilizing it via the oxidative Fenton reaction to enable the capture of the cellulose component of woody resources through cazymes [16]. Both mechanisms are non-specific and extracellular [17] suggesting they may be adapted to textile bioremediation.

Previous works, dating back almost a century, have identified various fungi that can degrade natural fibres such as cotton and wool, although these were conducted over short timescales and did not demonstrate complete bioremediation [7,8,18]. Likewise, initial steps of nylon degradation by white rot fungi have also been documented in laboratory conditions, using agar plates and sterile nylon discs as growth media [19,20]. Dye biodegradation has been the subject of many studies, in the context of wastewater treatment. Azo dyes, such as Reactive Black 5 (RB5) used in this study, are widely used but known to be toxic, mutagenic, and carcinogenic in waste [21]. This background, combined with the ecology and enzymatic capacity of wood decay fungi, suggested they had unexploited bioremediation potential. The hypothesis underpinning this research was that wood decay fungi would be able to grow on semi-synthetic textiles, with both white rot and brown rot fungi as potential candidates. The trial textiles used contained cotton, bamboo viscose, and an increasing percentage of elastane, and were dyed with RB5. The objectives were (i) to develop an agar-free microcosm that would be scalable for industrial use; (ii) to assess the ability of the fungi to bioremediate a commonly used dye from the textile; (iii) to analyze the volatile metabolome released during growth as a preliminary safety assessment of the process. The results of the study suggest wood decay fungi are worth further investigation as textile bioremediation agents.

2. Materials and Methods

2.1. Fungal Culture and Colonization of Woodchips

The strains used in this study were *Serpula himantioides* (MUCL38935, from the Belgian Co-ordinated Collections of Microorganisms) and *Hypholoma fasciculare* (HfGTWV2, from Professor Lynne Boddy at Cardiff University, Wales). Fungal strains were maintained on 2% malt extract agar (MEA, malt extract Oxoid, UK, agar Thermo Fisher Scientific, Waltham, MA, USA). 2 cm^3 blocks of Scots Pine (*Pinus sylvestris*, sourced from Portswood Timber Supplies, Southampton, UK) and European Beech (*Fagus sylvatica*, sourced from John Harrison, Wrexham, UK) were chipped using hammer and chisel, then autoclaved and colonized as in [22] with *S. himantioides* and *H. fasciculare* respectively. These were incubated at 25 °C for 4 weeks in the dark, to allow the fungi to fully colonize the wood chips before they were added to the microcosm setup.

2.2. Microcosm Setup

A layer of sterile perlite (Westland Horticulture Ltd., Dungannon, UK) was placed in the bottom of each sterile 250 mL polypropylene pot, and approximately 15 mL of sterile dH$_2$O was added. Using a sterile needle, 5 holes were poked in the side of each pot and were then covered with micropore tape (Wilko, Worksop, UK) to allow airflow. The three trial textiles were sourced from Bamboo Clothing Ltd. All contained the same plant-derived yarn—a blend of 70% bamboo viscose and 30% cotton—with either 0%, 4%, or 12% elastane, and all were black as a result of being dyed with Reactive Black 5 (RB5). The trial textiles were washed in a domestic washing machine using a non-biological washing powder 5 times at 30 °C with a 1000 rpm spin cycle, and line dried outside in between each wash. Three pieces of 3 cm × 3 cm textile were added to each pot. Pre-colonised wood chips were scraped to remove agar, then placed on top of the textile, in the centre of the pot. All pots were sealed, labelled, and incubated (LMS Cooled Incubator Model 600, LMS, Sevenoaks, UK) in the dark at 25 °C. They were checked regularly for water content and 5 mL sterile dH$_2$O added to the perlite if the fabric felt dry. The set up process is summarized in Figure 2.

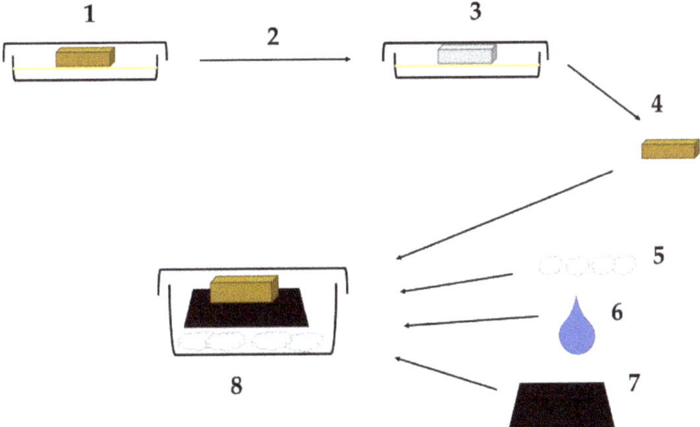

Figure 2. (**1**) the fungi were inoculated onto malt extract agar with wood chips. (**2**) the cultures were incubated at 25 °C for 4 weeks in the dark. (**3**) the fungus colonized the wood chips. (**4**) the colonized

wood chip was removed from the culture and the excess agar scraped off. (**5**) sterile perlite added to polypropylene container. (**6**) 15 mL sterile water added to perlite. (**7**) 3 squares of pre-washed trial textile added. (**8**) scraped colonized wood block placed on top of trial textile. The microcosm was incubated in the dark at 25 °C.

2.3. Imaging

At certain time points, the textiles were imaged using scanning electron microscopy (SEM, Zeiss Evo 50, Zeiss, Germany). The time points captured the colonization of the textile by the fungus (3 months), a mid-way time point (5 months), and the end of the trial (8 months). Textile samples were affixed to titanium stubs and then placed into a sputter coater (Polaron SC7640 Sputter Coater, Quorum Technologies, Laughton, UK). One side of the textile sample was coated with gold/palladium, with the machine set to 2 kV for 2 min at 20 mA, with Argon pumped into the chamber and maintained at a pressure of 10^{-7} mbar/PA. Each sample was placed into the chamber of the SEM, with the beam set to 5 kV. Images were taken at various magnifications, ranging from 35 to 2500.

2.4. Determination of Dye Content

2.4.1. Thermogravimetric Analysis

10 mg of textile was weighed into an alumina crucible for loading into a Thermogravimetric Analyzer (Mettler Toledo, Leicester, UK). The starting temperature was set to 25 °C, and the machine was programmed to increase to 1000 °C in increments of 10 °C per minute. Results were recorded as graphs recording mass change over time in the STARe software V10.00 (Mettler Toledo, Leicester, UK). Reactive Black 5 (RB5, Merck, Germany) was used as a control.

2.4.2. Dye Extraction Essay

The method developed by [23] was used to assay the dye content of the textile. Textile samples with no fungi on them were used as a control. RB5 dye was used as a standard, with a range from 1 mg/L to 10 mg/L. Prior to taking absorbance measurements, the samples were all diluted 1:4 in 1.5% (*v*/*v*) aqueous sodium hydroxide, as the undiluted samples exceeded an absorbance reading of 1. Samples were measured on a Unicam Helios Epsilon spectrophotometer (Thermo Fisher Scientific, Waltham, MA, USA).

2.5. Gas Chromatography Mass Spectrometry of Volatiles

A StableFlex™ 2 cm SPME fiber (Supelco, St. Louis, MO, USA) was used to capture volatiles released from an opened microcosm for one hour before injection into an Agilent 6890N GC with 5973N Inert MSD and 7683 Injector (Agilent, Santa Clara, CA, USA). The GCMS methodology started at 80 °C and ramped to a maximum of 320 °C with the temperature increasing at a rate of 10 °C per minute—for a total run time of 29 min. Controls with the different textiles and the relevant woods were used and any peak found in these controls was eliminated from further analysis. Peak data was collected from Total Ion Chromatograms (TICs) generated using Chemstation (Agilent, Santa Clara, CA, USA), and potential compound matches were collected from the NIST 2.0 database (National Institute of Standards and Technology, Gaithersburg, MD, USA) for each peak. Quality cutoff point was 60, and only compounds that were at or above 60 were recorded. Controls of textile and wood were also analysed, with compounds present in controls being removed from the experimental data sets. Compounds were then manually curated using the following databases to identify any known functions or health hazards associated with them: MetaCyc Metabolic Pathway Database version 20 [24]; PhytoHub version 1.4 [25]; the Yeast Metabolome Database version 2.0 [26,27]; Golm Metabolome Database [28]; the Metabolomics Workbench (https://www.metabolomicsworkbench.org/search/index.php, accessed between May and June 2022) [29]; and PubChem [30] for hazards and toxicity.

2.6. Bioinformatic Analysis of Genomes

Genomic data were obtained from the Joint Genomes Initiative (JGI) website (Home—*Serpula himantioides* (*S. lacrymans* var shastensis) MUCL38935 v1.0 (doe.gov) and Home—*Hypholoma sublateritium* v1.0 (doe.gov)) [31] using the standard annotations and search terms 'laccase' and 'peroxidase'. All results were included in the analysis.

3. Results

3.1. Fungal Growth on Textiles

Neither fungal species used in this project had previously shown the ability to grow on textiles. Wood chips pre-colonized with either the white rot fungus *Hypholoma fasciculare* (HfGTWV2) or brown rot fungus *Serpula himantioides* (MUCL38935) were placed on textile and incubated in the dark at 25 °C for 8 months. The *S. himantioides* microcosms did not show any discolouration (Figure 3). The growth exhibited on all textiles was patchy, with some thin cord formation but predominantly a mycelial network that appeared to weave through the textile at times. This growth pattern has been previously seen in *S. himantioides* when grown on soil with wood resource units added (Skrede, unpublished), suggesting that the microcosm set up here provides similar nutrient options and foraging patterns to soil with sparse lignocellulose availability.

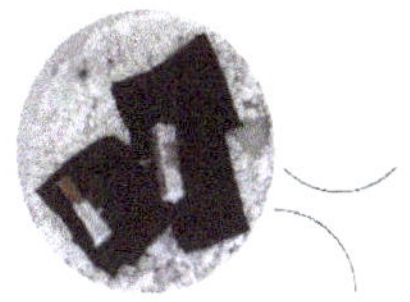

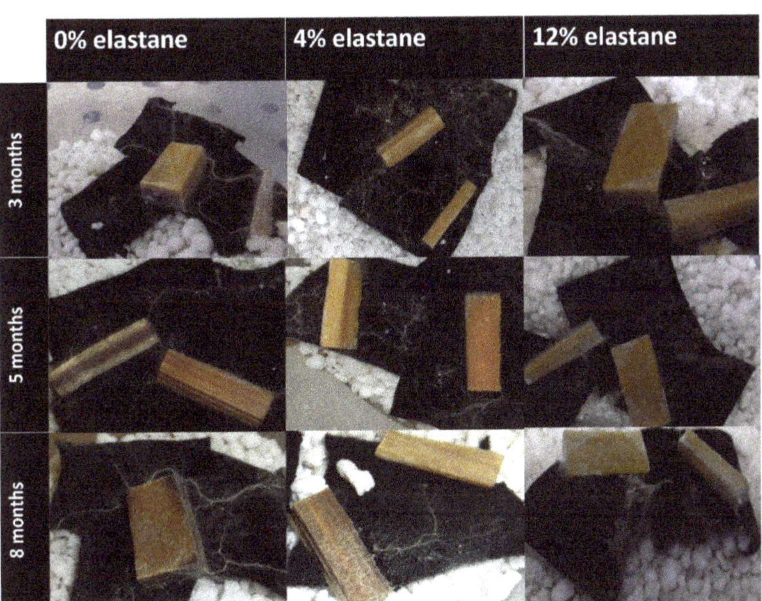

Figure 3. *S. himantioides* growing on a range of textiles, imaged at (top right, starting point) then at 3 months, 5 months, and 8 months incubation.

Within 3 months, both species had grown across the textile but the *H. fasciculare* microcosms also exhibited a marked colour change. This grew more pronounced over time, and interestingly, was more obvious on 0% elastane textiles compared with both the 4% and 12% textiles (Figure 4). The growth pattern shown here is typical of all the microcosms, and the dense concentric rings produced by tangential anastomoses of mycelial growth seen in the agar culture of *H. fasciculare* were not produced on textile. Rather, the growth on all textiles closely echoed that described by [32] as the textile appeared to be thoroughly colonized with mycelia, with thick connecting cords extending across the textile resource from the wood inoculum. The presence of these foraging cords as a hyphal network were particularly noticeable on the 12% elastane textiles (Figure 4). As shown below, the fungus was able to metabolize dye more easily from 0% and 4% elastane textiles (see Figure 6) and it is suggested that this pattern of increased cord development on 12% elastane may be due to more extensive resource capture as the textile was the key nutrient source in these micrososms, rather than the dye. As wood decay fungi access nutrition from solid organic resources, spatial capture is equivalent to nutrient capture [33], and *H. fasciculare* is well-adapted as a forest floor competitor to enact this rapidly. At no point did *H. fasciculare* extend foraging cords beyond the boundaries of the textile, suggesting adequate nutrition was available within the textile microcosm.

Figure 4. *H. fasciculare* growing on a range of textiles, imaged at (top right, starting point) then at 3 months, 5 months, and 8 months incubation.

Scanning Electron Microscopy (SEM) imaging of the textiles and individual fibres was undertaken to give a baseline and means of identifying any fibre preference the fungi may have displayed. The images are included in the supplementary information (Figure S1). Imaging a sample of colonized textiles by SEM revealed hyphal growth on all fibre types, with denser growth in the *H. fasciculare* microcosms (Figure 5) which confirmed the light microscopy findings. The images shown are a representative sample showing exemplars of the different structures. Figure 5a,b,d were taken at 8 months, but Figure 5c is from 5 months as no crystals were detected in *H. fasciculare* samples at 8 months. The hyphal morphologies seen in both species appeared normal compared with others' images [34], with some patchy structures (Figure 5b) likely to be fungal metabolites that have this sheet-like appearance under SEM conditions [35]. Crystalline structures (Figure 5c,d) were seen on all textiles at 3 months for both species. For *H. fasciculare*, they were also found on all three textiles at 5 months but were not visualized in the samples taken at 8 months. For *S. himantioides*, crystals were additionally seen on 0% and 4% at 5 months, and 4% and 12% at 8 months, suggesting that they may be present on all textiles at all time points for this fungus and increased sampling may have demonstrated this.

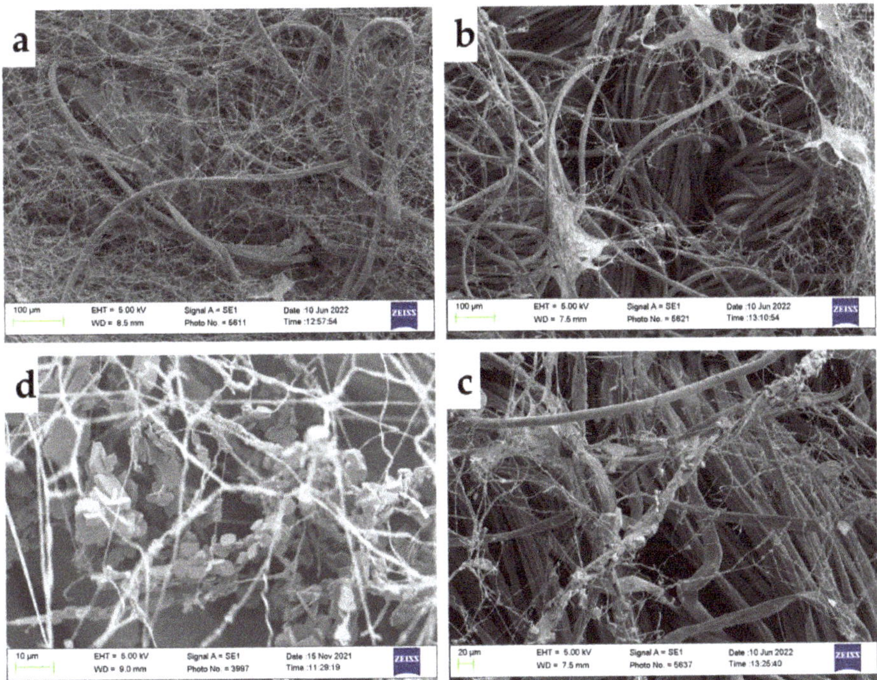

Figure 5. SEM images on a representative sample of microcosm textiles show the variety of structures seen. Clockwise from top left: (**a**) *H. fasciculare* on 12% elastane at 8 months incubation, (**b**) *S. himantioides* on 0% elastane at 8 months incubation, (**c**) *S. himantioides* on 4% elastane at 8 months incubation, (**d**) *H. fasciculare* on 4% elastane at 5 months incubation.

3.2. Quantification of Dye Remediation

Two methods of quantification were trialled. The unsuccessful method is reported here to enable others to benefit from the protocol development that contributed to this study.

3.2.1. Thermogravimetric Analysis (TGA)

For TGA to be a useful method in this study each fibre type and the dye would need a significantly different thermal decomposition point, that would be detectable when plotted as mass change over time. The textile used was dyed with Reactive Black 5 (RB5), a common black clothing dye. The 0% elastane textile was compared with undyed bamboo viscose-cotton but little difference was noted. RB5 powder was analysed in isolation but rather than showing a distinct 'step' in mass loss, there was a gradual loss over time. Elastane alone was compared with both 0% and 12% textiles, but again, there were no distinct mass changes. These data are shown in Figure S2 and demonstrate that TGA was not a suitable method for assessing the remediation of the dye in this context.

3.2.2. Dye Extraction and Spectrophotometric Quantification

The method that gave reproducible results was based on a protocol by [23]. The incubated textiles were compared with control samples from the same garments. All experiments were done in triplicate and the data presented in Figure 6 are the means.

At both 5 months and 8 months of incubation, the samples from *H. fasciculare* demonstrated a significant decrease in dye saturation as had been noted on visual examination of the microcosms. This was evident in all textile types with over 80% dye lost, although the 12% elastane textile showed a slightly lower level of dye remediation than the 4% and 0% textiles. The *S. himantioides* microcosm textiles were visually unchanged from the black starting point, but dye extraction and quantification showed that the 12% textile at 5 and 8 months incubation had a small but significant (*p*-value 0.001) drop in dye content compared with the controls.

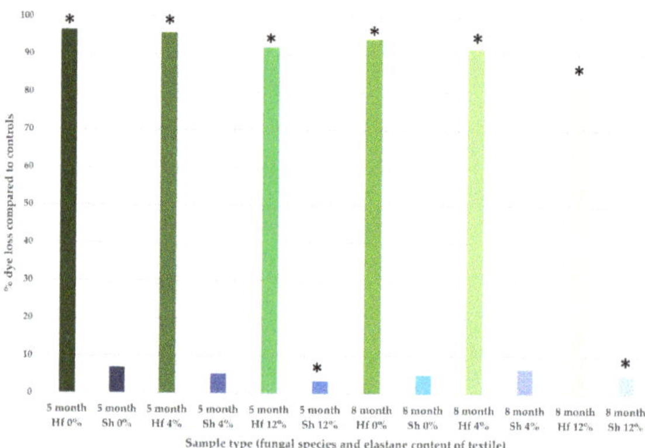

Figure 6. The dye loss from textiles at 5 and 8 months incubation with either *H. fasciculare* (Hf, shown in green) or *S. himantioides* (Sh, shown in blue), compared with the original textile. Percentages refer to elastane content. Each column shows the mean of three triplicate microcosms. The stars indicate samples that were significantly different from the control (as determined by paired two-tailed *t*-test).

3.3. Bioinformatic Analysis of Enzyme Capacity for Dye Remediation

Bioinformatic analysis of enzyme candidates for dye remediation offered a route to understanding the disparity in dye loss from the textiles between the two fungal species. The genome for *H. fasciculare* was unavailable at the time of writing, but that of a close relative, *Hypholoma sublaterium*, was used as a proxy to compare the potential enzymatic degradative capability of the two fungal species used. Two groups of key secreted enzymes, the laccases and peroxidases, known to be involved in the breakdown of complex phenolic compounds as part of the wood decay ecology, were analyzed. As seen in Figure 7, the *Hypholoma* genus appears to have far greater numbers of laccases and peroxidases which may account for the disparity in dye remediation seen here. It is acknowledged that many other enzymes may also be involved in this process and that *H. fasciculare* may differ from *H. sublaterium* in enzyme complement.

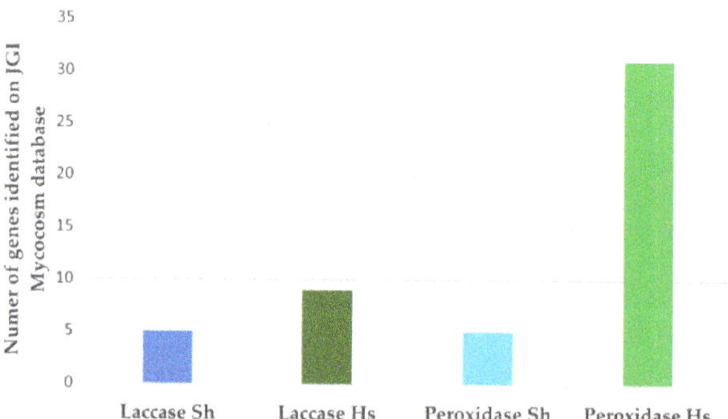

Figure 7. The laccase and peroxidase complement of *S. himantioides* (Sh, shown in blue) and *H. sublaterium* (Hs, shown in green). Data were obtained from the Joint Genomes Initiative website (Home—*Serpula himantioides* (*S. lacrymans* var shastensis) MUCL38935 v1.0 (doe.gov) and Home—*Hypholoma sublateritium* v1.0 (doe.gov)) [31].

3.4. Gas Chromatography Mass Spectrometry (GCMS) Analysis of the Volatile Metabolome

The volatile metabolome was captured and analysed by GCMS at three timepoints during the study, with uncolonized wood and textile used as controls to exclude volatiles released from the substrates. The GCMS spectra showed a change in volatile profile between the fungi, between different textiles, and over time. The majority of the compounds identified had no previously published connection to fungal metabolomics which made interpretation of function or role challenging (tables of tentatively identified compounds are included in Table S1). Metabolites of interest with identified function in at least two replicates of both fungal microcosms were butylated hydroxytoluene (a known lipophilic fungal metabolite with antioxidant properties) and dodecane (a medium chain alkane with known involvement in the biosynthesis of cuticular wax in vascular plants). Phytane (a diterpenoid specialized metabolite) was identified in at least two *H. fasciculare* 4% elastane microcosms, with 1,5,9-trimethyl-1,5,9-cyclododecatriene (a known plant volatile) released in at least two of the *S. himantioides* 4% elastane microcosms. Farnesane, a sesquiterpenoid specialized metabolite, was also identified in at least one microcosm of each fungus, suggesting that terpenoid metabolism may play a role in dye metabolism or textile degradation. There were different timescales and longevity for the production of these metabolites. For example, butylated hydroxytoluene was seen in *H. fasciculare* at 3 months on 4% elastane, whereas dodecane was seen in *H. fasciculare* at both 3 and 5 months incubation but only on 12% elastane. Butylated hydroxytoluene was detected in *S. himantioides* microcosms on all three textiles at 3 months incubation, but dodecane was only detected on 4% elastane at 3 months. The variability of the volatile metabolome over time and between samples was noted, being particularly pertinent at 8 months of incubation. At this time point, although volatiles were produced, no compounds were identified in more than one biological replicate in any of the conditions for either fungus. This high level of variation was not entirely unexpected given the few volatiles consistently identified in more than one replicate at earlier time points but is certainly an interesting avenue to explore further to understand why this variation may exist.

4. Discussion

The number of textile items produced globally has doubled in 15 years to more than 100 billion units, and when discarded, 73% of items are incinerated or sent to landfill [36]. As at least 60% of clothing contains synthetic fibres [37], it is not surprising that the fashion industry causes significant environmental and societal damage throughout the lifecycle. New research into alternative sources for textiles and methods of disposal is urgently needed.

This project signifies the first demonstration of fungal growth on textiles with elastane, with the development of agar-free microcosms successfully maintaining fungal culture for over 6 months. There were clear differences noted in the extent of hyphal growth and particularly cord formation in *H. fasciculare* on the different textiles (Figure 4). It was suggested this may be due to differential nutrient metabolism based on the elastane content, and further work is needed to develop a method for quantification of the relative textile components (cotton, bamboo viscose, elastane, and RB5 dye). This would enable a greater understanding of which specific components of the microcosm resources are being utilized by the fungi over time. Development of a protein extraction protocol for this microcosm set up has proved challenging, but successful capture of the proteome would enable important insights into the processes and metabolic pathways the fungi are using to maintain growth on the different textiles, and potentially provide greater knowledge of preferential resource use.

As far as the authors are aware, this is also the first demonstration of dye bioremediation directly from textiles by any fungus. This was not unexpected as other white rot fungi have previously been shown to be capable of RB5 degradation (e.g., *Trametes hirsuta* [38]) but in solution, rather than from textile. The results shown here demonstrate the biodegradation of the dye rather than the adsorption reported by others [21] as the hyphae remained white throughout the experiment (Figures 3 and 4). *H. fasciculare* demonstrated some impact of elastane content on dye removal as the 12% textile contained more dye at the 8 month incubation point than the 4% or 0% textiles (Figure 6). It is unclear whether this is due to the different chemical interactions of the dye with natural and synthetic textiles having an impact on the bioavailability of the fungus to metabolise the dye. In contrast, the 12% elastane textile appears to facilitate *S. himantioides* ability to utilize the dye. At both 5 and 8 month timepoints, there is a small but significant decrease in the amount of dye remaining in the textile compared to the control (Figure 6). This may suggest a different mechanism or underlying process by which these two fungi approach dyed textiles, which would be in keeping with their differential ecologies.

While this project did not attempt identification of the crystals seen on SEM images of both fungi, they bear significant structural similarity to whewellite, calcium oxalate monohydrate, which has been previously shown to be a fungal metabolite of *S. himantioides* [39] and had a very similar arrangement on the hyphae compared with whewellite crystal production by *Perenniporia meridionalis* [40]. Calcium oxalate crystals are also known to be produced during an interaction between *H. fasciculare* and another fungus (although the producing organism was not identified in that study) and may be associated with a fungal stress response [41].

The volatile metabolome analyses suggested heterogeneous textile samples may be challenging for predictability and identification, but the presence of elastane while altering the volatiles produced did not cause a significant release of known hazardous volatiles during the lifespan of the incubations used here. The identified metabolites had few hazard symbols associated with them in PubChem [30] but some were designated as irritants or health hazards which may have implications for ensuring scaled up applications are appropriately managed. Similar to others, the degradation products produced in these microcosms were not easily characterized but their potential toxicity is likely to be low based on others work. White rot fungi have been shown to oxidise RB5, although it is thought that where the appearance of RB5 changes to a light red colour as it does in the microcosms in this work, this is indicative of the continued presence of metabolites containing azo bonds and naphthalene rings. In silico docking experiments have suggested laccases as being of paramount importance in the biodegradation of RB5 [38] although peroxidases have also been suggested as promising enzymes to explore [42]. Where a fungal biodegradation mechanism for RB5 has been elucidated and the end point metabolites captured, there has been a demonstration of limited toxicity to plant seeds [21,43] and reduced cytotoxicity to human cell lines [38].

The findings presented here are promising, but some significant challenges remain to be able to demonstrate mycoremediation as a sustainable option for industrial scale removal of textile pollution. The first is that long timescales, beyond the life span of many research funding options, would be needed to demonstrate complete degradation. It is currently unknown whether the synthetic materials in the textile would be metabolized, and if they were, it is not known what the breakdown products may be. It is possible that generation of microplastics would be the unwelcome endpoint. If mycoremediation can be demonstrated at a laboratory scale, the final challenge would be scaling this up to be a viable option at an industrial scale. While the experimental work described here has been designed with scale up in mind, there would still be challenges of increased production of volatiles (whether harmful or not), the increase in temperature in large composting facilities may impact fungal growth and metabolism, and the risk of contamination. Recognizing the possible challenges and realizing the potential of mycoremediation for textile waste remain a promising field for further exploration.

Supplementary Materials: The following supporting information can be downloaded at: https://www.mdpi.com/article/10.3390/jof9060661/s1, Figure S1. SEM images of individual fibres to identify fibre types within the fabric. (a) shows a single woven thread of the bamboo viscose and cotton mix, with a total width of 230.7 μm. (b) shows individual fibres of cotton (left) and bamboo viscose (right) from the thread, with widths of 29.8 μm and 12.99 μm respectively. (c,e) show elastane fibres identified within the woven fabric structure seen in (g). (f) shows a clearer example of cotton vs. bamboo viscose fibres in the fabric, the bamboo viscose having striations and the cotton being less uniform. (h) shows some of the damage to the individual fibres seen in the washed fabric. These are representative images for the washed fabric controls, from several that were taken. Figure S2. Aggregate graph of thermogravimetric analysis (TGA) curves plotting the percentage of initial mass against temperature in °C for RB5 dye (black line), fabric with 12% elastane (red line), fabric with no elastane content (blue line) and the undyed bamboo viscose-cotton fibres (green line). However, no obvious difference in the decomposition of the 0% elastane fabric sample and the undyed bamboo viscose and the cotton mixture was noted, making it difficult to determine any dye loss via this method (blue and green lines). A sample of pure RB5 powder was analysed in the same way, and rather than showing a clear 'step' where a large mass loss occurred due to decomposition, it was more of a gradual loss over time. This confirmed that it would have been difficult to quantify dye loss via TGA (black line). Table S1. The volatile metabolome captured from microcosms at 3, 5, and 8-month timepoints on 0, 4, and 12% elastane fabrics. The NIST database was used for identification with a Quality Cutoff of 60. The metabolites included here were found in at least two biological replicates at that timepoint and on that fabric.

Author Contributions: Conceptualization, S.C.M.; Data curation, R.H. and S.C.M.; Formal analysis, S.C.M.; Funding acquisition, S.C.M.; Investigation, R.H.; Methodology, R.H. and S.C.M.; Project administration, S.C.M.; Resources, S.C.M.; Supervision, S.C.M.; Visualization, R.H.; Writing—original draft, R.H.; Writing—review & editing, S.C.M. All authors have read and agreed to the published version of the manuscript.

Funding: SCM thanks Applied Microbiology International for a New Lecturer Research Grant (grant number 04-200540) that bought much of the equipment used for this research.

Data Availability Statement: We confirm we are happy to supply the original GCMS.RAW files if requested.

Acknowledgments: We thank Bamboo Clothing Ltd. for partially funding this research and supplying textiles.

Conflicts of Interest: The authors declare no conflict of interest.

References

1. Wang, L.; Li, Y.; He, W. The energy footprint of China's textile industry: Perspectives from decoupling and decomposition analysis. *Energies* **2017**, *10*, 1461. [CrossRef]
2. Stone, C.; Windsor, F.M.; Munday, M.; Durance, I. Natural or synthetic—How global trends in textile usage threaten freshwater environments. *Sci. Total Environ.* **2020**, *718*, 134689. [CrossRef]
3. De Falco, F.; di Pace, E.; Cocca, M.; Avella, M. The contribution of washing processes of synthetic clothes to microplastic pollution. *Sci. Rep.* **2019**, *9*, 6633. [CrossRef] [PubMed]
4. Niinimäki, K.; Peters, G.; Dahlbo, H.; Perry, P.; Rissanen, T.; Gwilt, A. The environmental price of fast fashion. *Nat. Rev. Earth Environ.* **2020**, *1*, 189–200. [CrossRef]
5. Bick, R.; Halsey, E.; Ekenga, C.C. The global environmental injustice of fast fashion. *Environ. Health* **2018**, *17*, 92. [CrossRef]
6. Papamichael, I.; Chatziparaskeva, G.; Pedreno, J.N.; Voukkali, I.; Almendro Candel, M.B.; Zorpas, A.A. Building a new mind set in tomorrow fashion development through circular strategy models in the framework of waste management. *Curr. Opin. Green Sustain. Chem.* **2022**, *36*, 100638. [CrossRef]
7. Smith, G. 8—The Identification of Fungi Causing Mildew in Cotton Goods: The Genus *Aspergillus*. *J. Text. Inst. Trans.* **1928**, *19*, T92–T100. [CrossRef]
8. Marsh, P.B.; Bollenbacher, K. The Fungi Concerned in Fiber Deterioration. *Text. Res. J.* **1949**, *19*, 313–324. [CrossRef]
9. Liu, J.; Liang, J.; Ding, J.; Zhang, G.; Zeng, X.; Yang, Q.; Zhu, B.; Gao, W. Microfiber pollution: An ongoing major environmental issue related to the sustainable development of textile and clothing industry. *Environ. Dev. Sustain.* **2021**, *23*, 11240–11256. [CrossRef]
10. Haslinger, S.; Hummel, M.; Anghelescu-Hakala, A.; Määttänen, M.; Sixta, H. Upcycling of cotton polyester blended textile waste to new man-made cellulose fibers. *Waste Manag.* **2019**, *97*, 88–96. [CrossRef]

11. Lynch, J.M.; Moffat, A.J. Bioremediation—Prospects for the future application of innovative applied biological research. *Ann. Appl. Biol.* **2005**, *146*, 217–221. [CrossRef]
12. Viswanath, B.; Rajesh, B.; Janardhan, A.; Kumar, A.P.; Narasimha, G. Fungal Laccases and Their Applications in Bioremediation. *Enzym. Res.* **2014**, *2014*, 163242. [CrossRef]
13. Singh, R.K.; Tripathi, R.; Ranjan, A.; Srivastava, A.K. Fungi as potential candidates for bioremediation. In *Abatement of Environmental Pollutants*; Elsevier: Amsterdam, The Netherlands, 2020; pp. 177–191.
14. Noman, E.; Al-Gheethi, A.; Mohamed, R.M.S.R.; Talip, B.A. Myco-Remediation of Xenobiotic Organic Compounds for a Sustainable Environment: A Critical Review. *Top. Curr. Chem.* **2019**, *377*, 17. [CrossRef]
15. Chatterjee, S.; Sarma, M.K.; Deb, U.; Steinhauser, G.; Walther, C.; Gupta, D.K. Mushrooms: From nutrition to mycoremediation. *Environ. Sci. Pollut. Res.* **2017**, *24*, 19480–19493. [CrossRef]
16. Eastwood, D.C.; Floudas, D.; Binder, M.; Majcherczyk, A.; Schneider, P.; Aerts, A.; Asiegbu, F.O.; Baker, S.E.; Barry, K.; Bendiksby, M.; et al. The Plant Cell Wall–Decomposing Machinery Underlies the Functional Diversity of Forest Fungi. *Science* **2011**, *333*, 762–765. [CrossRef]
17. Saraswat, P.; Yadav, K.; Gupta, A.; Prasad, M.; Ranjan, R. Physiological and molecular basis for remediation of polyaromatic hydrocarbons. In *Handbook of Bioremediation*; Elsevier: Amsterdam, The Netherlands, 2021; pp. 535–550.
18. Błyskal, B. *Gymnoascus arxii*'s potential in deteriorating woollen textiles dyed with natural and synthetic dyes. *Int. Biodeterior. Biodegrad.* **2014**, *86*, 349–357. [CrossRef]
19. Deguchi, T.; Kakezawa, M.; Nishida, T. Nylon biodegradation by lignin-degrading fungi. *Appl. Environ. Microbiol.* **1997**, *63*, 329–331. [CrossRef]
20. Friedrich, J.; Zalar, P.; Mohorčič, M.; Klun, U.; Kržan, A. Ability of fungi to degrade synthetic polymer nylon-6. *Chemosphere* **2007**, *67*, 2089–2095. [CrossRef]
21. Pereira de Almeida, A.; Macrae, A.; Dias Ribeiro, B.; Pires do Nascimento, R. Decolorization and Detoxification of Different Azo Dyes by *Phanerochaete chrysosporium* ME-446 under Submerged Fermentation. *Braz. J. Microbiol.* **2021**, *52*, 727–738. [CrossRef]
22. Moody, S.C.; Dudley, E.; Hiscox, J.; Boddy, L.; Eastwood, D.C. Interdependence of primary metabolism and xenobiotic mitigation characterizes the proteome of *Bjerkandera adusta* during wood decomposition. *Appl. Environ. Microbiol.* **2018**, *84*, e01401–e01417. [CrossRef]
23. Home, J.M.; Dudley, R.J. Thin-layer chromatography of dyes extracted from cellulosic fibres. *Forensic Sci. Int.* **1981**, *17*, 71–78. [CrossRef]
24. Caspi, R.; Billington, R.; Keseler, I.M.; Kothari, A.; Krummenacker, M.; Midford, P.E.; Ong, W.K.; Paley, S.; Subhraveti, P.; Karp, P.D. The MetaCyc database of metabolic pathways and enzymes—A 2019 update. *Nucleic Acids Res.* **2020**, *48*, D445–D453. [CrossRef] [PubMed]
25. Bento da Silva, A.; Giacomoni, F.; Pavot, B.; Fillâtre, Y.; Rothwell, J.A.; Bartolomé Sualdea, B.; Veyrat, C.; Garcia-Villalba, R.; Gladine, C.; Kopec, R.; et al. PhytoHub V1.4: A new release for the online database dedicated to food phytochemicals and their human metabolites. In Proceedings of the International Conference on Food Bioactives & Health, Norwich, UK, 13 September 2016.
26. Jewison, T.; Knox, C.; Neveu, V.; Djoumbou, Y.; Guo, A.C.; Lee, J.; Liu, P.; Mandal, R.; Krishnamurthy, R.; Sinelnikov, I.; et al. YMDB: The Yeast Metabolome Database. *Nucleic Acids Res.* **2012**, *40*, D815–D820. [CrossRef] [PubMed]
27. Ramirez-Gaona, M.; Marcu, A.; Pon, A.; Guo, A.C.; Sajed, T.; Wishart, N.A.; Karu, N.; Djoumbou Feunang, Y.; Arndt, D.; Wishart, D.S. YMDB 2.0: A significantly expanded version of the yeast metabolome database. *Nucleic Acids Res.* **2017**, *45*, D440–D445. [CrossRef] [PubMed]
28. Hummel, J.; Selbig, J.; Walther, D.; Kopka, J. The Golm Metabolome Database: A database for GC-MS based metabolite profiling. In *Metabolomics: A Powerful Tool in Systems Biology*; Springer: Berlin/Heidelberg, Germany, 2007; pp. 75–95.
29. The Metabolomics Workbench. Available online: https://www.metabolomicsworkbench.org/search/index.php (accessed on 15 May 2022).
30. Kim, S.; Chen, J.; Cheng, T.; Gindulyte, A.; He, J.; He, S.; Li, Q.; Shoemaker, B.A.; Thiessen, P.A.; Yu, B.; et al. PubChem in 2021: New data content and improved web interfaces. *Nucleic Acids Res.* **2021**, *49*, D1388–D1395. [CrossRef]
31. Grigoriev, I.V.; Nordberg, H.; Shabalov, I.; Aerts, A.; Cantor, M.; Goodstein, D.; Kuo, A.; Minovitsky, S.; Nikitin, R.; Ohm, R.A.; et al. The Genome Portal of the Department of Energy Joint Genome Institute. *Nucleic Acids Res.* **2012**, *40*, D26–D32. [CrossRef]
32. Dowson, C.G.; Springham, P.; Rayner Adm Boddy, L. Resource relationships of foraging mycelial systems of *Phanerochaete velutina* and *Hypholoma fasciculare* in soil. *New Phytol.* **1989**, *111*, 501–509. [CrossRef]
33. Boddy, L. Interspecific combative interactions between wood-decaying basidiomycetes. *FEMS Microbiol. Ecol.* **2000**, *31*, 185–194. [CrossRef]
34. Salem, M.Z.M. EDX measurements and SEM examination of surface of some imported woods inoculated by three mold fungi. *Measurement* **2016**, *86*, 301–309. [CrossRef]
35. Anh, D.H.; Dumri, K.; Yen, L.T.H.; Punyodom, W. The earth-star basidiomycetous mushroom Astraeus odoratus produces polyhydroxyalkanoates during cultivation on malt extract. *Arch. Microbiol.* **2023**, *205*, 34. [CrossRef]
36. A New Textiles Economy: Redesigning Fashion's Future. Ellen MacArthur Foundation. 2017. Available online: http://www.ellenmacarthurfoundation.org/publications (accessed on 4 April 2023).

37. Schumacher, K.; Forster, A. Facilitating a Circular Economy for Textiles Workshop Report. 2022. Available online: https://nvlpubs.nist.gov/nistpubs/SpecialPublications/NIST.SP.1500-207.pdf (accessed on 22 April 2023).
38. Alam, R.; Mahmood, R.A.; Islam, S.; Ardiati, F.C.; Solihat, N.N.; Alam, M.B.; Lee, S.H.; Yanto, D.H.Y.; Kim, S. Understanding the biodegradation pathways of azo dyes by immobilized white-rot fungus, Trametes hirsuta D7, using UPLC-PDA-FTICR MS supported by in silico simulations and toxicity assessment. *Chemosphere* **2023**, *313*, 137505. [CrossRef]
39. Gadd, G.M.; Bahri-Esfahani, J.; Li, Q.; Rhee, Y.J.; Wei, Z.; Fomina, M.; Liang, X. Oxalate production by fungi: Significance in geomycology, biodeterioration and bioremediation. In *Fungal Biology Reviews*; Elsevier Ltd.: Amsterdam, The Netherlands, 2014; Volume 28, pp. 36–55.
40. Girometta, C.; Zeffiro, A.; Malagodi, M.; Savino, E.; Doria, E.; Nielsen, E.; Buttafava, A.; Dondi, D. Pretreatment of alfalfa stems by wood decay fungus *Perenniporia meridionalis* improves cellulose degradation and minimizes the use of chemicals. *Cellulose* **2017**, *24*, 3803–3813. [CrossRef]
41. Baptista, P.; Guedes de Pinho, P.; Moreira, N.; Malheiro, R.; Reis, F.; Padrão, J.; Tavares, R.; Lino-Neto, T. In vitro interactions between the ectomycorrhizal *Pisolithus tinctorius* and the saprotroph *Hypholoma fasciculare* fungi: Morphological aspects and volatile production. *Mycology* **2021**, *12*, 216–229. [CrossRef]
42. Singh, R.L.; Singh, P.K.; Singh, R.P. Enzymatic decolorization and degradation of azo dyes—A review. In *International Biodeterioration and Biodegradation*; Elsevier Ltd.: Amsterdam, The Netherlands, 2015; Volume 104, pp. 21–31.
43. Gul, R.; Sharma, P.; Kumar, R.; Umar, A.; Ibrahim, A.A.; Alhamami, M.A.M.; Jaswal, V.S.; Kumar, M.; Dixit, A.; Baskoutas, S. A sustainable approach to the degradation of dyes by fungal species isolated from industrial wastewaters: Performance, parametric optimization, kinetics and degradation mechanism. *Environ. Res.* **2023**, *216*, 114407. [CrossRef] [PubMed]

Journal of
Fungi

Article

Assessment of Tannin Tolerant Non-*Saccharomyces* Yeasts Isolated from *Miang* for Production of Health-Targeted Beverage Using *Miang* Processing Byproducts

Pratthana Kodchasee [1], Nattanicha Pharin [1], Nakarin Suwannarach [2], Kridsada Unban [3], Chalermpong Saenjum [4], Apinun Kanpiengjai [5], Dipayan Sakar [6], Kalidas Shetty [6], Martin Zarnkow [7] and Chartchai Khanongnuch [1,8,*]

[1] Division of Biotechnology, School of Agro-Industry, Faculty of Agro-Industry, Chiang Mai University, Chiang Mai 50100, Thailand
[2] Research Center of Microbial Diversity and Sustainable Utilization, Chiang Mai University, Chiang Mai 50200, Thailand
[3] Division of Food Science and Technology, School of Agro-Industry, Faculty of Agro-Industry, Chiang Mai University, Chiang Mai 50100, Thailand
[4] Department of Pharmaceutical Science, Faculty of Pharmacy, Chiang Mai University, Chiang Mai 50200, Thailand
[5] Division of Biochemistry and Biochemical Innovation, Department of Chemistry, Faculty of Science, Chiang Mai University, Chiang Mai 50200, Thailand
[6] Department of Plant Sciences, North Dakota State University, Fargo, ND 58108, USA
[7] Research Center Weihenstephan for Brewing and Food Quality, Technische Universität München, Alte Akademie 3, 85354 Freising, Germany
[8] Research Center for Multidisciplinary Approaches to Miang, Chiang Mai University, Mueang, Chiang Mai 50200, Thailand
* Correspondence: chartchai.k@cmu.ac.th; Tel.: +66-53-948-261

Citation: Kodchasee, P.; Pharin, N.; Suwannarach, N.; Unban, K.; Saenjum, C.; Kanpiengjai, A.; Sakar, D.; Shetty, K.; Zarnkow, M.; Khanongnuch, C. Assessment of Tannin Tolerant Non-*Saccharomyces* Yeasts Isolated from *Miang* for Production of Health-Targeted Beverage Using *Miang* Processing Byproducts. *J. Fungi* 2023, 9, 165. https://doi.org/10.3390/jof9020165

Academic Editor: Baojun Xu

Received: 21 December 2022
Revised: 18 January 2023
Accepted: 20 January 2023
Published: 27 January 2023

Abstract: This research demonstrated an excellent potential approach for utilizing *Miang* fermentation broth (MF-broth), a liquid residual byproduct from the *Miang* fermentation process as a health-targeted beverage. One hundred and twenty yeast strains isolated from *Miang* samples were screened for their potential to ferment MF-broth and four isolates, P2, P3, P7 and P9 were selected, based on the characteristics of low alcoholic production, probiotic properties, and tannin tolerance. Based on a D1/D2 rDNA sequence analysis, P2 and P7 were identified to be *Wikerhamomyces anomalus*, while P3 and P9 were *Cyberlindnera rhodanensis*. Based on the production of unique volatile organic compounds (VOCs), *W. anomalus* P2 and *C. rhodanensis* P3 were selected for evaluation of MF-broth fermentation via the single culture fermentation (SF) and co-fermentation (CF) in combination with *Saccharomyces cerevisiae* TISTR 5088. All selected yeasts showed a capability for growth with 6 to 7 log CFU/mL and the average pH value range of 3.91–4.09. The ethanol content of the fermented MF-broth ranged between 11.56 ± 0.00 and 24.91 ± 0.01 g/L after 120 h fermentation, which is categorized as a low alcoholic beverage. Acetic, citric, glucuronic, lactic, succinic, oxalic and gallic acids slightly increased from initial levels in MF-broth, whereas the bioactive compounds and antioxidant activity were retained. The fermented MF-broth showed distinct VOCs profiles between the yeast groups. High titer of isoamyl alcohol was found in all treatments fermented with *S. cerevisiae* TISTR 5088 and *W. anomalus* P2. Meanwhile, *C. rhodanensis* P3 fermented products showed a higher quantity of ester groups, ethyl acetate and isoamyl acetate in both SF and CF. The results of this study confirmed the high possibilities of utilizing MF-broth residual byproduct in for development of health-targeted beverages using the selected non-*Saccharomyces* yeast.

Keywords: non-*Saccharomyces* yeast; *Miang*; byproduct valorization; healthy beverage; fermentation

1. Introduction

Miang is an ethnic fermented tea (*Camellia sinensis* var. *assamica*) from northern Thailand, which is fermented naturally by mixed microorganisms. This fermented tea is a unique traditional food product known as chewing tea or eating tea. It is believed to provide health benefits due to its specific bioactive compounds with antioxidant and antimicrobial activities [1]. The natural microbial communities involved in biotransformation processes, including lactic acid bacteria, yeast, *Bacillus* spp. and filamentous fungi, have been proposed as the key microbes in the successive process of *Miang* fermentation [2–4]. These biotransformation processes cause the unique strong aromatic flavor and taste of this fermented food. *Miang* fermentation process can be divided into two types, the non-filamentous fungi growth-based process (NFP) and filamentous fungi growth-based process (FFP). The NFP used the steamed young tea leaves as raw material in fermentation without the requirement for the growth of filamentous fungi, while the FFP used the steamed mature tea leaves as raw materials and required for the growth of filamentous fungi, as one important step involved in the fermentation process and product quality [1]. The FFP begins with the aerobic growth of filamentous fungus and subsequently leads to anaerobic fermentation. In local practice, anaerobic fermentation is performed either by a semi-submerged fermentation (SSF) or submerged fermentation (SMF) [3]. Once fermentation is completed, *Miang* products are harvested for the market, while the *Miang* fermentation broth (MF-broth) residual byproduct is not utilized for any other use and discharged as liquid waste byproduct. Kodchasee et al. [3] confirmed that some bioactive constituents, such as the organic acids, catechin and catechin-related compounds remained in this liquid waste after the *Miang* FFP. Yeast species are key microbes in *Miang* fermentation processes, yeast species have been suggested to contribute the unique aroma and flavor in *Miang* products. Some of the non-*Saccharomyces* yeast species have been reported as the microbial population found in *Miang* products, including *Candida ethanolica*, *C. boidinii*, *Cyberlindnera rhodanensis*, *Debaryomyces hansenii*, *Pichia manshurica*, and *Wikerhamomyces anomalus*. These non-*Saccharomyces* yeasts are tannin tolerant and have been suggested to be involved in the development of flavors and qualities of *Miang* [2]. Furthermore, it has been suggested that yeast species found in *Miang* fermentation processes are obtained from the natural environment of the *Miang* production area, such as non-*Saccharomyces* yeast from tea flowers that may be associated with *Miang* fermentation [2,5]. Some previous studies described the potential application in beverage production processes of using non-*Saccharomyces* yeasts *Cyberlindnera rhodanensis* and *Wikerhamomyces anomalus*, due to their ability to produce a special aroma. *W. anomalus* is recognized as a good producer of fruity esters, such as acetate esters, ethyl acetate, isoamyl acetate and 2-phenylethyl acetate [6]. As a result, these species were explored for flavor enhancement in several fermented beverage products [7]. Furthermore, the *Cyberlindnera* species has been also reported for producing high levels of acetate esters, isoamyl, ethyl and 2-phenylethyl acetate, especially *C. saturnus* (formerly *Williopsis saturnus*), *C. mrakii* (formerly *W. saturnus* var. *mrakii*) and *C. subsufficiens* [8,9].

Among the varieties of yeast, *Saccharomyces cerevisiae* has been recognized for playing an important role in the alcoholic beverage industry because of its ability to ferment sugar into ethanol and other metabolic products [10,11]. However, there are some limitations in fermentation using *S. cerevisiae*, such as low levels of beneficial flavors and odor development, therefore leading to for the need to find a non-*Saccharomyces* yeast species for improving the flavor and odor forming compounds during the fermentation process. Several non-*Saccharomyces* yeasts have gained high interest due to their capability for β-glucosidase production during alcoholic beverage fermentation and this enzyme is reported to be involved in aroma formation through the catalytic activity on the glycosidic bonds of monoterpenes or other glycosides, and subsequently release the aromatic components in fermented beverages, such as in wine and black tea [12]. *Dekkera bruxellensis*, *Hanseniaspora uvarum*, *Metschnikowia pulcherrima*, *Pichia kudriavzevii* and *Wickerhamomyces anomalus* (formerly known as *P. anomala*) have been used to improve aromatic properties of wine by their release of glucosidase [13,14]. The characteristics of non-*Saccharomyces* yeast is a lower alcohol production but it results in a higher aromatic compound production, compared to *Saccharomyces* yeast and therefore is attractive for the design of consumer-targeted healthy beverages [15]. Moreover, the use of non-*Saccharomyces* yeasts in many aspects of fermentation, such as in inhibiting the growth of unwanted microorganisms, detoxifying mycotoxin and increasing the bioactive compounds have been also reported [16]. Generally, non-*Saccharomyces* yeasts can be found in nature, such as in ripened fruit or flowers, as well as in beverages and fermented foods, such as kombucha, kefir and *Miang* (fermented tea) [2,17].

The fermented tea-beverages produced, based on either green or black tea substrates, such as kombucha and low-alcoholic beverages, have recently been increasing in popularity among the health-conscious consumers and the increased demand is leading to the development of beverages derived from tea substrates as well. Tea wines, the low-alcohol beverages made by adding the sugar or juice of tea, followed by fermentation with yeast, has also received of interest. The benefits of the tea derived compounds in these healthy beverages, such as antioxidant and antimicrobial properties and tea catechins have been important targets for value addition [18]. Furthermore, low alcoholic beverages also provide potential health benefits, such as lowering cholesterol and increasing high-density lipoprotein (HDL), which contributes to the prevention of heart disease and lowering the risks of alcohol-related illnesses [17,19]. The attractiveness of low-alcohol healthy drink products derived from tea supports the conceptual idea for the development of this potential health-targeted beverage using MF-broth, a waste byproduct from the *Miang* fermentation process, as the primary substrate. The rationale for this approach is both the design of a new healthy beverage and also the concurrent advancement of zero waste technology contributing to carbon sequestration through the utilization of MF-broth byproducts.

This study therefore describes the use of an MF-broth byproduct, together with the newly isolated indigenous species of non-*Saccharomyces* yeast from *Miang* in the development of a potential low-alcoholic healthy beverage. This product process has the potential to allow for the improvement of product quality through the enhancement of flavors and enrichment of bioactive constituents by the co-culturing of these yeast isolates.

2. Materials and Methods

2.1. Strains, Media and Culture Conditions

A total of 120 yeast strains used in this study and previously isolated from *Miang* samples by Kanpiengjai et al. [2] were maintained in 25% (*v/v*) glycerol and kept at $-80\,^{\circ}$C as a stock culture at the Laboratory of Microbial Resources Development and Enzyme Technology, Faculty of Agro-Industry, Chiang Mai University. The reference strain *S. cerevisiae* TISTR 5088 was purchased from the culture collection of the Thailand Institute of Scientific and Technological Research (TISTR). The yeast strains were routinely grown at $30\,^{\circ}$C in yeast peptone dextrose (YPD) broth consisting of 1% (*w/v*) yeast extract, 2% (*w/v*) peptone and 2% (*w/v*) glucose.

2.2. Screening of the Yeasts for their Ethanol Producing Capability

A single colony of each of the 110 yeast isolates was inoculated into a Durham tube containing 10 mL YPD broth and statically incubated at $30\,^{\circ}$C and observed for gas formation following 24 h and 48 h cultivation. The yeast isolates capable of gas formation were selected and further investigated for glucose consumption and yield of ethanol production. The ethanol and glucose concentrations from the selected yeast isolates growing in the culture broth were analyzed by high-performance liquid chromatography (HPLC), as described by Kodchasee et al. [3].

2.3. Probiotic Characterization and Tannin-Tolerance Test

The selected ethanol producing yeasts were investigated for their probiotic characteristics, which included the ability to grow at $37\,^{\circ}$C, the resistance to grow in bile salt and low pH conditions, hemolytic activity and the adhesion capacity of cells [20]. Briefly, yeast cells were incubated in a simulated gastric juice containing 0.35% (*w/v*) pepsin at pH 2.0 and 3.0, and 0.3% (*w/v*) bile salt in phosphate buffered saline (PBS buffer) at pH 7 (PBS buffer pH 7 as control). The treated suspension was incubated at 37 °C for 3 h. Viable cell counts were determined by plating on yeast malt (YM) agar at 37 °C for 24 h, and the survival percentage was calculated as follows: survival (%) = [final (logCFU/mL)/control (logCFU/mL)] × 100. Hemolytic activity was determined by inoculating the yeast strain on a blood agar plate containing 5% defibrinated sheep blood and incubated at 37 °C for 72 h. The development of a clear zone of hydrolysis surrounding the colonies was observed and classified [21]. The cell surface hydrophobicity was tested in toluene (a nonpolar solvent). The yeast suspension was prepared in PBS to an optical density of 600 nm (OD600) = 1 (A0). Then, the volume of toluene was added into the yeast cell suspension at a ratio of 1:1 and the two-phase system was mixed for 5 min. Following 1 h of incubation at 37 °C, OD600 of the cell suspension was measured (A1) and the microbial adhesion to solvents (MATS) percentage was calculated as follows: MATS (%) = [(A0 − A1)/A0] × 100. Isolates with MATS above 50% were considered as hydrophobic [22]. The tannin-tolerance of the selected yeast isolates was also investigated on YM agar containing tannin, according to the method described by Kanpiengjai et al. [2]. A single colony of yeast isolate was picked up and spiked on YM agar with 10, 30 and 50 g/L tannin. The growth of the yeast isolates was observed after incubation at 30 °C for 3 days. The characteristics of low alcoholic producing properties, probiotic properties and tannin tolerance were used as the criteria for the selection of the proper yeast strain used in the healthy beverage fermentation using *Miang* fermentation broth residual byproducts.

2.4. Molecular Identification and Carbon Assimilation Profiles

The selected yeasts were identified, based on the morphological characteristics, biochemical tests, and the presence of D1/D2 region of the 26S rRNA gene. Extraction of genomic DNA was performed as described by Kanpiengjai et al. [2] and the PCR was performed using two universal primer pairs NL1 (5′-GCATATCAATAA GCGGAGGAAAAG-3′) and NL4 (5′-GGTCCGTGTTTCAAGACGG-3′). The 26S rRNA gene were sequenced and analyzed using the BLAST algorithm of GenBank (http://www.ncbi.nlm.nih.gov/blast, accessed on 25 November 2022) and were deposited under the accession numbers OL63634 (P2 strain), OL636341 (P3 strain), OL636342 (P7 strain) and ON460286 (P9 strain). Carbon assimilation was performed with API ID 32C strips (Biomérieux, Craponne, France) and carbon fermentation was analyzed and compared, according to a standard yeast identification method [23]. The samples were evaluated visually for turbidity in wells, differentiating the positive (+), negative (−) and weak (w) growths.

2.5. Analysis of the Volatile Compounds

The selected yeast strains were investigated for their ability to produce volatile compounds (VOCs) during cultivation on YM slants at 30 °C in the dark, following the method described by Suwannarach et al. [24]. The VOCs in the air space above each yeast colony were captured for 45 min at 25 ± 2 °C using a solid-phase microextraction (SPME) fiber, composed of 50/30 divinylbenzene/carboxen on polydimethyl-siloxane on a stable flex fiber (Supelco, Bellefonte, PA, USA). The adsorbed fiber was placed into the splitless injection port of a gas chromatograph (Agilent Technologies Inc, Palo Alto, CA, USA) equipped with a DB-WAX capillary column (0.25 mm × 30 m I.D × 0.25 μm film thickness, Supelco, Bellefonte, PA, USA). The column temperature was set to 40 °C for 2 min, then to 200 °C for 5 min. At an initial column head pressure of 60 kPa, the carrier gas was ultra-high purity helium. Prior to trapping the volatiles, the fiber was cleaned at 250 °C for 57 min under a flow of helium gas. The adsorbed volatiles were introduced into the gas chromatography (GC) interfaced with a mass spectrometer over a 30-s injection duration (Agilent 5975C, Agilent Technologies Inc., Santa Clara, CA, USA). With ionization energy of 70 eV, the mass detector was operated in the electron impact mode. The GCMS solution was used to collect and process the data (Agilent). The volatiles were identified by comparison with a mass spectra library (NIST) and the RI of the pure standard compounds were confirmed by GC-MS (https://webbook.nist.gov/chemistry/ accessed on 8 March 2022) [24].

2.6. MF (Miang Fermentation) Broth Fermentation

MF-broth, the remaining byproduct liquid waste from the submerged fermentation process in *Miang* production, was collected from the FFP *Miang* producing plant in Phrae province, Thailand. The starter cultures of two selected strains of non-*Saccharomyces* yeast species (*Wikerhamomyces anomalus* P2 and *Cyberlindnera rhodanensis* P3) and *S. cerevisiae* TISTR 5088 were prepared by cultivating the yeast strains in 100 mL YM broth at 30 °C for 24 h on a 150-rpm rotary shaker. The cells were harvested by centrifugation at 6000 rpm for 5 min at 4 °C, washed twice and re-suspended in a sterile solution of 0.85% NaCl (*w/v*) and used as the starter for fermentation. Glucose was added into the MF-broth to achieve the final concentration of 5% (*w/v*) glucose in 300 mL in a 500 mL flask and sterilized at 121 °C for 20 min. Seven fermentative groups consisting of three single microbe fermentations (SF) of *S. cerevisiae* TISTR 5088, *W. anomalus* P2 and *C. rhodanensis* P3 and four co-fermentations (CF) of *S. cerevisiae* TISTR 5088 + *W. anomalus* P2, *S. cerevisiae* TISTR 5088 + *C. rhodanensis* P3, *W. anomalus* P2 + *C. rhodanensis* P3, *S. cerevisiae* TISTR 5088 + *W. anomalus* P2 + *C. rhodanensis* P3 were conducted and a non-inoculated sterile broth was used as a control. The fermentation was carried out in a 500 mL Erlenmeyer flask with 5% (*v/v*) starter containing a final inoculated concentration of approximately 6.5 logCFU/mL *S. cerevisiae* and non-*Saccharomyces* yeast cultures at a 1:1 ratio in the *Miang* byproduct fermented broth and incubated at 121 °C for 5 days. Samples were taken at 0, 12, 24, 36, 48, 72, 96 and 120 h to determine the viable cell counts and pH. The bioactive compound in the fermented broth was determined by the method described by Kodchasee et al. [3]. The remaining sugars, organic acids and ethanol concentrations were analyzed by HPLC. Volatile compounds analyses (VOCs) were performed using headspace vials with the sample of the final fermentation time (120 h).

2.7. Enumeration of the Microbes and the pH Measurement

The freshly fermented sample (1 mL) was transferred into 9 mL of 0.85% (*w/v*) NaCl solution and serially diluted with a 10-fold dilution. A volume of 10 µL of 10^{-2} to 10^{-5} dilutions were dropped on YM agar and incubated at 30 °C for 24 h. The viable cell counts of the yeast cultures were observed and presented as the log of a colony forming unit per milliliter (log CFU/mL). Measurement of the pH was performed by a Metrohm 744 pH meter with a glass electrode (Metrohm Co. Ltd., Herisa, Switzerland).

2.8. Analysis of Glucose, Ethanol, Organic Acids, Catechins and Caffeine by HPLC

Glucose, ethanol, and organic acid were analyzed using HPLC, according to Kodchasee et al. [3]. Briefly, a solution of 10 µL of each sample was filtered through a 0.45 µm filter paper (Whatman Inc., Clifton, NJ, USA), and injected into the HPLC system (Agilent 1000 series, Agilent Technologies Inc., Palo Alto, CA, USA) equipped with a 150 × 7.80 mm Rezex ROA organic acid H+ (8%) column (Phenomenex, Torrance, CA, USA). A mobile phase of 0.005 M H_2SO_4 was used for the glucose and ethanol analyses via a refractive index (RI) detector with a flow rate of 0.6 mL/min and 60 °C temperature. The concentrations of glucose and ethanol were calculated, based on the area peak corresponding with their standard retention time (RT).

The organic acids analysis was carried out with the same column using an isocratic elution of 50% (A) 0.05 M H_2SO_4 and 50% (B), 2% acetonitrile with a flow rate of 0.5 mL/min; temperature 40 °C, with a UV detector at the wavelength of 210 nm. The concentration of organic acids was calculated, based on the retention time (RT) corresponding to standards (acetic, citric, formic, gallic, glucuronic, lactic, malic, oxalic, succinic, and tartaric acids).

Catechins and caffeine were analyzed using a C18 column (250 × 4.6 mm, PhenomenexGemini, Torrance, CA, USA) with the HPLC conditions as follows; UV detector;

mobile phase (A) 0.1% acetic acid in acetonitrile and mobile phase (B) 0.1% acetic acid in deionized (DI) water at a flow rate of 1.0 mL/min at 20 °C and injection volume of 10 μL. The concentrations of catechins and caffeine were calculated, based on their RT, compared with the standards that included epigallocatechin gallate (EGCG), epicatechin gallate (ECG), gallocatechin gallate (GCG), epigallocatechin (EGC), epicatechin (EC), catechins (C), gallocatechin (GC) and caffeine (Sigma-Aldrich, St. Louis, MO, USA) [25].

2.9. Determination of the Bioactive Compounds

The bioactive compounds containing total phenolic (TP), total tannin (TT) and total flavonoids (TF) contents were determined following the methods described by Kodchasee et al. [3]. TP was measured according to a modified Folin–Ciocalteu method. Briefly, 200 μL of fermented MF-broth was added into a glass test tube (12 × 100 mm) containing 200 μL of 2 M Folin–Ciocalteu reagent and mixed using a vortex mixer (Vortex Genie 2, Scientific Industries, Bohemia, NY, USA). Then, 1.8 mL of deionized water (DI water) was added, and the reaction mixture was incubated at ambient temperature for 3 min. Then, 400 μL of 10% (*w/v*) sodium carbonate was added. The volume was adjusted using DI water to 4 mL and incubated in the dark at ambient temperature for 1 h. The absorbance of the blue solution was measured at 725 nm using the spectrophotometer. The total phenolic content was determined from the calibration curve using gallic acid as the reference standard (0–200 mg/L of gallic acid).

TT was determined by the Folin–Ciocalteu method and using polyvinylpolypyrrolidone (PVPP) to separate the tannins from other phenols. Briefly, 1 mL of the MF-broth sample was mixed with 1 mL of 10% (*w/v*) PVPP, vortexed and incubated at 4 °C for 15 min. Then, it was centrifuged at 3000× *g*, at 4 °C for 10 min. The remaining TP of the PVPP precipitated supernatant was measured with the Folin–Ciocalteu reagent and TT was estimated using the formula: TT = TP − PVPP precipitation.

TF contents were determined using the aluminum chloride colorimetric method. Briefly, 250 μL *Miang* extract was mixed with 250 μL 10% (*w/v*) aluminum nitrate and 50 μL 1 M potassium acetate. Then, 1.65 mL 80% (*v/v*) methanol was added to the reaction mixture and incubated in the dark at ambient temperature for 40 min. The absorbance was measured at 415 nm. The content of the flavonoids was determined from the calibration curve using the quercetin equivalent as the reference standard (0–250 mg/L of quercetin equivalent).

Antioxidant activity was based on 1,1-diphenyl-2-picrylhydrazyl (DPPH) (Sigma-Aldrich, St. Louis, MO, USA) free radical assay [26]. Briefly, 100 μL of diluted sample in DI water (10 μL/mL) was mixed with 400 μL of 0.15 mM DPPH in 80% methanol and incubated in the dark at ambient temperature for 30 min. The absorbance was measured at 517 nm, the radical scavenging percentage was calculated against a blank using the following equation: inhibition (%) = (1 − (B/A)) × 100, where A is the absorbance of the mixture without sample, and B is the absorbance of the mixture containing the sample.

2.10. β-Glucosidase Activity Assay

β-Glucosidase activity was assayed by measuring the amount of p-nitrophenol (pNP) liberated from *p*-nitrophenyl-β-D-glucoside (pNPG) as substrate [16]. First, 0.125 mL MF-broth was mixed with 0.125 mL of a 4 mM solution of pNPG in 0.1 M citrate–phosphate buffer (pH 5.0). The reaction mixture was incubated at 30 °C for 20 min and subsequently 0.250 mL of 50 mM sodium carbonate (Na_2CO_3) was added to stop the reaction. The absorbance of the final reaction mixture was measured at 405 nm by a spectrophotometer (Metertech SP-8001 UV/Visible Spectrophotometer, Metertech Inc., Taipei, Taiwan). The enzyme activity unit (U) was expressed as the ability to liberate 1 μmole pNP from the substrate solution/min.

2.11. Statistical Analysis

All experiments were performed in triplicate and the data are represented as the mean ± standard deviation. Statistical analyses were conducted using IBM SPSS statistics 23.0 software package (SPSS Inc., Chicago, IL, USA) employing the Tukey multiple range tests at a significance level of $p < 0.05$. The principal component analysis (PCA), hierarchical cluster analysis and heat-maps were performed through GraphPad PRISM 9 (GraphPad Software, La Jolla, CA, USA) and SPSS software, respectively.

3. Results and Discussion

3.1. Screening of Yeast with Potential Characteristics

Among 120 yeast strains, 24 isolates showed a gas formation (Figure 1a) and the ethanol production in the culture broth were confirmed by HPLC, in comparison to the efficient ethanol producing yeast *Saccharomyces cerevisiae* TISTR 5088 (Figure 1b). All 24 gas producing yeasts showed a variety of both ethanol concentration and sugar consumption. Most of the sucrose (20 g/L) was consumed by the isolates P2, P3, P7 and P9, similar to *S. cerevisiae* TISTR 5088, but the lower ethanol yields were observed from the yeast isolates P2, P3, P7 and P9 at 2.5, 3.3, 3.1 and 2.7 g/g substrate, respectively, while the ethanol yield of 0.46 ± 0.02 g ethanol/g glucose (9.21 ± 0.35 g/L) was produced by *S. cerevisiae* TISTR 5088. Aligned with the purpose of this research, yeast isolates P2, P3, P7 and P9 were selected due to potential of high glucose consumption (>19.90 g/L) and low ethanol production, which were approximately 50% of the ethanol yield observed for the ethanol producer as *S. cerevisiae* TISTR 5088.

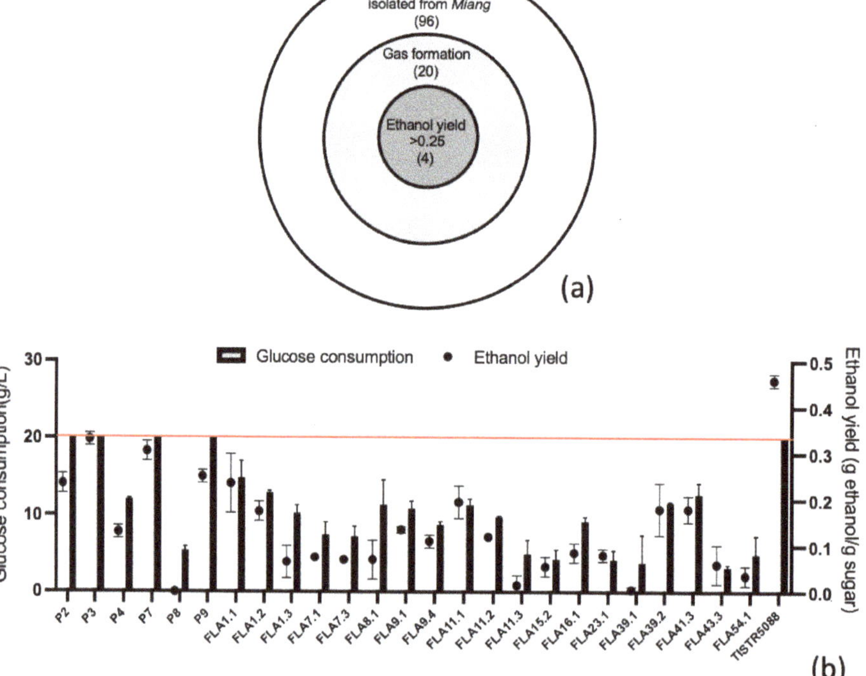

Figure 1. Screening for the ethanol producing yeast isolated from *Miang*. The graphical ratio of a total screened yeast number, in comparison to the gas-forming and ethanol-producing isolates (**a**). Glucose consumption (g/L) and ethanol yield (g/g sugar consumption) of 24 yeast isolates and *S. cerevisiae* TISTR 5088 (control) after 3 days of fermentation in YPD medium at 30 °C (**b**).

The tolerance to both a low pH condition and bile salts were used to evaluate the probiotic characteristics. The selected isolates showed a high tolerance to pepsin at pH 2.0 after 3 h exposure, with a survival rate of 83.9 to 99.9% (Table 1), whereas the survival rate of bile salt was 66.6 to 91.4%. The highest probiotic potential was in isolate P2, which demonstrated the highest percentage of survival rate, both in bile salt and low pH conditions, followed by isolates P3 and P9, respectively. However, the strains are required to survive against the natural barriers of the gastro-intestinal tract, including body temperature, a low pH in the stomach and high bile acid concentration at the duodenum in the small intestine [22]. The hydrophobicity of the yeasts was assessed indirectly as the ability to interact with chloroform and the results showed the hydrophobicity index ranged from 31.14 to 52.31%. None of the yeast strains showed β-hemolysis after 5 days incubation (Figure 2a). Several reports have previously described the probiotic properties of non-*Saccharomyces* yeast species in beverage and food production [10,27–30]. This result shows the additional advantage to the potential health benefit using MF-broth. Although, *C. rhodanensis* were reported to be associated with clinical bovine mastitis [31], but the *C. rhodanensis* strains investigated in this study are negative for a hemolytic response. In addition to the edible source of the yeast isolate, the negative hemolytic test indicates the safety status for *C. rhodanensis* used in this study [32].

Table 1. Probiotic characteristics of four selected yeast strains.

Isolate	Survival Rate (%)			Microbial Adhesion to Solvents (%)
	Bile Salt	pH2	pH3	
W. anomalus P2	91.42 ± 0.10 [a]	91.76 ± 0.18 [b]	96.40 ± 0.08 [b]	43.81 ± 0.89 [c]
C. rhodanensis P3	70.28 ± 0.81 [d]	83.88 ± 0.15 [c]	99.71 ± 0.09 [a]	39.45 ± 0.54 [d]
W. anomalus P7	66.61 ± 0.68 [e]	84.67 ± 0.11 [c]	99.94 ± 0.08 [a]	31.14 ± 0.62 [e]
C. rhodanensis P9	75.51 ± 0.12 [c]	91.86 ± 0.23 [b]	95.21 ± 0.12 [b]	52.31 ± 0.94 [b]
S. cerevisiae TISTR 5088	85.59 ± 0.66 [b]	95.93 ± 0.01 [a]	99.61 ± 0.01 [a]	63.77 ± 1.02 [a]

Note: [a–e] Different letters within the same column indicate a statistically significant difference ($p < 0.05$).

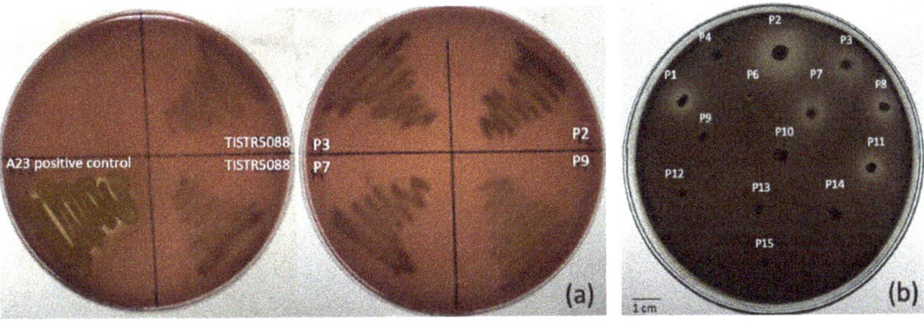

Figure 2. Hemolytic activity of yeast isolates P2, P3, P7 and P9 on blood agar at 30 °C for 7 days (**a**) and growth of the yeast isolate on YMA containing 50 g/L tannin at 37 °C for 3 days (**b**).

Tannin in *Miang* products is likely to influence the microbial growth and development throughout the fermentation process. Therefore, the tannin tolerance capability was investigated to predict whether the yeast could grow in MF-broth. Isolates P2, P3, P7 and P9 were cultured on YM agar with varied tannin concentrations of 10, 30 and 50 g/L, respectively. All isolates could grow in all tannin concentrations and formed clear zones around their colonies, indicating that they have a tannin degrading system initiated by the key enzyme as tannase (Figure 2b). The previous discovery of the cell associated tannase (CAT) with high capability in gallate and catechin biotransformation have been reported from yeasts isolated from *Miang*, *C. rhodanensis* A22.3 and *C. rhodanensis* A45.3 [33]. As a result, we expected that the yeast species obtained from the *Miang* sample would be able to grow in

MF-broth containing tannins and increase the antioxidant activity because of the efficient tannin and catechin biotransformation caused by CATs.

3.2. Molecular Identification and Carbon Assimilation Profiles

Based on the result of the D1/D2 region nucleotide sequence analysis, the isolates P3 and P7 were identified to be *Cyberlindera rhodanensis*, while the isolates P2 and P9 were *Wickerhamomyces anomalus* (Figure 3). The results of the analysis of the biochemical characteristics of the yeasts are also summarized in Table 2. The carbon fermentation and assimilation analysis found that sucrose and glucose utilization were positive in all strains. Meanwhile, the assimilation of cellobiose, galactose, raffinose and xylose were different in *C. rhodanensis* P3 and P7 and *W. anomalous* P2 and P9. Interestingly, *C. rhodanensis* showed positive results for the assimilation of cellobiose, a known substrate of β-glucosidase, which is an enzyme involved in the aromatic compound formation in wine [13]. The several non-*Saccharomyces* yeasts isolated from fermented food were examined for their ability to improve wine fragrance during fermentation. From these studies, non-*Saccharomyces* yeasts were recognized for their low ethanol fermentation rate, in comparison to *S. cerevisiae*. *W. anomalus* has previously been investigated in the production of low alcohol with high levels of esters, such as in kiwi wine [34], muscat bailey wine [13], apple wine [35] and Baijiu [36]. The members of the genus *Cyberlindnera*, as well as *C. subsufficiens*, *C. mrakii*, *C. jadinii*, *C. fabianii* and *C. saturnus* [9,37–39] were studied on the generation of a fruity aroma in non-alcoholic beer, but *C. rhodanensis* has not been reported for use in fermented beverages. Therefore, two of the selected non-*Saccharomyces* yeasts *C. rhodanensis* isolated from a traditional fermented tea of FFP *Miang* are attractive to elucidate and develop for further applications, especially in the development of the value-added beverages from the waste byproduct produced during the *Miang* fermentation process.

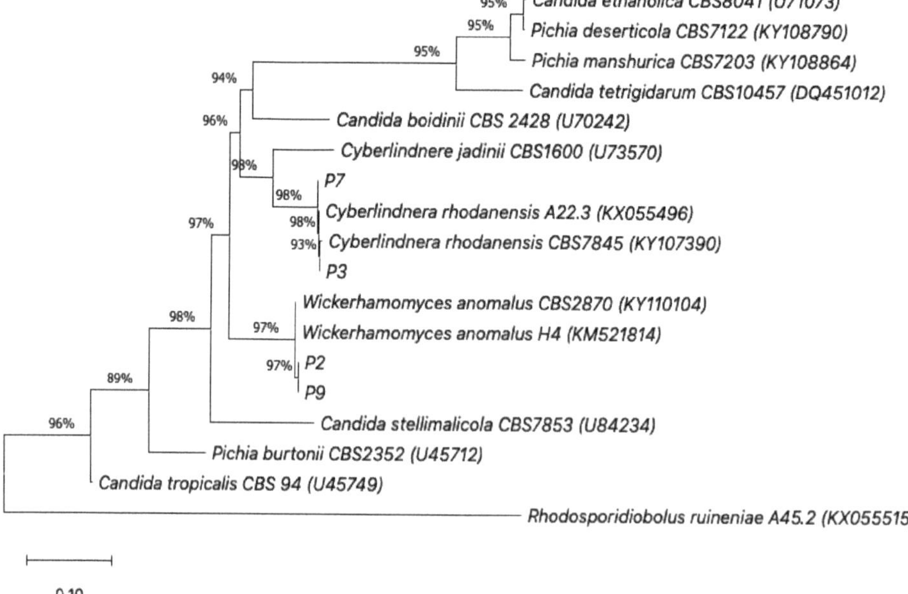

Figure 3. Phylogenetic tree analysis of the yeast *Cyberlindera rhodanensis* P3 and P7, *Wikerhamomyces anomalus* P2 and P9, based on particle 26S rDNA D1/D2 region sequence using the neighbor-joining method. Bootstrap values >50% (based on 1000 replication) are given at the branch points. The scale bar shows a patristic distance of 0.10.

Table 2. Carbon fermentation and assimilation characteristics of the four selected yeast strains.

Characteristic	Strains				Characteristic	Strains			
	P2	P3	P7	P9		P2	P3	P7	P9
Assimilation of Carbon					**Assimilation of Carbon**				
D-galactose	+	-	-	+	Potassium gluconate	-	w	w	-
Cyacloheximide (actidione)	-	-	-	-	Levulinic acid	-	-	-	-
D-saccharose (sucrose)	+	+	+	+	D-Glucose	+	+	+	+
N-Acetyl-glucosamine	+	-	-	+	L-Sorbose	-	w	w	-
Lactic acid	+	+	+	+	Glucosamine	-	-	-	-
L-arabinose	-	-	-	-	Esculin	+	+	+	+
D-Cellobiose	-	+	+	-	**Fermentation**				
D-Raffinose	+	-	-	+	Cellobiose	-	+	+	-
D-Maltose	+	+	+	+	D-glucose	+	+	+	+
D-Trehalose	-	+	+	-	D-galactose	+	-	-	+
Potassium 2 ketogluconate	-	-	-	-	Maltose	+	w	w	+
Methyl-αD-glucopyranoside	+	-	-	+	Sucrose	+	+	+	+
D-Mannitol	+	-	-	+	Lactose	-	-	-	-
D-Lactose (bovine origin)	+	-	-	+	Trehalose	-	-	-	-
Inositol	-	-	-	-	Raffinose	+	-	-	+
D-Sorbitol	+	+	+	+	Xylose	-	+	+	-
D-Xylose	+	+	+	+	**Other Growth Characteristics**				
D-Ribose	+	-	-	+	50% glucose yeast extract	+	+	+	+
Glycerol	+	+	+	+	60% glucose yeast extract	+	+	+	+
L-Rhamnose	-	+	+	-	10% NaCl	-	-	-	-
Palatinose	+	+	+	+	Growth at 20 °C	+	+	+	+
Erythritol	+	-	-	+	Growth at 25 °C	+	+	+	+
D-Melibiose	-	-	-	-	Growth at 30 °C	+	+	+	+
Sodium glucuronate	-	-	-	-	Growth at 37 °C	+	+	+	+
D-Melezitose	+	+	+	+	Growth at 45 °C	-	-	-	-

Note: +, positive; -, negative; w, weak positive.

The analysis of VOCs produced by *S. cerevisiae* TISTR 5088, *C. rhodanensis* P3 and P7 and *W. anomalous* P2 and P9 was performed by the headspace SPME-GC/MS technique that indicated that higher alcohols were the main VOCs of *S. cerevisiae*, while acetate esters were the main esters of the non-*Saccharomyces* yeast species. *S. cerevisiae* and *W. anomalus* P2 and P9 produced at least five VOCs on YM slant agar, while *C. rhodanensis* P3 and P7 produced 10 types of VOCs. The VOCs were analyzed by comparing their mass spectral characteristics to those in the NIST database (Figure 4). Isoamyl alcohol (63.54% of total area) was the most abundant aroma compound generated by *S. cerevisiae*, followed by 2-phenylmethyl ethanol and isoamyl acetate. Whereas acetic acid ethyl ester was the most abundant VOC in the culture of *W. anomalus* P2 and P9, followed by 1-butanol- 3-methyl acetate, 2-phennethyl acetate, 2-phenylmethyl ethanol and 2-phenylmethyl ethanol. 1-Butanol-3-methyl acetate was the most abundant VOC produced by *C. rhodanensis* P3 and P7, followed by acetic acid ethyl ester, 2-phennethyl acetate, 2-phenylmethyl ethanol and 2-phenylmethyl ethanol. Furthermore, 2-furyl methyl acetate, benzyl acetate and 2-propenoic acid were found in the fermented broth of *C. rhodanensis* P3 and P7 but not in *S. cerevisiae* or *W. anomalus* P2 and P9. However, esters are a type of volatile compound produced by yeast cells during alcoholic fermentation, and various studies have demonstrated that their presence is related to wine quality. Ethyl acetate and isoamyl acetate are the most important esters in wine as both are associated with fruity wine/apple-like sweet pear drop and sweet banana ripe flavors, respectively [40]. Since the strains of *C. rhodanensis* P3 and P7 and *W. anomalus* P2 and P9 produced significant amounts of ethyl acetate and isoamyl acetate, we decided to select the representative yeasts *C. rhodanensis* P3 and *W. anomalus* P2 for further studies, as both isolates produced the highest ester in each species.

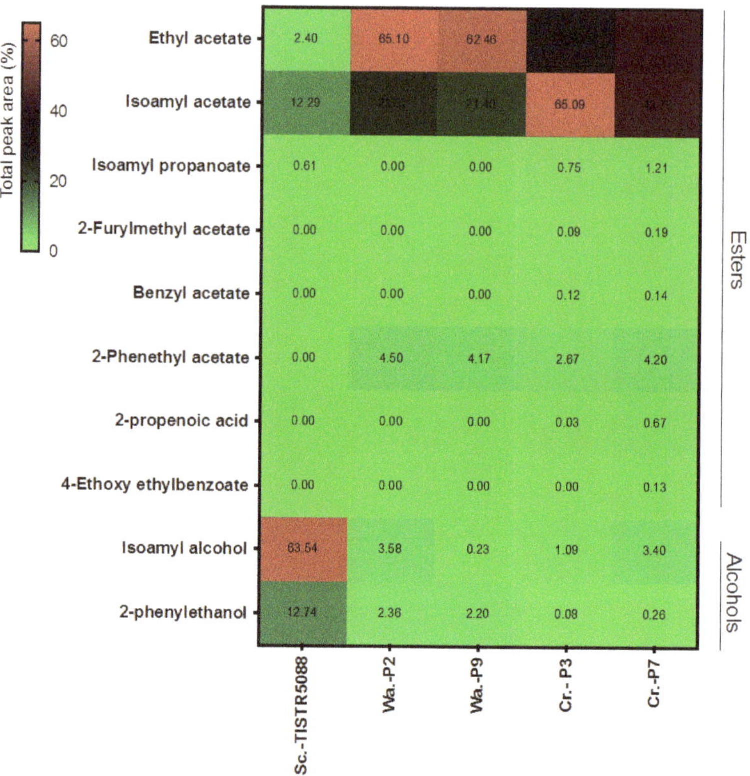

Figure 4. Heatmap using the total peak area (%) of VOCs produced with *S. cerevisiae* TISTR 5088, *W. anomalus* (P2, P9) and *C. rhodanensis* (P3, P7) on YM agar at 30 °C for 2 days.

3.3. Yeast Growth Dynamic and pH Changes in MF (Miang Fermentation) Broth

The growth dynamics of *S. cerevisiae* TISTR 5088 and non-*Saccharomyces* yeast species in the respective single fermentation (SF) and co-fermentation (CF) of MF-broth fermentation are shown in Figure 5. The control MF-broth did not have any microbial growth, while the culture of non-*Saccharomyces* species P2 and P3 showed the increase of the viable yeast number log 6.80 ± 0.03 and log 6.61 ± 0.03 CFU/mL, respectively. In CF, the growth of the yeast population showed a similar trend in all treatments and reached the approximate level of log 6.60 ± 0.02 CFU/mL at 120 h fermentation. The pH of MF-broth slightly decreased after 120 h of fermentation in the range of 4.04 ± 0.03 to 3.88 ± 0.01, which is very close to the pH of the initial MF-broth (Figure 5). The MF-broth fermentation inoculated with non-*Saccharomyces* yeast species exhibited a lower growth than that of *S. cerevisiae* TISTR 5088, which might be due to the numerous parameters involved, such as the cultivability loss of non-*Saccharomyces* strains in alcoholic fermentation, the nitrogen limitation, low oxygen availability and inhibition by increased ethanol, as well as extrinsic factors, such as SO_2 [41,42]. However, when *S. cerevisiae* and non-*Saccharomyces* yeasts were inoculated in a 1:1 ratio, the amount of yeast was slightly reduced in SF, indicating a mutual inhibition under the current conditions, especially *W. anomalus*, which has a reported killer activity against *S. cerevisiae*. To minimize the technological issues caused by sluggish or incomplete alcoholic fermentation, the compatibility of the selected killer *W. anomalus* strains with the primary microbial agents involved in wine production must be investigated throughout the selection stage [6,7,43].

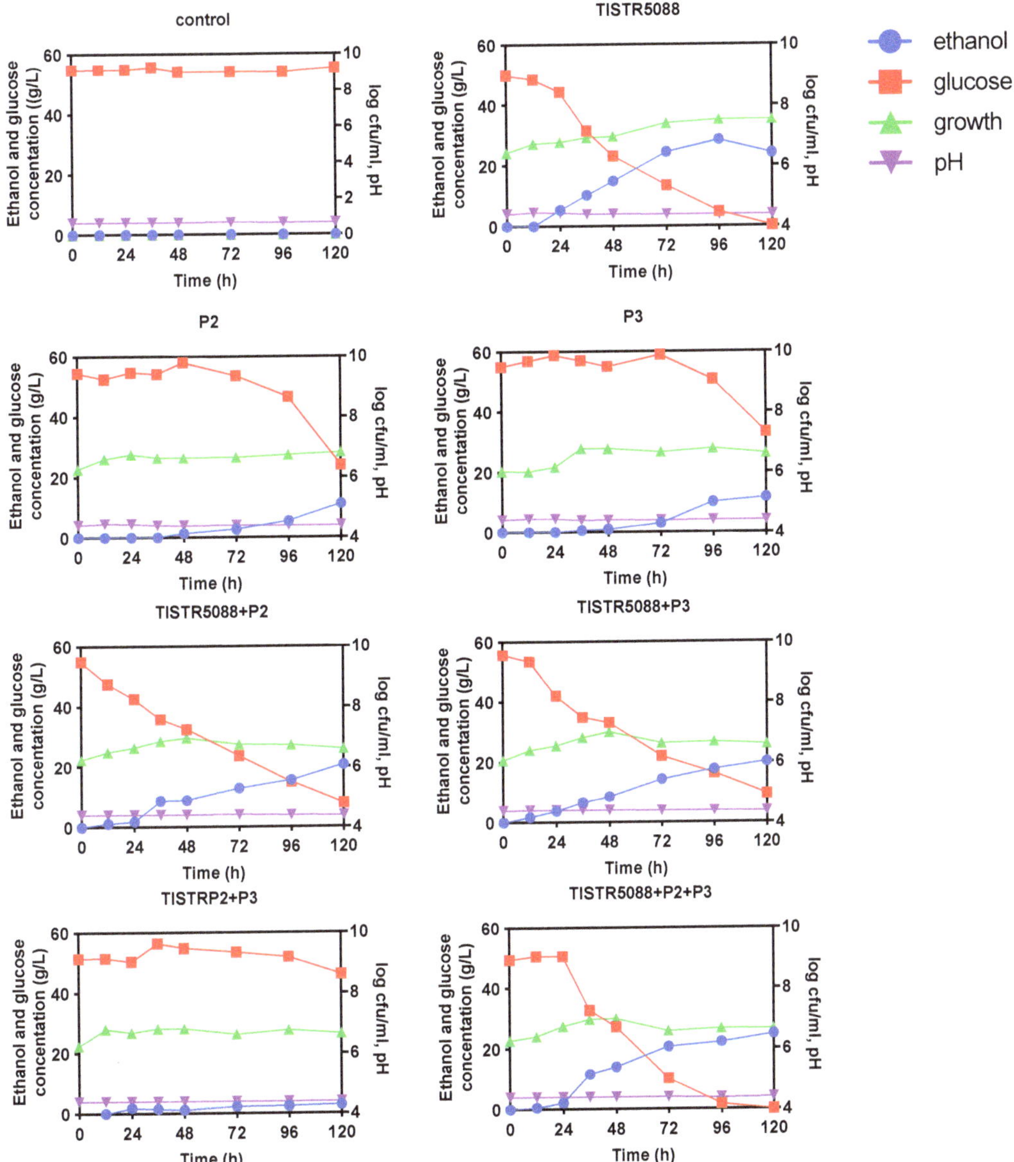

Figure 5. Cell growth, glucose, and ethanol concentration profiles during the fermentation of MF-broth fermented by *S. cerevisiae* TISTR 5088, *W. anomalus* P2 and *C. rhodanensis* P3, with single and co-fermentation at 30 °C for 120 h.

3.4. Changes in the Sugar Consumption, Ethanol Production and Organic Acid Formation

The glucose concentration in the original MF-broth was 3.56 ± 0.10 g/L, which occurred via the microbial degradation of tea leaf polysaccharides during *Miang* fermentation. This original sugar content is low for supporting yeast fermentation. The external source of glucose was added up to the final concentration of 5% (*w/v*) as the main carbon source for yeast growth and conversion to alcohol. The initial glucose concentration was at 53.21 ± 0.48 g/L. In SF, the glucose consumption ranged around 20–50 g/L at 120 h

(Figure 5). *S. cerevisiae* TISTR 5088 showed a high glucose consumption, since a significant low residual concentration of glucose (0.19 ± 0.77 g/L) was detected (Table 3), followed by *C. rhodanensis* P3 and *W. anomalus* P2. In case of CF, a range of varied glucose concentrations (5 to 50 g/L) was observed, co-fermentation with *S. cerevisiae* TISTR 5088 clearly showed a higher glucose consumption than the SF of *C. rhodanensis* P3 and *W. anomalus* P2. The combination of *S. cerevisiae* TISTR 5088 with either *C. rhodanensis* P3 or *W. anomalus* P2, in CF, showed the almost complete consumption of glucose during the final period of fermentation.

In SF, the ethanol production by *S. cerevisiae* TISTR 5088 was observed after 24 h and reached the maximum at 96 h with the highest ethanol concentrations of 23.67 g/L (Figure 5). Whereas the ethanol productions by *C. rhodanensis* P3 and *W. anomalus* P2 were observed after 48 h and showed a final ethanol concentration of 11.03 and 11.56 g/L, respectively, at 120 h (Table 3). The CF with *S. cerevisiae* TISTR 5088 showed a more rapid ethanol production after 12 h, both in response to *S. cerevisiae* TISTR 5088 + *W. anomalus* P2 and *S. cerevisiae* TISTR 5088 + *C. rhodanensis* P3, and an almost same level of ethanol production was observed (20.74 ± 0.21 and 20.14 ± 0.43 g/L, respectively), while *S. cerevisiae* TISTR 5088 + *W. anomalus* P2 + *C. rhodanensis* P3 produced the highest ethanol concentration at 24.91 ± 0.76 g/L. Moreover, CF of *W. anomalus* P2 + *C. rhodanensis* P3 produced a low amount of ethanol at only 2.92 ± 0.05 g/L. The greater ethanol production of *S. cerevisiae* TISTR 5088 in SF, when compared to *W. anomalus* P2 and *C. rhodanensis* P3 is probably caused from the weak metabolic response of non-*Saccharomyces* yeasts in consumption or the utilization of the nutrients in the culture broth. Furthermore, it has been reported that the low level of oxygen in the fermenting activity of this yeast leads to an increase in the cell biomass and a decrease in ethanol yield, which is a strategy that leads to the reduction of the ethanol level in wine produced by the co-culture with *S. cerevisiae* [44,45]. Therefore, in CF, the ethanol content might mainly come from *S. cerevisiae* TISTR 5088. Furthermore, the sugar content control in MF-broth fermented with non-*Saccharomyces* yeast species resulted in a low ethanol product that might be the strategy for the development of health-targeted low alcohol beverages. Several studies indicated the potential health benefits from the production of low alcoholic beverages, such as lowering cholesterol and increasing high-density lipoprotein (HDL), which helps to prevent heart disease and the risks of alcohol-related illnesses [19]. The genus *Cyberlindnera*, as well as *C. subsufficiens*, *C. mrakii*, *C. jadinii*, *C. fabianii* and *C. saturnus*, have been found to produce low-alcoholic beers in both mono and co-fermentation with *S. cerevisiae* [9,38,39,46]. In addition, *W. anomalus* was targeted to be used in alcohol reduction in fermented beverages. Furthermore, a low ethanol producing non-*Saccharomyces* yeast has been also targeted for use in the high level formation of aromatic compounds, such as esters, higher alcohols and fatty acids [44,47].

Organic acids in traditional MF-broth residual byproducts have been quantified and include acetic, citric, glucuronic, gallic, oxalic and succinic acids (Table 3), which were similar to the organic acids found in the *Miang* sample [3,48]. These acids were assumed to be released from the *Miang* sample during fermentation. Consistent with the previous reports on tea products, green tea and black tea were the best precursors for organic acids of the kombucha culture, such as acetic acid and glucuronic acid, respectively [49].

Table 3. Physicochemical properties of fermented *Miang* broth with single and co-cultures of *S. cerevisiae* TISTR 5088, *W. anomalus* P2 and *C. rhodanensis* P3 at 120 h after fermentation.

Property	*Miang* Broth Original	Control (0 h)	Following 120 h Fermentation							
			Control	5088	P2	P3	5088 + P2	5088 + P3	P2 + P3	5088 + P2 + P3
Glucose (g/L)	3.56 ± 0.02	53.21 ± 0.02 [h]	53.27 ± 0.12 [h]	0.19 ± 0.01 [b]	23.86 ± 0.45 [e]	33.19 ± 0.32 [f]	7.93 ± 0.12 [c]	9.50 ± 0.21 [d]	46.20 ± 1.09 [g]	0.00 ± 0.00 [a]
Ethanol (g/L)	0.00 ± 0.00 [f]	0.00 ± 0.00 [f]	0.00 ± 0.00 [f]	23.67 ± 0.70 [b]	11.03 ± 0.07 [d]	11.56 ± 0.00 [d]	20.74 ± 0.0 [c]	20.15 ± 0.50 [c]	2.93 ± 0.0 [e]	24.91 ± 0.0 [a]
pH	4.06 ± 0.0 [a]	4.1 ± 0.0 [a]	4.08 ± 0.03 [a]	3.9 ± 0.04 [a]	3.94 ± 0.00 [a]	4.04 ± 0.06 [a]	3.91 ± 0.03 [a]	3.97 ± 0.0 [a]	4.02 ± 0.02 [a]	4.02 ± 0.0 [a]
Organic acid										
Acetic acid (g/L)	3.04 ± 0.15 [a]	2.24 ± 0.12 [d]	1.91 ± 001 [f]	1.37 ± 0.15 [f]	2.03 ± 0.12 [e]	1.72 ± 0.02 [f]	3.09 ± 0.34 [a]	2.64 ± 0.25 [b]	1.48 ± 0.12 [g]	2.42 ± 0.18 [d]
Citric acid (g/L)	2.79 ± 0.05 [a]	1.17 ± 0.09 [de]	1.22 ± 0.2 [de]	1.21 ± 0.05 [de]	1.16 ± 0.05 [f]	0.93 ± 0.01 [f]	1.47 ± 0.01 [b]	1.17 ± 0.02 [de]	1.28 ± 0.03 [cd]	1.38 ± 0.02 [bc]
Glucuronic acid (g/L)	1.70 ± 0.15 [b]	1.20 ± 0.01 [e]	1.31 ± 0.2 [de]	1.39 ± 0.03 [d]	2.02 ± 0.23 [a]	1.57 ± 0.04 [c]	2.04 ± 0.04 [a]	1.75 ± 0.02 [b]	1.29 ± 0.01 [de]	1.94 ± 0.01 [a]
Lactic acid (g/L)	1.90 ± 0.01 [a]	1.10 ± 0.03 [d]	1.08 ± 0.05 [d]	1.38 ± 0.03 [bc]	1.31 ± 0.01 [c]	1.08 ± 0.02 [d]	1.44 ± 0.05 [b]	1.31 ± 0.01 [bc]	1.37 ± 0.03 [bc]	1.45 ± 0.03 [b]
Oxalic acid (g/L)	0.22 ± 0.02 [a]	0.13 ± 0.02 [bc]	0.13 ± 0.01 [bc]	0.15 ± 0.01 [bc]	0.17 ± 0.04 [ab]	0.10 ± 0.01 [c]	0.16 ± 0.01 [bc]	0.13 ± 0.03 [bc]	0.15 ± 0.01 [bc]	0.14 ± 0.01 [bc]
Succinic acid (g/L)	2.59 ± 0.06 [a]	1.61 ± 0.05 [de]	1.81 ± 0.14 [c]	1.08 ± 0.07 [g]	1.45 ± 0.01 [f]	1.19 ± 0.01 [g]	1.65 ± 0.04 [de]	1.55 ± -0.05 [ef]	1.76 ± 0.05 [cd]	2.02 ± 0.01 [b]
Bioactive compounds										
Total polyphenol (g/L)	7.86 ± 0.2 [a]	7.70 ± 0.98 [a]	6.61 ± 0.28 [b]	7.97 ± 0.09 [a]	8.15 ± 0.015 [a]	7.98 ± 0.06 [a]	7.99 ± 0.12 [a]	8.27 ± 0.31 [a]	7.24 ± 0.02 [b]	6.54 ± 0.07 [b]
Total tannin (g/L)	6.26 ± 0.0 [b]	6.73 ± 0.21 [ab]	6.09 ± 0.35 [b]	7.07 ± 0.28 [ab]	6.96 ± 0.09 [ab]	7.12 ± 0.15 [ab]	7.71 ± 0.06 [a]	7.91 ± 0.12 [a]	6.94 ± 0.31 [ab]	6.00 ± 0.02 [ab]
Total flavonoid (g/L)	0.30 ± 0.01 [a]	0.27 ± 0.09 [a]	0.26 ± 0.23 [a]	0.36 ± 0.08 [a]	0.37 ± 0.13 [a]	0.37 ± 0.07 [a]	0.23 ± 0.01 [a]	0.20 ± 0.02 [a]	0.20 ± 0.03 [a]	0.24 ± 0.01 [a]
DPPH free radical scavenging (%)	67.91 ± 0.06 [a]	66.15 ± 2.69 [a]	67.28 ± 0.90 [a]	68.93 ± 1.47 [a]	71.38 ± 1.22 [a]	69.69 ± 2.35 [a]	70.23 ± 3.78 [a]	71.43 ± 1.70 [a]	69.49 ± 1.60 [a]	70.23 ± 1.13 [a]
Total catechin (g/L)	2.22 ± 0.01 [a]	1.4 ± 0.02 [d]	1.24 ± 0.01 [e]	1.22 ± 0.01 [e]	1.23 ± 0.01 [e]	1.64 ± 0.00 [b]	1.43 ± 0.00 [d]	1.53 ± 0.01 [d]	1.64 ± 0.00 [b]	1.60 ± 0.01 [b]
Catechin (g/L)	1.02 ± 0.01 [a]	0.53 ± 0.02 [f]	0.51 ± 0.01 [f]	0.56 ± 0.03 [e]	0.56 ± 0.03 [e]	0.77 ± 0.02 [b]	0.66 ± 0.03 [d]	0.71 ± 0.12 [c]	0.75 ± 0.03 [b]	0.71 ± 0.02 [c]
Gallocatechin (g/L)	0.74 ± 0.03 [a]	0.44 ± 0.03 [b]	0.35 ± 0.02 [c]	0.25 ± 0.03 [e]	0.24 ± 0.04 [e]	0.33 ± 0.03 [c]	0.26 ± 0.01 [e]	0.26 ± 0.03 [e]	0.30 ± 0.01 [d]	0.30 ± 0.00 [d]
Epigallocatechin (g/L)	0.13 ± 0.01 [f]	0.17 ± 0.00 [f]	0.13 ± 0.01 [f]	0.15 ± 0.01 [e]	0.16 ± 0.00 [de]	0.20 ± 0.01 [c]	0.19 ± 0.01 [c]	0.21 ± 0.01 [b]	0.21 ± 0.02 [b]	0.26 ± 0.01 [a]
Epicatechin (g/L)	0.15 ± 0.01 [d]	0.20 ± 0.00 [c]	0.20 ± 0.01 [c]	0.21 ± 0.01 [c]	0.21 ± 0.01 [c]	0.26 ± 0.01 [b]	0.24 ± 0.01 [b]	0.25 ± 0.01 [b]	0.29 ± 0.01 [a]	0.24 ± 0.01 [b]
Epigallocatechin gallate (g/L)	0.18 ± 0.01 [a]	0.06 ± 0.001 [de]	0.05 ± 0.00 [e]	0.05 ± 0.00 [e]	0.06 ± 0.01 [de]	0.08 ± 0.00 [cd]	0.08 ± 0.02 [bc]	0.10 ± 0.02 [b]	0.09 ± 0.01 [bc]	0.09 ± 0.00 [bc]
Gallic acid (g/L)	0.13 ± 0.01 [a]	0.05 ± 0.01 [b]	0.05 ± 0.01 [b]	0.06 ± 0.01 [b]	0.06 ± 0.01 [b]	0.06 ± 0.00 [b]	0.05 ± 0.00 [b]	0.06 ± 0.00 [b]	0.06 ± 0.00 [b]	0.06 ± 0.00 [b]
Pyrogallol (g/L)	0.55 ± 0.01 [a]	0.32 ± 0.01 [c]	0.42 ± 0.01 [b]	0.33 ± 0.02 [c]	0.31 ± 0.01 [de]	0.32 ± 0.02 [cd]	0.29 ± 0.02 [de]	0.28 ± 0.01 [e]	0.31 ± 0.01 [cd]	0.32 ± 0.01 [cd]
Caffeine (g/L)	0.86 ± 0.02 [a]	0.55 ± 0.02 [ef]	0.58 ± 0.02 [cd]	0.59 ± 0.01 [bc]	0.58 ± 0.02 [cd]	0.61 ± 0.02 [b]	0.53 ± 0.01 [f]	0.58 ± 0.02 [cd]	0.57 ± 0.02 [cd]	0.55 ± 0.03 [de]

Note: [a–g] Different letters within the same row indicate a statistically significant difference ($p < 0.05$).

During fermentation, acetic acid was found as the main organic acid in MF-broth byproduct fermentation followed by succinic, glucuronic, lactic, citric, gallic acid and oxalic acid, respectively. Figure 6 indicates the fluctuation of the total acidity during fermentation. Interestingly, CF fermentation significantly increased the level of acetic acid and glucuronic acid levels (Table 3), which are special organic acids that are reported to be associated with health benefits in kombucha [50]. The increased acetic acid content might be due to the synthesis by yeast metabolically, is derived through the phosphogluconate pathway, acetate kinase pathway and citric acid metabolism [34]. In *S. cerevisiae*, a direct relationship has been established between glycerol and acetic acid production during fermentation [51]. Whereas, *W. anomalus* has been reported to produce a high level of acetic acid, which is one factor that is useful for inhibiting other microbial processes that occur during wine fermentation with *S. cerevisiae* [52]. However, acetic acid is the main acid responsible for the wine fault termed volatile acidity and is a contributor taste ingredient that gives fermented foods their typical strong, pungent and vinegar flavor [44,53]. A glucuronic acid increase detected in MF-broth might be due to the oxidative pathway converting glucose to gluconic acid with yeast [54]. Glucuronic acid is one of the most important components found in kombucha tea that is a vitamin C precursor and plays an important role in the formation of glycosaminoglycans which allows for the detoxifying function via conjugation [55]. Other organic acids found in MF-broth fermentation, including citric, succinic, and oxalic acids are important intermediates in the Krebs cycle. The change in the content of these compounds could be attributed to the yeast cell catabolism [51,52]. According to studies, succinic acid possesses the bitterness and saltiness, and the reduced succinic acid content after fermentation may have the advantages in improving wine taste.

A PCA was used to analyze the results of products, in terms of glucose, ethanol and organic acids that remained after fermentation (Figure 7). The total variance was 78.09%, PC1 (46.57%) and PC2 (31.52%). Glucose, pH and gallic acids had been represented at PC1, whereas ethanol and organic acids were revealed at PC2. Regarding the relevance of the PC score, the PC1 projects the sample of the original MF-broth (0h), control, P3, and P2 + P3. PC2 was differentiated among samples of TISIR 5088, P2, TISIR 5088 + P2, TISIR 5088 + P3 and TISIR 5088 + P2 + P3. A clear separation of the fermented MF-broth obtained with the *S. cerevisiae* TISTR5088 was strongly related to ethanol production. In contrast, glucose concentration was observed in the unfermented sample of the initial MF-broth and the control at 120 h. The final important product, organic acids correlated with the original MF-broth that might come from what remains of the biotransformation during the FFP *Miang* fermentation, except gallic acid that was most associated with *C. rhodanensis* P3.

3.5. Phenolic Compounds and Antioxidant Activity

Phenolic compounds, which are natural antioxidants, have markedly generated interest in *Miang* production due to their potential beneficial effects on humans [3,4,26]. Interestingly, the MF-broth residual byproduct is similar to the extraction content of fermented leaves with water during the anaerobic fermentation, which causes the water-soluble phenolic compounds to be solubilized after release from the *Miang* leaves. Therefore, the fermented MF-broth byproduct was expected to have the same benefits as other types of fermented tea beverages as well. The overall perception of the experimental analysis with the yeast inoculation in fermentation revealed that TP, TT and TF were similar to the initial control. TP values ranged from 6.61 ± 0.28 to 8.27 ± 0.31 g/L, TT values ranged from 6.00 ± 0.02 to 7.91 ± 0.12 g/L and TF values ranged from 0.20 ± 0.03 to 0.36 ± 0.08 g/L (Figure 8). The antioxidant capacity of SF and CF were similar, where 10 μL of each sample exhibited DPPH scavenging abilities at a range of $68.93 \pm 1.47\%$ to $71.43 \pm 1.7\%$, which showed no difference between the initial MF-broth and after 120 h fermentation. The bioactive molecules in the broth might be the result of biotransformation in the SMF process, in which microbes convert macronutrients, such as carbohydrates, proteins and lipids into bioactive compounds and dissolve them in aqueous solutions during fermentation [4]. This study added only glucose as a carbon source without other substrates for yeast to grow in and convert into ethanol. As a result, there may be insufficient precursors in MF-broth fermentation to be converted into the bioactive components. However, by comparing the bioactive compounds in the fermented *Miang*, it was found to be close to or higher than other fermented tea beverages. For example, in kombucha made from black, white and red teas, the TP content after 7 days of fermentation, was at 2.19 ± 0.21, 2.05 ± 0.30, and 2.70 ± 0.40 g/L, respectively, whereas the DPPH scavenging ability was 70.63 ± 0.53%, 79.13 ± 0.93% and 77.37 ± 0.80% [56]. It is expected that by using the benefits of tea containing catechins, gallic acid and other antioxidative compounds, the fermented MF-broth could be developed into a health-targeted beverage, similar to other tea wines, such as Oolong tea and pear wine [18].

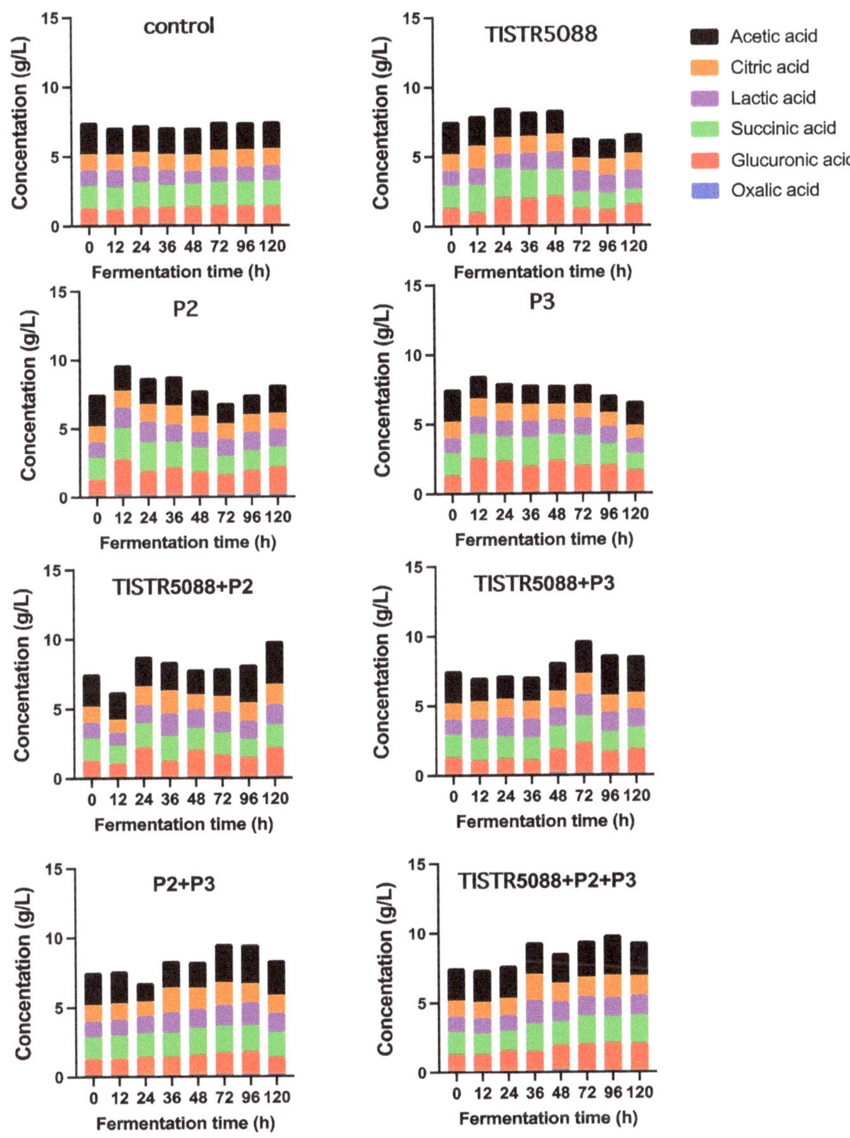

Figure 6. Organic acid profiles during fermentation of MF-broth fermented by *S. cerevisiae* TISTR 5088, *W. anomalus* P2 and *C. rhodanensis* P3, with single and co-fermentation at 30 °C for 120 h.

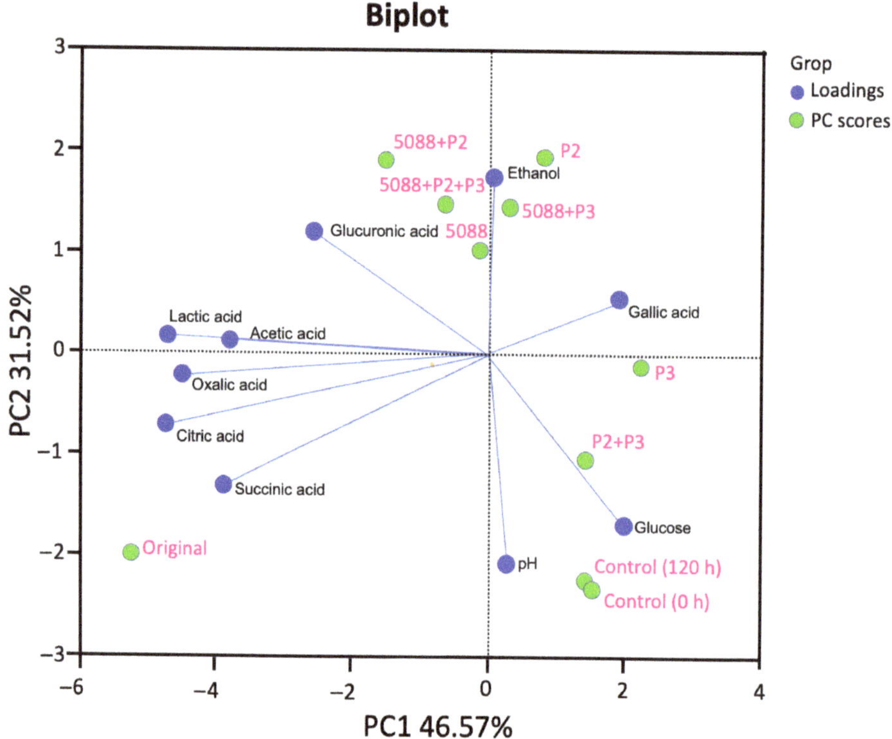

Figure 7. PCA biplot of the organic acids, glucose, ethanol, and pH variables in the original and fermented MF-broth by *S. cerevisiae* TISTR 5088, *W. anomalus* P2 and *C. rhodanensis* P3, with single and co-fermentation at 30 °C for 120 h.

Tea phenolic compounds were identified and quantified in MF-broth products, including catechins, gallic acid, pyrogallol and caffeine. In the original MF-broth C, GC, EGC, EC and EGCG were confirmed and this was similar to that in *Miang* FFP [3]. However, following the fermentation of MF-broth byproduct substrate, the tea polyphenols were significantly decreased, except EC and EGC (Table 3). Considering the amount of each substance, C increased while GC decreased in all fermentation samples. EC, EGC and EGCG were stable, except for the samples inoculated with *C. rhodanensis* P3, in the single and co-cultured fermentations, and showed a higher content of EC and EGC, compared to the original MF-broth (Table 3). The increase in EC and EGC might be due to the hydrolysis of gallate polyphenols (EGCG) by esterase or tannase from the yeast [36]. However, the reduction of the total catechins in all fermented samples might be from the biotransformation by the yeast. That may lead to the contribution of polyphenols to MF-broth and simultaneously the reduction of the bitter and astringent tastes of MF-broth.

3.6. β-Glucosidase Activity in MF-Broth Fermentation

Changes of β-Glucosidase activity in MF-broth were observed and it was found that the increase in enzyme activity was observed in all fermentation broths, except the non-inoculated treatment (control). The SF fermented MF-broth with *S. cerevisiae* TISTR 5088 produced β-glucosidase at 24.51 ± 1.42 to 46.09 ± 1.73 mU/mL, whereas non-*Saccharomyces W. anomalus* P2 and *C. rhodanensis* P3 produced β-glucosidase in the range of 63.47 ± 5.12 to 89.86 ± 5.31 mU/mL and 20.78 ± 1.42 to 196.68 ± 3.55 mU/mL, respectively. *C. rhodanensis* P3 showed the highest β-glucosidase activity of 196.68 ± 3.55 mU/mL at 24 h fermentation.

In the case of the CF fermentation, all treatments showed the activity of β-glucosidase at 12 h to 48 h except TISTR 5088 + P3, which was continuous to 120 h (Figure 9). TISTR 5088 + P3 showed the highest activity of 127.56 ± 1.24 mU/mL at 24 h fermentation and this ranged from 23.35 ± 4.09 to 127.56 ± 1.24 mU/mL, followed by TISTR 5088 + P2 (40.84 ± 0.89 to 97.40 ± 4.44 mU/mL), P2 + P3 (17.59 ± 2.67 to 82.94 ± 1.78 mU/mL) and TISTR 5088 + P2 + P3 (1.63 ± 0.31 to 49.64 ± 1.71 mU/mL), respectively. However, the decrease of β-glucosidase levels after 24 h was observed (Figure 5). This might be due to an increase in ethanol levels and high acid, as the previous research found that high acidity, high glucose and high ethanol concentrations affected the function of this enzyme [57]. However, MF-broth fermentation with *C. rhodanensis* P3 showed remarkable enzyme activities in both SF and CF fermentations, in which the functional activity of β-glucosidase is hydrolyzing the glycosidic bonds from various aglycone structures into monoterpenes, which are precursors of aromatic compounds [13,14]. Consequently, previous research investigated the use of β-glucosidase produced from *Cyberlindnera* to develop an aroma in fermented beverages and tea [9,37,39,58]. Wang et al. [58] found that *C. saturuns* var. *mrakii* NCYC 2251 β-glycosides enhanced the tea aroma in green tea slurry by hydrolyzing glycoside precursors (β-glucosides and β-primeverosides) such as geranyl, linalyl, benzyl and 2-phenylethyl glucosides. Furthermore, the benefits of β-glycosides in improving the beverage flavor were also revealed in the *W. anomalus* species, and certain strains were reported to have β-glycosidase, that is multi-functionally active under common oenological conditions. Thus, the enzyme might have multiple applications in fermentation, such as enhancing the sensory and bioactive component concentrations by splitting the glycosylated precursors [6,7]. Due to the reasons mentioned above, β-glucosidase from both non-*Saccharomyces* strains used in this experiment possibly plays an important role in odor and aroma formation, leading to the formation of special quality characteristics in fermented MF-broth byproducts, which requires further research.

3.7. Volatile Aromatic Compounds in MF-Broth Fermentation

Figure 10 shows the volatile compounds associated with the metabolic activities of *S. cerevisiae* TISTR 5088, *W. anomalus* P2 and *C. rhodanensis* P3 and their co-culture during fermentation and the original MF-broth from the byproduct. In total, 36 volatile compounds were found in both the unfermented and fermented samples, indicating that alcohol, ester, and terpenes were quantitatively the major group of VOCs in MF-broth byproduct fermentation. A heatmap with hierarchical clustering analysis was used to determine whether the volatile flavor compounds associated with each sample could stimulate the sample clustering, based on the Pearson correlation. Cluster A, which represented the control, had limited effect on the aroma component profile and a significant concentration effect of vitispirane, as shown in Figure 10. The original MF-broth byproduct and yeast-associated fermented samples produced increasing levels of esters, acids and alcohols in Cluster B. Cluster C gathered yeast and co-culturing effects that produced more esters, acids, and alcohols after 120 h of fermentation.

The original MF-broth VOCs were considered the byproduct of SMF *Miang* biotransformation, with microorganisms probably converting the tea aromatic precursors, including carotenoids, lipids, glycosides and other compounds, as previously reported [59,60]. As a result, 22 aromatic substances were detected in the original MF-broth, accounting for 70.16% of the total aroma content. Linalool and methyl salicylate were the main VOCs in that broth, followed by linalool oxide, (Z)-3-hexen-1-ol, benzyl alcohol, ethyl acetate, isoamyl acetate and other substances. Following the preparation of the MF-broth substrate with 5% (*w/v*) glucose, the VOCs profiles changed with linalool, methyl salicylate, ethyl acetate and isoamyl acetate decreasing with the production process. In contrast, new compounds, such as vitispirane, β-ocimene, limonene and α-terpinene, were discovered in the MF-broth substrate. Vitispirane had a high level of VOCs, which has been described as floral, fruity, woody or reminiscent of eucalyptus [61]. Compared to this, the further fermented MF-broth byproduct with yeast was significantly different. Higher levels of flavor intensity (total peak area) were discovered at 120 h fermentation with co-culture, when compared to the unfermented and fermented MF-broth byproducts. This result demonstrated the synergistic effect of co-culturing on the production of aroma compounds during MF-broth fermentation.

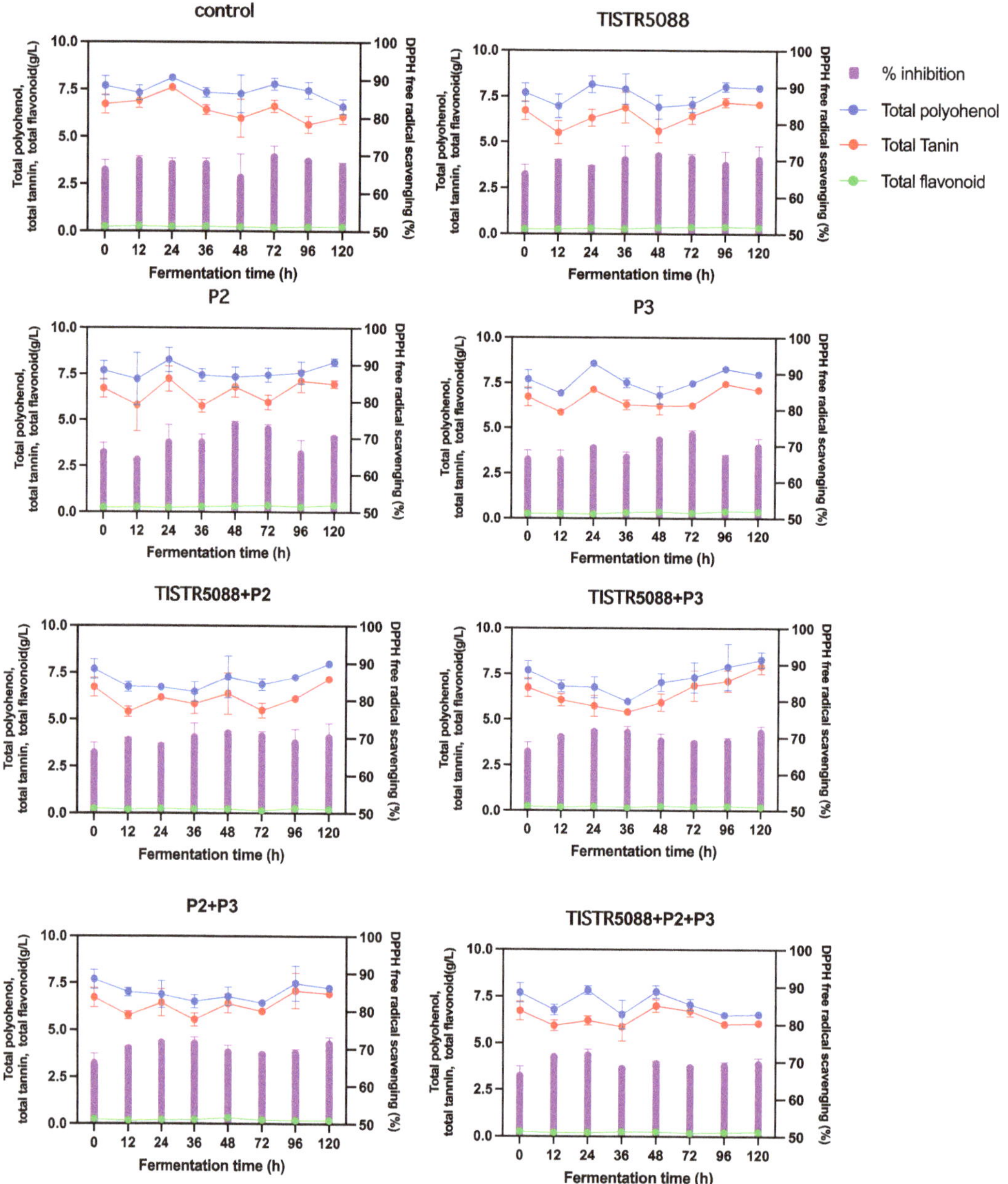

Figure 8. Bioactive compound and antioxidant profiles during the fermentation of MF-broth fermented by *S. cerevisiae* TISTR 5088, *W. anomalus* P2 and *C. rhodanensis* P3, with single and co-fermentation.

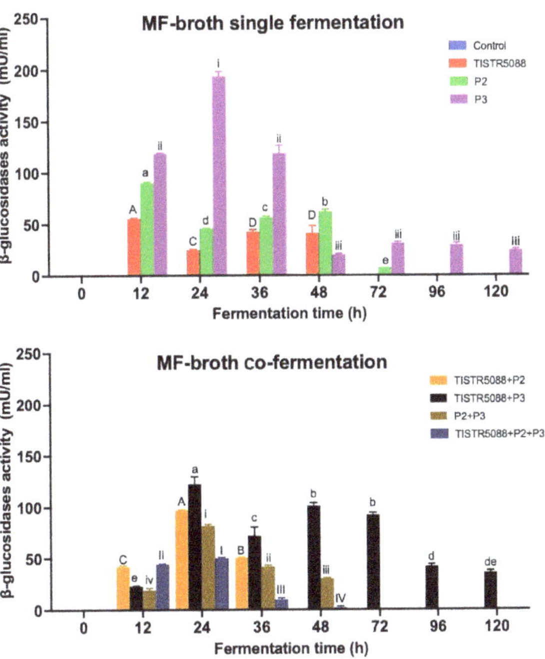

Figure 9. β-glucosidase activity of non-inoculated (control), *S. cerevisiae* TISTR 5088, *W. anomalus* P2 and *C. rhodanensis* P3 during MF- broth fermentation.

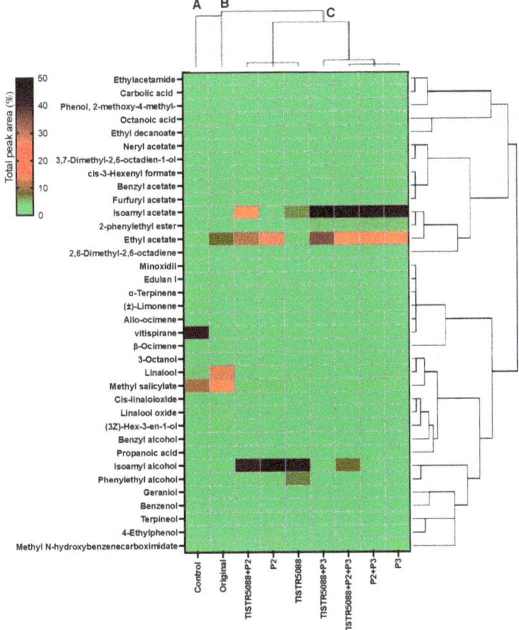

Figure 10. Heatmap and hierarchical cluster analysis representation corresponding to the 36 volatile compounds of original and fermented MF-broth inoculated with *S. cerevisiae* TISTR 5088, *W. anomalus* P2 and *C. rhodanensis* P3 at 30 °C for 120 h.

The fermented MF-broth was clearly in the hierarchical clustering that separated the between-group yeast providing alcohols and ester (Figure 10). In the alcohol group, isoamyl alcohol exhibited high levels in all fermentations with *S. cerevisiae* TISTR 5088 and *W. anomalus* P2, in SF and CF. Meanwhile, *C. rhodanensis* P3 represented a high amount of ester group, ethyl acetate and isoamyl acetate, which were remarkable in both of SF and CF. During fermentation, yeast uses the Ehrlich pathway to convert amino acids or sugars into isobutanol, isoamyl alcohol and 2-phenylethanol [34,62–64]. Dzialo et al. [62] suggested isoamyl alcohol and 2-phenylethanol were the major fuel alcohols found in alcoholic beverages. In this study, MF-broth fermentation revealed the presence of isoamyl alcohol and 2-phenylethanol, with alcohol produced at a higher level in both the SF and CF with *S. cerevisiae* TISTR 5088 and *W. anomalus* P2, which was consistent with previous studies [34,36,43,65]. However, although in the original MF-broth, a high quantity of (Z)-3-hexen-1-ol and benzyl alcohol (known as tea leaf alcohol) was found, after fermentation that compound was reduced by the yeast [66].

Non-*Saccharomyces* yeast strains are known as the efficient producers of esters in fermented foods and beverages. The earlier result of VOCs formed by *W. anomalus* P2 and *C. rhodanensis* P3, the confirmed ethyl acetate (fruity aroma), isoamyl acetate (banana aroma) and 2-phenethyl acetate (2PA), have been described as good characteristics and pleasant aromatic properties to wine [67,68]. In this investigation, the higher ester in fermented MF-broth was generated with *C. rhodanensis* P3, followed by *W. anomalus* P2, in SF and CF. Isoamyl acetate and ethyl acetate were dominant esters found in all fermented MF-broths, which were inoculated with *C. rhodanensis* P3. Furthermore, in *C. rhodanensis* P3, benzyl acetate, 2-phenethyl acetate, furfuryl acetate, ethyl decanoate and neryl acetate were discovered, all of which were not found in the other samples without P3. Different ester characteristics in MF-broth might offer a unique odor in each sample. Several reports confirmed that esters contribute the most to the aroma of alcoholic beverages. The principal contribution to the typical fruity scents of the fermentation aroma is a mixture of ethyl caproate and ethyl caprylate (apple-like aroma), while isoamyl acetate imparts a banana-like aroma and 2-phenylethyl acetate contributes a fruity and flowery flavor [8,69,70]. However, *Cyberlindnera* yeast has been reported to produce high concentrations of acetate esters, in particular isoamyl acetate, ethyl acetate and 2-phenylethyl acetate [37]. In particular, the *Cyberlindnera* species is known for its high ester production, which was shown in the studies with *C. saturnus*, *C. mrakii* and *C. subsufficiens*. Furthermore, it has been suggested to use yeasts with a high production of flavor compounds (i.e., esters, higher alcohols) to mask the wort-like flavor of non-alcoholic beer, cider, and wine [69,71,72]. This study is the first report on the VOCs' formation by *C. rhodanensis* in a fermented beverage.

Terpenes, volatile phenols, and other esterified VOCs were also found in fermented MF-broth. For example, the terpene groups found were β-ocimene, linalool oxide, terpineol and geraniol, whereas the volatile phenols found were benzenol and 4-ethylphenol. In general, terpenes belong to the primary aroma group that contribute to desirable descriptors, such as flowery, honey and citrus notes. Linalool and terpineol have low odor thresholds and offer floral aromas [13,57]. Meanwhile, after 120 h of fermentation, several VOCs, such as methyl salicylate, (Z)-3-hexen-1-ol and linalool, were dramatically reduced. The flavor characteristic of MF-broth increased as fermentation proceeded. Following 120 h of fermentation, more acetate esters and acids were generated but there were fewer terpenoid alcohols than in the initial MF-broth. Terpenoid alcohols can be liberated from glycosides by β-glucosidase or produced via yeast during fermentation, resulting in their accumulation in fermented samples. Meanwhile, terpenoid alcohols can be isomerized into other terpenoids or converted into corresponding acetate esters. For example, methyl salicylate, (Z)-3-hexen-1-ol, 2-phenylethanol and benzyl alcohol were released with glucosides, and the alcohols can be further transferred into esters by yeast [58,73].

The PCA of VOCs evaluated the relationship between VOCs and the fermentation method. The first two principal components (PCs) accounted for 75.03% of the total variance, of which PC1 and PC2 accounted for 55.20% and 19.83%, respectively. The fermented MF-broth was divided into four groups, based on their similarity on the score plot (Figure 11). Group 1 was positioned in the lower right quadrant and was fermented with P3, P2 + P3 and TISTR 5088 + P3, which produced more acetate esters, particularly ethyl acetate, isoamyl acetate, 2-phenylethyl acetate, furfuryl acetate, benzyl acetate, cis-3-hexenyl formate and neryl acetate, and terpenes, such as 2,6-dimethyl-2,6-octadiene, β-ocimene and 3,7-dimethyl-2,6-octadien-1-ol. Meanwhile, TISTR 5088 + P2 and TISTR 5088 + P2 + P3 were located in the upper right quadrant (Group 2) due to the higher levels of octanoic acid, ethyl caprate and N-ethyl acetamide synthesis. The fermented MF-broth with TISTR 5088, P2, and original formed Group 3, corresponded with high alcohols and terpenes, particularly isoamyl alcohol, phenylethyl alcohol, benzenol, 3-octanol linalool, (Z)-3-hexen-1-ol, geraniol, benzyl alcohol, linalool oxide and cis-linalool. The control was located in the lower left quadrant, distant from the sequential fermentations, which were characterized by methyl salicylate, 4-ethylphenol, terpineol, limonene, α-terpinene, allo-ocimene and vitispirane. According to the PCA results, the yeast fermentation had a significant impact on the aroma profile of MF-broth. Therefore, this study indicates the performance of two VOCs groups on the properties of the fermented MF-broth, including the formation of higher alcohols by *S. cerevisiae* TISTR 5088 and *W. anomalus* P2, and higher esters by *C. rhodanensis* P3, and clearly showed the VOCs and correlation of yeast group between unfermented and fermented MF-broth. The results of this experiment demonstrated the ability of non-*Saccharomyces* yeast in the improvement of fermented MF-broth odor.

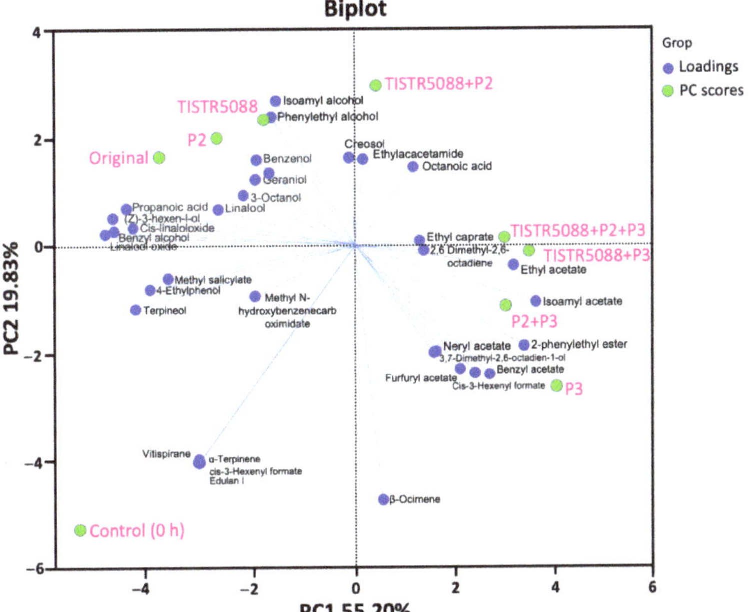

Figure 11. PCA biplot of the volatile compounds in the original and fermented MF-broth by *S. cerevisiae* TISTR 5088, *W. anomalus* P2 and *C. rhodanensis* P3, with single and co-fermentation at 120 h.

As aforementioned, although the original MF-broth benefits from partial bioactivity during the *Miang* fermentation processes, such as organic acids, catechins and other substances. The original MF-broth, Moreover, has a disagreeable odor for consumption. As a result, it was interesting to search for opportunities to strengthen the flavor while also providing additional benefits. In this research, we selected non-*Saccharomyces* that exhibited a distinct odor for the MF-broth to enhance the scent and add bioactive compounds, as well as the advantages from a low-alcohol beverage. We expect this study will be advantageous in the development of healthy drinks and add value to the MF-broth. Future iterations of this research will be the sensory evaluation for consumer acceptance and commercial development.

4. Conclusions

The findings of this study demonstrated the potential for using the waste byproduct as MF-broth residual byproduct substrate in the development of health-targeted beverages using non-*Saccharomyces* yeast species. Organic acids that were important in fermentation include acetic, citric, glucuronic, lactic, succinic, oxalic and gallic acids. The bioactive compounds and antioxidant activity were similar with the initial fermentation stage. β-Glucosidase was found to be involved in the quality improvement of fermented MF-broth by enhancing the volatile alcohols and esters compounds. Thirty-six volatile compounds were detected in both the unfermented and fermented samples, indicating that alcohol, ester, and terpenes were the most abundant VOCs in the MF-broth byproduct fermentation. The PCA and hierarchical clustering analysis clearly showed the VOCs and the correlation of the yeast group that classified Cluster A (control), Cluster B (original MF-broth) and Cluster C evaluated the yeast and co-culturing that created more esters, acids, and alcohols.

Author Contributions: Conceptualization, P.K. and C.K.; methodology and formal analysis, P.K., N.S., A.K., K.U., C.S. and M.Z.; investigation, P.K., N.P., N.S., K.U., D.S. and C.K.; writing—original draft preparation, P.K. and C.K.; writing—review and editing, P.K., K.S. and C.K.; supervision, C.K. All authors have read and agreed to the published version of the manuscript.

Funding: This manuscript was supported by Chiang Mai University, the National Research Council of Thailand (NRCT), and the Royal Golden Jubilee Ph.D. (RGJ) Programme (Grant No. PHD/0137/2560).

Institutional Review Board Statement: Not applicable.

Informed Consent Statement: Not applicable.

Data Availability Statement: Not applicable.

Acknowledgments: The authors are grateful to the Faculty of Agro-Industry, Chiang Mai University for access to the research facilities.

Conflicts of Interest: The authors declare no conflict of interest.

References

1. Khanongnuch, C.; Unban, K.; Kanpiengjai, A.; Saenjum, C. Recent research advances and ethno-botanical history of Miang, a traditional fermented tea (*Camellia sinensis* var. *assamica*) of northern Thailand. *J. Ethn. Foods* **2017**, *4*, 135–144. [CrossRef]
2. Kanpiengjai, A.; Chui-Chai, N.; Chaikaew, S.; Khanongnuch, C. Distribution of tannin-'tolerant yeasts isolated from Miang, a traditional fermented tea leaf (*Camellia sinensis* var. *assamica*) in northern Thailand. *Int. J. Food Microbiol.* **2016**, *238*, 121–131. [CrossRef]
3. Kodchasee, P.; Nain, K.; Abdullahi, A.D.; Unban, K.; Saenjum, C.; Shetty, K.; Khanongnuch, C. Microbial dynamics-linked properties and functional metabolites during Miang fermentation using the filamentous fungi growth-based process. *Food Biosci.* **2021**, *41*, 100998. [CrossRef]
4. Unban, K.; Khatthongngam, N.; Shetty, K.; Khanongnuch, C. Nutritional biotransformation in traditional fermented tea (*Miang*) from north Thailand and its impact on antioxidant and antimicrobial activities. *J. Food Sci. Technol.* **2019**, *56*, 2687–2699. [CrossRef]
5. Kanpiengjai, A.; Khanongnuch, C.; Lumyong, S.; Kummasook, A.; Kittibunchakul, S. Characterization of *Sporidiobolus ruineniae* A45. 2 Cultivated in Tannin substrate for use as a potential multifunctional probiotic yeast in aquaculture. *J. Fungi* **2020**, *6*, 378. [CrossRef] [PubMed]
6. Sabel, A.; Martens, S.; Petri, A.; König, H.; Claus, H. *Wickerhamomyces anomalus* AS1: A new strain with potential to improve wine aroma. *Ann. Microbiol.* **2014**, *64*, 483–491. [CrossRef]
7. Padilla, B.; Gil, J.V.; Manzanares, P. Challenges of the non-conventional yeast *Wickerhamomyces anomalus* in winemaking. *Fermentation* **2018**, *4*, 68. [CrossRef]
8. Akyereko, Y.G.; Wireko-Manu, F.D.; Alemawor, F.; Adzanyo, M. Effects of production methods on flavour characteristics of nonalcoholic wine. *J. Food Qual.* **2021**, *2021*, 3014793. [CrossRef]
9. Bellut, K.; Michel, M.; Zarnkow, M.; Hutzler, M.; Jacob, F.; Atzler, J.J.; Hoehnel, A.; Lynch, K.M.; Arendt, E.K. Screening and application of *Cyberlindnera* yeasts to produce a fruity, non-alcoholic beer. *Fermentation* **2019**, *5*, 103. [CrossRef]

10. Lara-Hidalgo, C.; Hernández-Sánchez, H.; Hernández-Rodríguez, C.; Dorantes-Álvarez, L. Yeasts in fermented foods and their probiotic potential. *Austin J. Nutr. Metab.* **2017**, *4*, 1045.

11. Moslehi-Jenabian, S.; Lindegaard, L.; Jespersen, L. Beneficial effects of probiotic and food borne yeasts on human health. *Nutrients* **2010**, *2*, 449–473. [CrossRef] [PubMed]

12. Swangkeaw, J.; Vichitphan, S.; Butzke, C.E.; Vichitphan, K. Characterization of β-glucosidases from *Hanseniaspora* sp. and *Pichia anomala* with potentially aroma-enhancing capabilities in juice and wine. *World J. Microbiol. Biotechnol.* **2011**, *27*, 423–430. [CrossRef]

13. Lee, S.-B.; Park, H.-D. Isolation and investigation of potential non-*Saccharomyces* yeasts to improve the volatile terpene compounds in Korean Muscat Bailey A wine. *Microorganisms* **2020**, *8*, 1552. [CrossRef] [PubMed]

14. Zhang, W.; Zhuo, X.; Hu, L.; Zhang, X. Effects of crude β-glucosidases from *Issatchenkia terricola*, *Pichia kudriavzevii*, *Metschnikowia pulcherrima* on the flavor complexity and characteristics of wines. *Microorganisms* **2020**, *8*, 953. [CrossRef]

15. Zhu, X.; Navarro, Y.; Mas, A.; Torija, M.-J.; Beltran, G. A rapid method for selecting non-*Saccharomyces* strains with a low ethanol yield. *Microorganisms* **2020**, *8*, 658. [CrossRef]

16. Han, X.; Qing, X.; Yang, S.; Li, R.; Zhan, J.; You, Y.; Huang, W. Study on the diversity of non-*Saccharomyces* yeasts in Chinese wine regions and their potential in improving wine aroma by β-glucosidase activity analyses. *Food Chem.* **2021**, *360*, 129886. [CrossRef]

17. Baschali, A.; Tsakalidou, E.; Kyriacou, A.; Karavasiloglou, N.; Matalas, A.-L. Traditional low-alcoholic and non-alcoholic fermented beverages consumed in European countries: A neglected food group. *Nutr. Res. Rev.* **2017**, *30*, 1–24. [CrossRef]

18. Li, Y.; Zhang, S.; Sun, Y. Measurement of catechin and gallic acid in tea wine with HPLC. *Saudi J. Biol. Sci.* **2020**, *27*, 214–221. [CrossRef]

19. Vasdev, S.; Gill, V.; Singal, P.K. Beneficial effect of low ethanol intake on the cardiovascular system: Possible biochemical mechanisms. *Vasc. Health Risk Manag.* **2006**, *2*, 263. [CrossRef]

20. Helmy, E.; Soliman, S.; Abdel-Ghany, T.M.; Ganash, M. Evaluation of potentially probiotic attributes of certain dairy yeast isolated from buffalo sweetened Karish cheese. *Heliyon* **2019**, *5*, e01649. [CrossRef]

21. Menezes, A.G.T.; Ramos, C.L.; Cenzi, G.; Melo, D.S.; Dias, D.R.; Schwan, R.F. Probiotic potential, antioxidant activity, and phytase production of indigenous yeasts isolated from indigenous fermented foods. *Probiotics Antimicrob. Proteins* **2020**, *12*, 280–288. [CrossRef]

22. Amorim, J.C.; Piccoli, R.H.; Duarte, W.F. Probiotic potential of yeasts isolated from pineapple and their use in the elaboration of potentially functional fermented beverages. *Food Res. Int.* **2018**, *107*, 518–527. [CrossRef] [PubMed]

23. Kurtzman, C.P.; Fell, J.W.; Boekhout, T.; Robert, V. Methods for isolation, phenotypic characterization and maintenance of yeasts. In *The Yeasts*; Elsevier: Amsterdam, The Netherlands, 2011; pp. 87–110.

24. Suwannarach, N.; Kaewyana, C.; Yodmeeklin, A.; Kumla, J.; Matsui, K.; Lumyong, S. Evaluation of *Muscodor cinnamomi* as an egg biofumigant for the reduction of microorganisms on eggshell surfaces and its effect on egg quality. *Int. J. Food Microbiol.* **2017**, *244*, 52–61. [CrossRef] [PubMed]

25. Wangkarn, S.; Grudpan, K.; Khanongnuch, C.; Pattananandecha, T.; Apichai, S.; Saenjum, C. Development of HPLC method for catechins and related compounds determination and standardization in Miang (Traditional Lanna fermented tea leaf in Northern Thailand). *Molecules* **2021**, *26*, 6052. [CrossRef] [PubMed]

26. Abdullahi, A.D.; Kodchasee, P.; Unban, K.; Pattananandecha, T.; Saenjum, C.; Kanpiengjai, A.; Shetty, K.; Khanongnuch, C. Comparison of phenolic contents and scavenging activities of Miang extracts derived from filamentous and non-filamentous fungi-based fermentation processes. *Antioxidants* **2021**, *10*, 1144. [CrossRef]

27. Canonico, L.; Zannini, E.; Ciani, M.; Comitini, F. Assessment of non-conventional yeasts with potential probiotic for protein-fortified craft beer production. *LWT* **2021**, *145*, 111361. [CrossRef]

28. Fernández-Pacheco, P.; Rosa, I.Z.; Arévalo-Villena, M.; Gomes, E.; Pérez, A.B. Study of potential probiotic and biotechnological properties of non-*Saccharomyces* yeasts from fruit Brazilian ecosystems. *Braz. J. Microbiol.* **2021**, *52*, 2129–2144. [CrossRef]

29. Senkarcinova, B.; Dias, I.A.G.; Nespor, J.; Branyik, T. Probiotic alcohol-free beer made with *Saccharomyces cerevisiae* var. *boulardii*. *LWT* **2019**, *100*, 362–367. [CrossRef]

30. Tamang, J.P.; Lama, S. Probiotic properties of yeasts in traditional fermented foods and beverages. *J. Appl. Microbiol.* **2022**, *132*, 3533–3542. [CrossRef]

31. Dalanezi, F.M.; da Paz, G.S.; Joaquim, S.F.; Guimarães, F.F.; Bosco, S.d.M.G.; Langoni, H. The first report of *Cyberlindnera rhodanensis* associated with clinical bovine mastitis. *J. Dairy Sci.* **2018**, *101*, 581–583. [CrossRef]

32. World Health Organization; Food and Agriculture Organization of the United Nations; World Organisation for Animal Health. *Monitoring Global Progress on Addressing Antimicrobial Resistance: Analysis Report of the Second Round of Results of AMR Country Self-Assessment Survey*; WHO: Geneva, Switzerland, 2018; Volume 2018.

33. Leangnim, N.; Aisara, J.; Unban, K.; Khanongnuch, C.; Kanpiengjai, A. Acid stable yeast cell-associated tannase with high capability in gallated catechin biotransformation. *Microorganisms* **2021**, *9*, 1418. [CrossRef] [PubMed]

34. Huang, J.; Wang, Y.; Ren, Y.; Wang, X.; Li, H.; Liu, Z.; Yue, T.; Gao, Z. Effect of inoculation method on the quality and nutritional characteristics of low-alcohol kiwi wine. *LWT* **2022**, *156*, 113049. [CrossRef]

35. Satora, P.; Tarko, T.; Sroka, P.; Blaszczyk, U. The influence of *Wickerhamomyces anomalus* killer yeast on the fermentation and chemical composition of apple wines. *FEMS Yeast Res.* **2014**, *14*, 729–740. [CrossRef] [PubMed]

36. Wang, R.; Sun, J.; Lassabliere, B.; Yu, B.; Liu, S.Q. Biotransformation of green tea (*Camellia sinensis*) by wine yeast *Saccharomyces cerevisiae*. *J. Food Sci.* **2020**, *85*, 306–315. [CrossRef] [PubMed]

37. Bellut, K.; Arendt, E.K. Chance and challenge: Non-*Saccharomyces* yeasts in nonalcoholic and low alcohol beer brewing–A review. *J. Am. Soc. Brew. Chem.* **2019**, *77*, 77–91. [CrossRef]

38. Methner, Y.; Hutzler, M.; Zarnkow, M.; Prowald, A.; Endres, F.; Jacob, F. Investigation of non-*Saccharomyces* yeast strains for their suitability for the production of non-alcoholic beers with novel flavor profiles. *J. Am. Soc. Brew. Chem.* **2022**, *80*, 341–355. [CrossRef]

39. Van Rijswijck, I.M.; van Mastrigt, O.; Pijffers, G.; Wolkers–Rooijackers, J.C.; Abee, T.; Zwietering, M.H.; Smid, E.J. Dynamic modelling of brewers' yeast and *Cyberlindnera fabianii* co-culture behaviour for steering fermentation performance. *Food Microbiol.* **2019**, *83*, 113–121. [CrossRef]

40. Longo, R.; Blackman, J.W.; Torley, P.J.; Rogiers, S.Y.; Schmidtke, L.M. Changes in volatile composition and sensory attributes of wines during alcohol content reduction. *J. Sci. Food Agric.* **2017**, *97*, 8–16. [CrossRef]

41. Contreras, A.; Hidalgo, C.; Schmidt, S.; Henschke, P.; Curtin, C.; Varela, C. The application of non-*Saccharomyces* yeast in fermentations with limited aeration as a strategy for the production of wine with reduced alcohol content. *Int. J. Food Microbiol.* **2015**, *205*, 7–15. [CrossRef] [PubMed]

42. Wang, C.; Mas, A.; Esteve-Zarzoso, B. The interaction between *Saccharomyces cerevisiae* and non-*Saccharomyces* yeast during alcoholic fermentation is species and strain specific. *Front. Microbiol.* **2016**, *7*, 502. [CrossRef]

43. Wang, W.; Fan, G.; Li, X.; Fu, Z.; Liang, X.; Sun, B. Application of *Wickerhamomyces anomalus* in simulated solid-state fermentation for Baijiu production: Changes of microbial community structure and flavor metabolism. *Front. Microbiol.* **2020**, *11*, 598758. [CrossRef] [PubMed]

44. Escalante, W.D.E. Perspectives and uses of non-*Saccharomyces* yeasts in fermented beverages. In *Frontiers and New Trends in the Science of Fermented Food and Beverages*; Books on Demand: Paris, France, 2019; p. 107.

45. Vilela, A. Modulating wine pleasantness throughout wine-yeast co-inoculation or sequential inoculation. *Fermentation* **2020**, *6*, 22. [CrossRef]

46. Methner, Y.; Hutzler, M.; Matoulková, D.; Jacob, F.; Michel, M. Screening for the brewing ability of different non-*Saccharomyces* yeasts. *Fermentation* **2019**, *5*, 101. [CrossRef]

47. Xu, A.; Xiao, Y.; He, Z.; Liu, J.; Wang, Y.; Gao, B.; Chang, J.; Zhu, D. Use of non-*Saccharomyces* yeast co-fermentation with *Saccharomyces cerevisiae* to Improve the polyphenol and volatile aroma compound contents in Nanfeng Tangerine cines. *J. Fungi* **2022**, *8*, 128. [CrossRef] [PubMed]

48. Unban, K.; Kodchasee, P.; Shetty, K.; Khanongnuch, C. Tannin-tolerant and extracellular tannase producing *Bacillus* isolated from traditional fermented tea leaves and their probiotic functional properties. *Foods* **2020**, *9*, 490. [CrossRef]

49. Jayabalan, R.; Marimuthu, S.; Swaminathan, K. Changes in content of organic acids and tea polyphenols during kombucha tea fermentation. *Food Chem.* **2007**, *102*, 392–398. [CrossRef]

50. Martínez Leal, J.; Valenzuela Suárez, L.; Jayabalan, R.; Huerta Oros, J.; Escalante-Aburto, A. A review on health benefits of kombucha nutritional compounds and metabolites. *CyTA-J. Food* **2018**, *16*, 390–399. [CrossRef]

51. Chidi, B.S.; Bauer, F.; Rossouw, D. Organic acid metabolism and the impact of fermentation practices on wine acidity: A review. *S. Afr. J. Enol. Vitic.* **2018**, *39*, 1–15. [CrossRef]

52. Wei, J.; Zhang, Y.; Yuan, Y.; Dai, L.; Yue, T. Characteristic fruit wine production via reciprocal selection of juice and non-*Saccharomyces* species. *Food Microbiol.* **2019**, *79*, 66–74. [CrossRef]

53. Deshmukh, G.; Manyar, H. Production pathways of acetic acid and its versatile applications in the food industry. In *Biomass*; IntechOpen: London, UK, 2020.

54. Li, Y.; Nguyen, T.T.H.; Jin, J.; Lim, J.; Lee, J.; Piao, M.; Mok, I.-K.; Kim, D. Brewing of glucuronic acid-enriched apple cider with enhanced antioxidant activities through the co-fermentation of yeast (*Saccharomyces cerevisiae* and *Pichia kudriavzevii*) and bacteria (*Lactobacillus plantarum*). *Food Sci. Biotechnol.* **2021**, *30*, 555–564. [CrossRef]

55. Jayabalan, R.; Subathradevi, P.; Marimuthu, S.; Sathishkumar, M.; Swaminathan, K. Changes in free-radical scavenging ability of kombucha tea during fermentation. *Food Chem.* **2008**, *109*, 227–234. [CrossRef] [PubMed]

56. Jakubczyk, K.; Kałduńska, J.; Kochman, J.; Janda, K. Chemical profile and antioxidant activity of the kombucha beverage derived from white, green, black and red tea. *Antioxidants* **2020**, *9*, 447. [CrossRef]

57. Lai, Y.-T.; Hsieh, C.-W.; Lo, Y.-C.; Liou, B.-K.; Lin, H.-W.; Hou, C.-Y.; Cheng, K.-C. Isolation and identification of aroma-producing non-*Saccharomyces* yeast strains and the enological characteristic comparison in wine making. *LWT* **2022**, *154*, 112653. [CrossRef]

58. Wang, R.; Sun, J.; Lassabliere, B.; Yu, B.; Liu, S.Q. β-Glucosidase activity of *Cyberlindnera* (*Williopsis*) *saturnus* var. *mrakii* NCYC 2251 and its fermentation effect on green tea aroma compounds. *LWT* **2021**, *151*, 112184. [CrossRef]

59. Ho, C.-T.; Zheng, X.; Li, S. Tea aroma formation. *Food Sci. Hum. Wellness* **2015**, *4*, 9–27. [CrossRef]

60. Yang, Z.; Baldermann, S.; Watanabe, N. Recent studies of the volatile compounds in tea. *Food Res. Int.* **2013**, *53*, 585–599. [CrossRef]

61. Oliveira, I.; Ferreira, V. Modulating fermentative, varietal and aging aromas of wine using non-*Saccharomyces* yeasts in a sequential inoculation approach. *Microorganisms* **2019**, *7*, 164. [CrossRef]

62. Dzialo, M.C.; Park, R.; Steensels, J.; Lievens, B.; Verstrepen, K.J. Physiology, ecology and industrial applications of aroma formation in yeast. *FEMS Microbiol. Rev.* **2017**, *41*, S95–S128. [CrossRef]

63. Hazelwood, L.A.; Daran, J.-M.; Van Maris, A.J.; Pronk, J.T.; Dickinson, J.R. The Ehrlich pathway for fusel alcohol production: A century of research on *Saccharomyces cerevisiae* metabolism. *Appl. Environ. Microbiol.* **2008**, *74*, 2259–2266. [CrossRef]
64. Walker, G.M.; Stewart, G.G. *Saccharomyces cerevisiae* in the production of fermented beverages. *Beverages* **2016**, *2*, 30. [CrossRef]
65. Zha, M.; Sun, B.; Wu, Y.; Yin, S.; Wang, C. Improving flavor metabolism of *Saccharomyces cerevisiae* by mixed culture with *Wickerhamomyces anomalus* for Chinese Baijiu making. *J. Biosci. Bioeng.* **2018**, *126*, 189–195. [CrossRef] [PubMed]
66. Jiang, Q.-x.; Li, L.-j.; Chen, F.; Rong, B.; Ni, H.; Zheng, F.-p. β-Glucosidase improve the aroma of the tea infusion made from a spray-dried Oolong tea instant. *LWT* **2022**, *159*, 113175. [CrossRef]
67. Carpena, M.; Fraga-Corral, M.; Otero, P.; Nogueira, R.A.; Garcia-Oliveira, P.; Prieto, M.A.; Simal-Gandara, J. Secondary aroma: Influence of wine microorganisms in their aroma profile. *Foods* **2020**, *10*, 51. [CrossRef] [PubMed]
68. Gutiérrez, A.; Boekhout, T.; Gojkovic, Z.; Katz, M. Evaluation of non-*Saccharomyces* yeasts in the fermentation of wine, beer and cider for the development of new beverages. *J. Inst. Brew.* **2018**, *124*, 389–402. [CrossRef]
69. Aung, M.T.; Lee, P.-R.; Yu, B.; Liu, S.-Q. Cider fermentation with three *Williopsis saturnus* yeast strains and volatile changes. *Ann. Microbiol.* **2015**, *65*, 921–928. [CrossRef]
70. Ravasio, D.; Carlin, S.; Boekhout, T.; Groenewald, M.; Vrhovsek, U.; Walther, A.; Wendland, J. Adding flavor to beverages with non-conventional yeasts. *Fermentation* **2018**, *4*, 15. [CrossRef]
71. Bellut, K.; Michel, M.; Zarnkow, M.; Hutzler, M.; Jacob, F.; De Schutter, D.P.; Daenen, L.; Lynch, K.M.; Zannini, E.; Arendt, E.K. Application of non-*Saccharomyces* yeasts isolated from kombucha in the production of alcohol-free beer. *Fermentation* **2018**, *4*, 66. [CrossRef]
72. Liu, S.-Q.; Quek, A.Y.H. Evaluation of beer fermentation with a novel yeast *Williopsis saturnus*. *Food Technol. Biotechnol.* **2016**, *54*, 403. [CrossRef]
73. López, M.C.; Mateo, J.J.; Maicas, S. Screening of β-glucosidase and β-xylosidase activities in four non-*Saccharomyces* yeast isolates. *J. Food Sci.* **2015**, *80*, C1696–C1704. [CrossRef]

Journal of
Fungi

Article

Isolation and Characterization of Ruminal Yeast Strain with Probiotic Potential and Its Effects on Growth Performance, Nutrients Digestibility, Rumen Fermentation and Microbiota of Hu Sheep

Yao Wang [1,2], Zihao Li [1,2], Wei Jin [1,2,*] and Shengyong Mao [1,2]

1   Ruminant Nutrition and Feed Engineering Technology Research Center, College of Animal Science and Technology, Nanjing Agricultural University, Nanjing 210095, China
2   Laboratory of Gastrointestinal Microbiology, National Center for International Research on Animal Gut Nutrition, College of Animal Science and Technology, Nanjing Agricultural University, Nanjing 210095, China
*   Correspondence: jinwei@njau.edu.cn

Abstract: Yeast strains are widely used in ruminant production. However, knowledge about the effects of rumen native yeasts on ruminants is limited. Therefore, this study aimed to obtain a rumen native yeast isolate and investigate its effects on growth performance, nutrient digestibility, rumen fermentation and microbiota in Hu sheep. Yeasts were isolated by picking up colonies from agar plates, and identified by sequencing the ITS sequences. One isolate belonging to *Pichia kudriavzevii* had the highest optical density among these isolates obtained. This isolate was prepared to perform an animal feeding trial. A randomized block design was used for the animal trial. Sixteen Hu sheep were randomly assigned to the control (CON, fed basal diet, $n = 8$) and treatment group (LPK, fed basal diet plus *P. kudriavzevii*, CFU = 8×10^9 head/d, $n = 8$). Sheep were housed individually and treated for 4 weeks. Compared to CON, LPK increased final body weight, nutrient digestibility and rumen acetate concentration and acetate-to-propionate ratio in sheep. The results of Illumina MiSeq PE 300 sequencing showed that LPK increased the relative abundance of lipolytic bacteria (*Anaerovibrio* spp. and *Pseudomonas* spp.) and probiotic bacteria (*Faecalibacterium* spp. and *Bifidobacterium* spp.). For rumen eukaryotes, LPK increased the genera associated with fiber degradation, including protozoan *Polyplastron* and fungus *Pichia*. Our results discovered that rumen native yeast isolate *P. kudriavzevii* might promote the digestion of fibers and lipids by modulating specific microbial populations with enhancing acetate-type fermentation.

Keywords: *Pichia kudriavzevii*; fiber degradation; rumen fermentation; microbial community composition; rumen native yeast

Citation: Wang, Y.; Li, Z.; Jin, W.; Mao, S. Isolation and Characterization of Ruminal Yeast Strain with Probiotic Potential and Its Effects on Growth Performance, Nutrients Digestibility, Rumen Fermentation and Microbiota of Hu Sheep. *J. Fungi* **2022**, *8*, 1260. https://doi.org/10.3390/jof8121260

Academic Editor: Baojun Xu

Received: 5 November 2022
Accepted: 26 November 2022
Published: 29 November 2022

Publisher's Note: MDPI stays neutral with regard to jurisdictional claims in published maps and institutional affiliations.

1. Introduction

The rumen contains a large number of microorganisms that ferment structural and non-structural carbohydrates to produce volatile fatty acids (VFAs) and synthesize microbial proteins for the host [1]. Microbial degradation and fermentation, which can enhance ruminal digestion and absorption capacity and nutrient metabolism, are of central importance in ruminant nutrition [2]. Even if ruminal microorganisms are inherently efficient, probiotics are often added during production to further improve rumen digestion, thus increasing economic benefit.

Yeast products are widely used as probiotic additives in ruminant production [3]. Previous studies have reported that live yeast can consume ruminal oxygen and decrease the redox potential, which favors the activity of anaerobic microorganisms [3,4]. Moreover, the addition of yeast provides nutrients (vitamins, peptides and growth factors) for the host, thereby stimulating the growth of ruminal microorganisms [5]. The regulation of yeast on ruminal microorganisms is reflected in increasing fiber-degrading and lactate-utilizing bacteria, which improves further rumen fermentation efficiency and subsequent production capacity [6,7]. In ruminant production, *Saccharomyces cerevisiae* is the most commonly used yeast additive [8]. However, Ishaq et al. reported no change in the proportion of *S. cerevisiae* in the rumen after feeding active *S. cerevisiae* to dairy cows, indicating that live *S. cerevisiae* may not serve a functional purpose [9]. Therefore, the effectiveness of probiotics depends on their adaptability to a specific ecosystem.

In this regard, potential probiotic yeasts can be selected from the rumen to enhance ruminants' adaptability [10]. The proportion of yeasts in the rumen may be more than 1×10^3 CFU/mL, which plays an important role in improving rumen functions [11]. In particular, *Pichia kudriavzevii*, as a rumen native yeast, accounts for the high relative abundance of ruminal yeasts [12,13]. Results in vitro experiments have revealed that, unlike *S. cerevisiae*, *Pichia kudriavzevii* survives well in the rumen environment [14,15]. Additionally, it possesses fiber-degrading enzymes and the ability to utilize lactate. Feeding rumen-derived *P. kudriavzevii* to dairy cows has recently been applied to increase milk production and feed conversion rates [16,17]. However, the previous studies only focused on milk performance, rumen bacteria and fungi in cows [16,17]. To date, the effects of rumen native *P. kudriavzevii* on growth performance, ruminal microorganisms (bacteria and eukaryote) and the fermentation parameters and digestibility of nutrients remain poorly understood in Hu sheep.

Herein, we isolated rumen native *P. kudriavzevii* from Hu sheep to investigate its effect on growth performance, ruminal fermentation, rumen bacteria and eukaryote community and digestibility of nutrients in Hu sheep. Our study brought new insights to the application of rumen-native yeast in future sheep production practice.

2. Materials and Methods

The protocols of this trial and the use of animals were approved by the Animal Care and Use Committee of Nanjing Agricultural University.

2.1. Isolating, Identifying and Characterizing Yeast Isolates

Fresh rumen fluid samples were collected to isolate yeasts from the rumen fistula of Hu sheep. The rumen fluid was 10 times gradually diluted from 10^{-1} to 10^{-5} with 0.75% sterile saline. About 200 μL of each of the dilutions was spread on yeast agar plates consisting (1000 mL) of yeast extract 5 g, peptone 10 g, glucose 20 g, and agar 14 g. Streptomycin sulfate (2000 U/mL) and Penicillin potassium (1600 U/mL) used to inhibit bacteria were purchased from Jilin Huamu Animal Health Products Co., Ltd. (Changchun, China). These agar plates were incubated at 30 °C for 48 h. In total, 94 colonies were picked up, purified (3 times) and morphologically identified [14]. Then, biomass production was evaluated in liquid medium (per 1000 mL contained yeast extract 10 g, peptone 20 g, glucose 20 g, adenine sulfate 0.03 g) at 30 °C for 48 h based on optical density (OD) at 600 nm. Fourteen isolates with smooth round colony and OD > 1.0 were finally selected. The growth curves of the 14 isolates were further measured in liquid medium at 39 °C, 150 rmp for 48 h to evaluate their growth potential in the rumen. Triplicate was set for each isolate. OD was measured at an interval of 4 h. To identify these isolates, internal transcribed spacer (ITS) regions of fungi were sequenced and BLAST in GenBank. The No. 8 isolate belonging to *P. kudriavzevii* had the highest OD value. Therefore, isolate 8 was used to perform animal feeding experiment. A Plant/Fungi DNA Isolation Kit (Omega Bio-Tek, Shanghai, China) was used to extract the yeast DNA following the instructions. PCR amplification was performed targeting ITS using the primers ITS1F/ITS4R [18]. The sequence of primer ITS1F was 5′-TCCGTAGGTGAACCTGCGG-3′, and ITS4R was 5′-TCCTCCGCTTATTGATATGC-3. PCR was performed in 50 μL reactions including 5× FastPfu buffer 10 μL, 2.5 mmol/L dNTPs 2 μL, each primer 1 μL, FastPfu Polymerase (AP221-01, Tran, Beijing, China) 0.5 μL. The procedure of PCR was 95 °C, 5 min; 95 °C for 30 s, 55 °C for 30 s, 72 °C for 45 s, 30 cycles; 72 °C for 10 min. The amplicons were sequenced using a 3730XL DNA Analyzer (Applied Biosystems, Foster, CA, USA).

Live yeast additive (4×10^8 CFU/g) was prepared using a freeze dryer (JK-FD-1N, Shanghai Jingke Scientific Instrument Co., Ltd., Shanghai, China). Prior to freeze-drying, 10% glycerine and 12% trehalose were mixed with the culture of isolate 8, which had been incubated at 30 °C, 48 h. The procedure of freeze-drying was −50 °C, 90 min; −40 °C, 60 min; −30 °C, 60 min; −20 °C, 60 min; −10 °C, 60 min; 0 °C, 180 min; 10 °C, 300 min; 20 °C, 180 min; 30 °C, 180 min; 30 °C, 300 min. Finally, 5000 g of live yeast additive was stored under seal at 4 °C until use.

2.2. Animal Trial

The animal trial was carried out from January to February 2021, at a sheep farm in Huzhou, China. Randomized block design was used in this study. Sixteen healthy Hu sheep (Age (mean ± SD) = 108 ± 5 d, Body weight = 28.32 ± 0.71 kg) were randomly assigned into CON and LPK groups. Sheep assigned to CON were fed a basal ration ($n = 8$). Sheep assigned to LPK were fed a basal ration plus live yeast additive (20 g/d/head, $n = 8$). All sheep were housed individually (pen, 0.9 m × 1.5 m), and fed twice at 08:00 am and 5:00 pm every day. Free access to water was given. Approximately 5% feed residual was allowed to reduce sorting. The adaptation period was 1 week, and the treatment period was 4 weeks. All sheep were fed pelleted rations. Live yeast was first mixed with a 10 g corn meal. Then, the mixture was dressed on the top of basal diet in the LPK group. In the CON group, only 10 g of corn meal was dressed on the top of basal diet. The basal ration was formulated according to NY/T 816-2004 (Ministry of Agriculture of China, 2004), which met the requirements of 300 g/d of weight gain for sheep weighing 30 kg (Table 1).

Table 1. Composition and nutrient levels of diets.

Items	Content
Ingredient (air-dried basis)	
Mushroom residue [1]	10.00
Maize	40.00
Soybean meal	10.00
Corn germ meal	10.00
Barley malt sprouts	5.50
Barley peel	10.00
Rice husk	6.00
Rice bran	3.50
NaCl	0.80
Dicalcium phosphate	0.80
Limestone meal	1.40
Mineral and vitamin mixture [2]	2.00
Total	100.00
Nutrient level (dry matter basis)	-
Metabolic energy (ME) [3] ($MJ \cdot kg^{-1}$)	11.01
Dry matter (DM), %	87.42
Crude protein (CP), %	14.00
Ether extract (EE), %	3.12
Neutral detergent fiber (NDF), %	43.16
Acid detergent fiber (ADF), %	23.55
Crude ash (Ash), %	9.54
Calcium (Ca), %	1.30
Phosphorus (P), %	0.40

[1] *Flammulina velutipes* residue: the remaining cultivation substrate after the harvest of *F. velutipes*, containing moisture 5.59%, CP 12.7%, NDF 45.2%, ADF 22.3%, EE 1.8%, and Ash 11.6%. [2] Mineral and vitamin mixture was composed of Fe, 3.5 g; Zn, 4.5 g; Cu, 2.0 g; Mn, 3.0 g; Co, 80 mg; I, 140 mg; Se, 40 mg; vitamin A, 550,000 IU; vitamin D, 48,000 IU; vitamin E, 2000 IU. [3] ME was the calculated value [19].

All sheep were weighed before morning feeding on day 1 and 28 during the treatment period. The feed given and orts were recorded every day. Average daily gain (body weight gain divided by no. of days) and feed efficiency (DMI/BW gain, kg/kg) were calculated.

2.3. Sample Collection

Feed samples were collected weekly. Rectum contents (fecal samples) were collected on day 26, 27 and 28. The mL sulfuric acid solution (concentration, 10%) was added to each 100 g fecal sample to fix nitrogen. Finally, all fecal samples from each sheep were mixed and stored at −20 °C.

Rumen contents were sampled on day 28 using an oral tube (Wuhan Kelibo Equipment Co., Ltd., Wuhan, China) at 4 h after morning feeding. In order to avoid saliva contamination, discard the first 50 mL of rumen content [20]. A portable pH meter (HI 9024C; HANNA Instruments, Woonsocket, RI) was used to determine rumen content. pH. Five mL of rumen content was preserved at −80 °C for subsequent microbiota assays. The remaining rumen content was filtrated through four layers of gauze. The filtrate was conserved at −20 °C for further measuring VFAs, ammonia nitrogen (NH_3-N) and microbial crude protein (MCP).

2.4. Chemical Analysis

The DM of feed and feces samples was measured [21]. The samples were firstly dried at 65 °C for 48 h, then dried at 105 °C for 3 h using an air-drying oven. CP was measured using a Kjeldahl apparatus (Kjeltec 8400, Shenzhen, China), with a pretreated process by copper sulfate and concentrated sulfuric acid [21]. EE was measured using a Soxhlet extractor with anhydrous ether as an extractant [21]. NDF and ADF were measured using a fiber analyzer (ANKOM200, ANKOM Technologies, Macedon, NY, USA) by washing with neutral detergent, anhydrous sodium sulfite and α-amylase, and acid detergent [22]. The total tract apparent digestibility (TAD) was measured using acid insoluble ash (AIA) as an interior label [23]. AIA was measured using a muffle furnace (Sx2-4-10N, Tianjin China) with a process treated by hydrochloric acid. The VFAs were measured using a GC-14B (Shimadzu, Shijota, Japan) [24]. Crotonic acid solution was mixed with samples as internal standard. Lactate was determined with sulfuric acid, copper sulfate and p-hydroxybiphenyl [25]. Ammonia nitrogen was determined using the indophenol method [26]. MCP was measured using a colorimetric method [27].

2.5. Microbial DNA Extraction, Sequencing and Data Analysis

Microbial DNA of the rumen content (200 mg) was extracted using an E.Z.N.A. Soil DNA kit (Omega Bio-Tek, USA) combining a bead-beating process. The bacterial sequencing was targeted on the V3 and V4 regions of 16S rRNA genes. The primers' sequences were 341F (5′-CCT AYG GGR BGC ASC AG-3′) and 806R (5′-GGA CTA CNN GGG TAT CTA AT-3′) [28]. Primers for amplifying eukaryotic 18S rRNA genes included TAReukaseFWD1 (5′-CCA GCA SCY GCG GTA ATT CC-3′) and TAReukREV3 (5′-ACT TTC GTT CTT GAT YRA-3′) [29]. PCR amplicons were sequenced on an Illumina MiSeq PE 300 platform in Shanghai BIOERON Biotech. Co., Ltd. (Shanghai, China).

Paired-end reads were merged using FLASH (version 1.2.7) [30]. Then, reads were assigned into amplicon sequence variants (ASVs) at 100% identical similarity using DADA2. The representative ASV sequences were identified by SILVA (138.1) and classified into specific databases of bacteria (16S rRNA sequencing), protozoa and fungi (18S rRNA sequencing). The abundance of microbial communities was tested for significance based on the method of the non-parametric test (Mann–Whiney U) using SPSS (v. 25). Alpha diversity (Chao 1 indices, ACE indices and Shannon indices) was calculated using QIIME2 [31]. The PCA was implemented in the R program (v. 3.6.1) based on Bray–Curtis distance. The analysis of similarities (ANOSIM) was determined for the difference between the CON and LPK groups. The correlations among *P. kudriavzevii*, VFAs' concentration and the microbial relative abundance were assessed by Spearman's test.

2.6. Statistical Analysis

The data of the growth performance, digestibility and rumen fermentation were analyzed with a general linear model (single variable) by SPSS (v. 25, SPSS Inc., Chicago,

IL, USA). Fixed factors were group and treatment. Covariant was the initial body weight of sheep. The data of microbes were analyzed by the nonparametric test (Kruskal–Wallis). The post hoc test was performed by independent samples test to evaluate the difference between any two groups. A significant level was indicated at $p < 0.05$.

3. Results

3.1. Growth Curve of Yeast Isolates

The growth curve of 14 yeast isolates was determined (Figure 1). From 8 to 24 h after inoculation, these isolates entered the logarithmic growth phase, then entered the stable growth phase after 32 h of inoculation. Isolate 8 had the fastest growth rate and highest OD value, and formed white smooth colonies with elliptical edges and a convex center. The sequence of the ITS region showed that isolate 8 belonged to *P. kudriavzevii*.

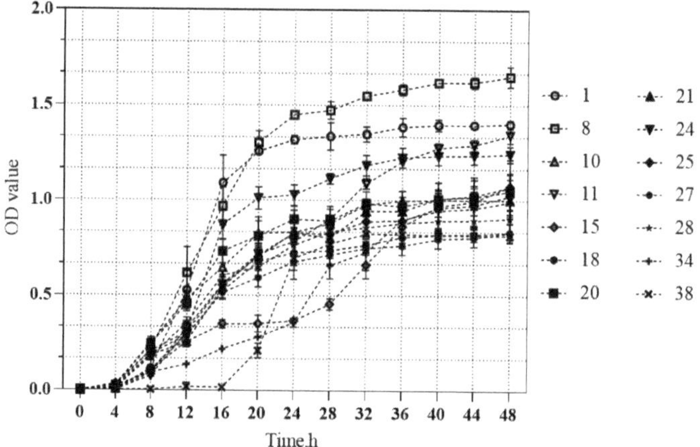

Figure 1. Growth curves of 14 rumen-derived yeast strains incubated for 48 h at 39 °C and 150 rmp in a culture-shaker under aerobic conditions.

3.2. Animal Performance and Nutrient Digestibility

LPK did not affect DMI ($p = 0.906$), initial BW ($p = 0.292$), ADG ($p = 0.906$) and feed efficiency ($p = 0.300$). The final BW was significantly increased in LPK compared to CON ($p = 0.033$, Table 2). The digestibility of DM ($p = 0.033$), EE ($p = 0.010$), NDF ($p = 0.003$) and ADF ($p = 0.005$) in LPK was higher than CON, but CP was not affected ($p = 0.121$, Table 3).

Table 2. Effects of live *Pichia kudriavzevii* on growth performance of Hu sheep [1].

Items	CON	LPK	*p*-Value
Dry matter intake, g/d	1519.06 ± 21.95	1551.13 ± 23.29	0.906
Initial body weight, kg	27.97 ± 0.99	28.66 ± 1.08	0.292
Final body weight, kg	34.90 ± 1.01	36.23 ± 0.93	0.033
Average daily gain, g/d	241.33 ± 19.31	266.83 ± 14.81	0.906
Feed efficiency	6.07 ± 0.45	5.81 ± 0.50	0.300

[1] The values are present as means ± SEM.

Table 3. Effects of live *Pichia kudriavzevii* on nutrient digestibility of Hu sheep [1] (%).

Items	CON	LPK	*p*-Value
Dry matter	60.62 ± 0.37	64.41 ± 1.46	0.033
Crude protein	60.50 ± 1.85	64.68 ± 1.16	0.121
Ether extract	58.08 ± 2.65	67.56 ± 1.41	0.010

Table 3. *Cont.*

Items	CON	LPK	*p*-Value
Neutral detergent fiber	44.38 ± 1.26	51.44 ± 1.28	0.003
Acid detergent fiber	37.65 ± 1.35	45.5 ± 1.63	0.005

[1] The values are present as means ± SEM.

3.3. Rumen Fermentation Parameters

The concentrations of acetate ($p < 0.001$), total VFAs ($p = 0.005$) and the acetate-to-propionate ratio ($p = 0.011$) were higher in the LPK group. However, ruminal pH ($p = 0.618$) and concentrations of NH_3-N ($p = 0.466$), lactate ($p = 0.988$), MCP ($p = 0.562$) and other VFAs ($p > 0.05$) were not affected by treatment (Table 4).

Table 4. Effects of live *Pichia kudriavzevii* on rumen fermentation parameters of Hu sheep [1].

Items	CON	LPK	*p*-Value
Rumen pH	6.30 ± 0.08	6.18 ± 0.39	0.618
Ammonia, mg/dL	19.23 ± 1.97	21.68 ± 2.27	0.466
Microbial crud protein, mg/dL	43.55 ± 0.04	38.69 ± 0.06	0.562
Lactate, mmol/L	0.26 ± 0.02	0.26 ± 0.02	0.988
Acetate, mmol/L	40.73 ± 1.67	52.08 ± 2.02	<0.001
Propionate, mmol/L	24.42 ± 2.15	20.94 ± 0.91	0.137
Isobutyrate, mmol/L	0.17 ± 0.05	0.29 ± 0.05	0.098
Butyrate, mmol/L	16.27 ± 1.26	17.36 ± 1.03	0.444
Isovalerate, mmol/L	0.47 ± 0.08	0.47 ± 0.05	0.997
Valerate, mmol/L	1.77 ± 0.20	1.56 ± 0.15	0.382
Acetate/propionate	1.76 ± 0.19	2.52 ± 0.14	0.011
Total volatile fatty acids, mmol/L	83.84 ± 2.58	92.69 ± 1.72	0.005

[1] The values are present as means ± SEM.

3.4. Rumen Microbial Community Diversity

PCoA analysis showed that LPK did not separate the rumen bacterial communities from CON. (ANOSIM: R = 0.082, $p = 0.174$; Figure 2A). However, there were clear segregation and dissimilarities for ruminal protozoa (ANOSIM: R = 0.423, $p = 0.001$; Figure 2B) and fungi (ANOSIM: R = 0.326, $p = 0.002$; Figure 2C).

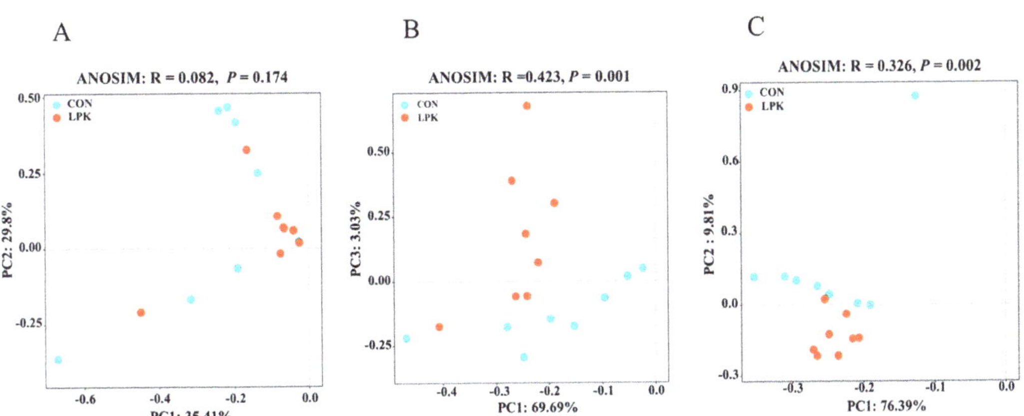

Figure 2. Principal coordinate analysis (PCoA) of rumen bacteria (**A**), protozoa (**B**), and fungi (**C**) based on operational taxon data. Control (CON) group (blue circle), live *Pichia kudriavzevii* (LPK) group (red circle).

The α-diversity was shown in Table 5. For bacteria, the richness was higher in LPK than CON (Observed ASVs, $p = 0.029$; Chao 1, $p = 0.030$). However, LPK did not affect the Shannon ($p = 0.110$) and Simpson index ($p = 0.213$). For eukaryotes, LPK did not affect the α-diversity.

Table 5. Effects of live *Pichia kudriavzevii* on microbial community diversity in the ruminal content of Hu sheep [1].

Item	CON	LPK	*p*-Value
Rumen bacteria			
Observed ASVs [2]	167.38 ± 17.58	230.75 ± 19.12	0.029
Coverage	1.00 ± 0.0002	1.00 ± 0.0002	0.255
Chao 1	170.38 ± 18.50	234.25 ± 18.93	0.030
Shannon	3.24 ± 0.26	3.80 ± 0.20	0.110
Simpson	0.15 ± 0.04	0.09 ± 0.02	0.213
Rumen protozoa			
Observed ASVs	11.00 ± 2.30	10.50 ± 1.09	0.847
Coverage	0.98 ± 0.01	0.98 ± 0.003	0.618
Chao 1	13.13 ± 3.01	11.38 ± 1.12	0.600
Shannon	1.41 ± 0.30	1.52 ± 0.16	0.740
Simpson	0.43 ± 0.11	0.34 ± 0.06	0.512
Rumen fungi			
Observed ASVs	24.38 ± 3.18	23.38 ± 1.53	0.781
Coverage	0.98 ± 0.01	0.98 ± 0.003	0.868
Chao 1	28.00 ± 4.53	29.00 ± 4.74	0.881
Shannon	2.20 ± 0.13	2.28 ± 0.08	0.597
Simpson	0.20 ± 0.03	0.16 ± 0.01	0.148

[1] The values are presented as means ± SEM; [2] ASVs = Amplicon Sequence Variants.

3.5. Rumen Microbiota Composition

The dominant phyla were Bacteroidetes, Firmicutes and Proteobacteria in the two groups, accounting for about 93% to 97% of the total phyla, respectively. Thirty dominant genera were > 0.5% in proportion. The first three dominant bacterial genera were *Ruminococcus*, *Selenomonas* and *Prevotella*. The relative abundance of *Anaerovibrio* ($p = 0.028$), *Pseudomonas* ($p = 0.028$), *Faecalibacterium* ($p = 0.001$) and *Bifidobacterium* ($p = 0.038$) was higher in LPK than CON. While the percentages of *Pseudoscardovia* ($p = 0.011$) and *Syntrophococcus* ($p = 0.035$) were lower in LPK (Figure 3B).

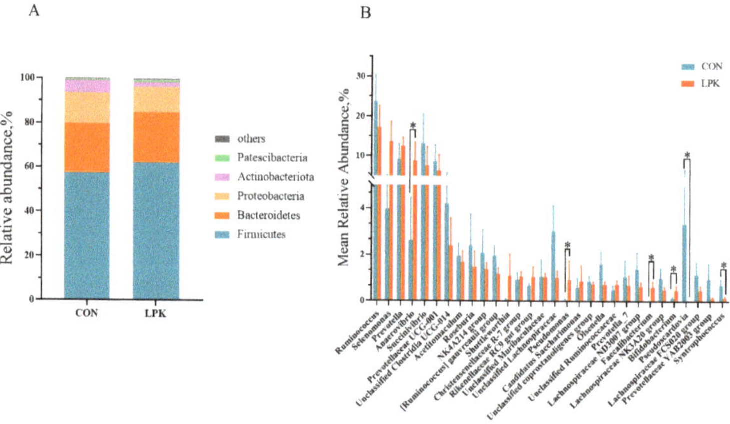

Figure 3. The bacterial community in the rumen of Hu sheep: (**A**) The mean relative abundance of phylum $\geq$ 0.5% in at least one group is present. (**B**) The mean relative abundance of genus $\geq$ 0.5% in at least one group is present. Error bars, SEM. *, $p < 0.05$. CON = control group; LPK = live *Pichia kudriavzevii* group.

All protozoa identified belonged to the phylum Ciliophora. At the genus level, Entodinium was the most abundant genus, accounting for 60.25% and 77.14% in LPK and CON, respectively. The relative abundances of *Polyplastron* (p = 0.002) and unclassified Trichostomatia (p = 0.004) were higher in LPK than CON, while the percentage of unclassified Hypotrichia was lower in LPK (p = 0.032, Figure 4B).

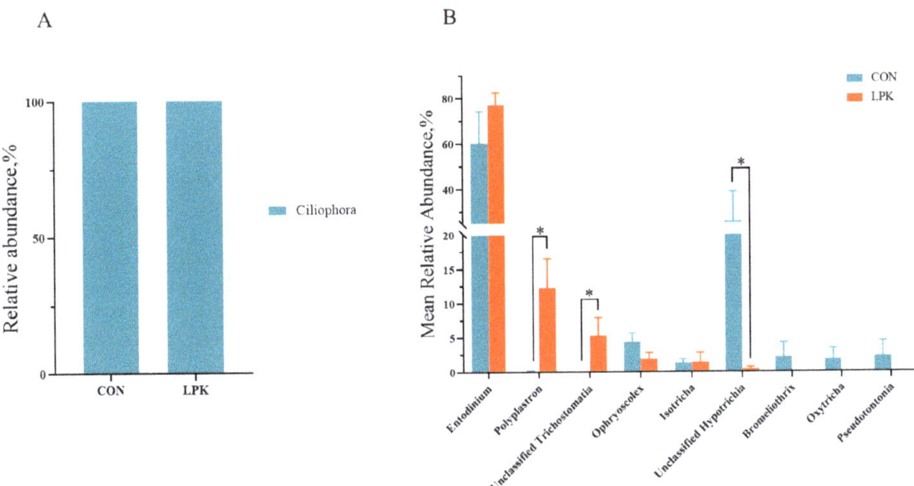

Figure 4. The rumen protozoa community in Hu sheep: (**A**) The mean relative abundance of phylum $\geq$ 1% in at least one group is present. (**B**) The mean relative abundance of genus $\geq$ 1% in at least one group is present. Error bars, SEM. * p < 0.05. CON = control group; LPK = live *Pichia kudriavzevii* group.

For the rumen fungi, the dominant phyla were Ascomycota and Basidiomycota, accounting for more than 99% of the total phyla. The relative abundance of Ascomycota was higher in LPK than CON (p = 0.003; Figure 5A), while Basidiomycota was lower in LPK (p = 0.003; Figure 5A). The dominant genera (the relative abundance > 1%) were *Kurtzmaniella Candidaclade*, *Pichia* and the *Clavispora-Candida clade*. The relative abundance of *Pichia* was higher in LPK than in CON (p = 0.012; Figure 5B).

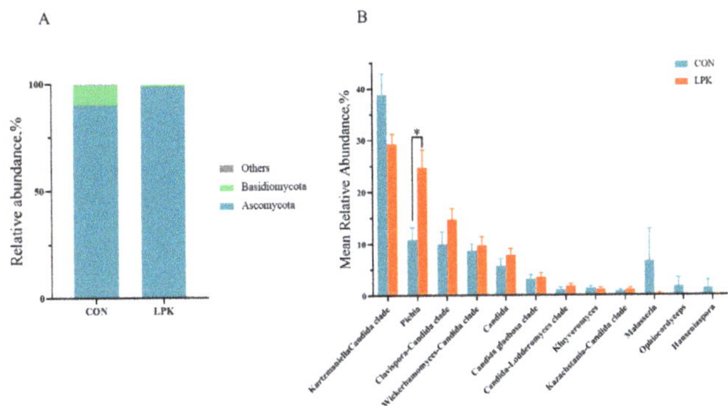

Figure 5. The rumen fungal community in Hu sheep: (**A**) The mean relative abundance of phylum $\geq$ 1% in at least one group is present. (**B**) The mean relative abundance of genus $\geq$ 1% in at least one group is present. Error bars, SEM. * p < 0.05. CON = control group; LPK = live *Pichia kudriavzevii* group.

3.6. Correlation Analysis

Spearman correlation analysis was performed among *P. kudriavzevii*, rumen VFAs and microorganisms (the top 10 bacteria, fungi and protozoa based on relative abundance at genus level). For the bacteria (Figure 6), the abundance of *Prevotella* was negatively correlated with the concentration of propionate (R = −0.65; p = 0.007) and positively correlated with the acetate-to-propionate ratio (R = 0.676; p = 0.005). *Selenomonas* (R = 0.673; p = 0.005) and *Anaerovibrio* (R = 0.682; p = 0.004) were positively correlated with the concentration of total VFA. Unclassified Clostridia UCG-014 had a positive correlation with butyrate concentration (R = 0.597; p = 0.016). For the protozoa, *Polyplastron* (R = 0.809; p < 0.001) was positively correlated with *P. kudriavzevii* and had a positive correlation with acetate concentration (R = 0.567; p = 0.022) and acetate-to-propionate ratio (R = 0.547; p = 0.028). *Oxytricha* was positively associated with the concentration of valerate (R = 0.199; p = 0.049), and *Acineta* was negatively associated with the acetate-to-propionate ratio (R = −0.524; p = 0.037). For the fungi, *P. kudriavzevii* showed a positive correlation with Pichia (R = 0.924; p < 0.001) and negatively correlated with *Malassezia* (R = −0.668; p = 0.005), whereas *Pichia* was positively linked with the acetate concentration (R = 0.661; p = 0.005) and acetate-to-propionate ratio (R = 0.595; p = 0.015). *Candida* was negatively correlated with the concentration of Valerate (R = −0.571; p = 0.020). There was a negative correlation between the *Kazachstania-Candida clade* and the Propionate concentration (R = −0.508; p = 0.044) (Figure 6).

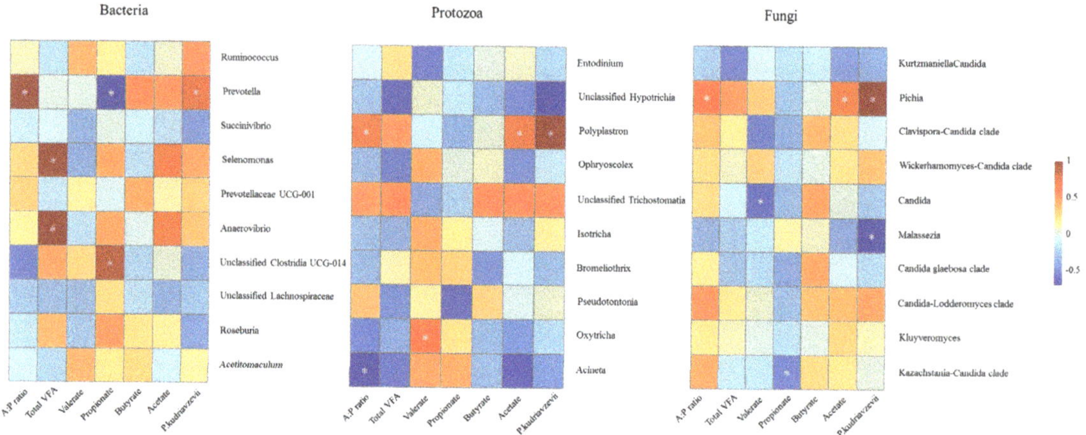

Figure 6. Spearman rank correlation analysis between rumen microbial community (top 10 relative abundance at the genus level of bacteria, protozoa and fungi), *Pichia kudriavzevii* and rumen volatile acids. Red, positive; blue, negative. * p < 0.05.

4. Discussion

In the present study, the function of rumen native yeast *P. kudriavzevii* was investigated through animal *in vivo* and *in vitro* experiments. It had been reported that the growth of yeast was affected by species, oxygen, nutrients, temperature and other aspects [32]. Under a healthy physiological state, the rumen temperature is maintained at 38.5 to 40 °C. Therefore, the biomass production capability of the obtained isolates was evaluated at 39 °C in the present study. Isolate 8 grew rapidly and had a relatively high OD value, suggesting that this isolate would grow faster and have greater biomass in rumen conditions. It was identical to the previous report by Suntara et al. [15]. Therefore, we predicted that isolate 8 (*Pichia kudriavzevii*) would maintain activity in the rumen environment and regulate rumen microbial composition.

Isolate 8 elevated the concentrations of acetate and total VFA, which were consistent with previous studies [33,34]. Since VFA produced by microbial fermentation on carbohydrates in the rumen provides more than 70% of the host's energy requirements [35], a higher concentration of total VFA would be expected to provide more energy to the body. An increased concentration of acetate may hint at an enhanced ability to digest structural carbohydrates [36]. This observation corresponded to the increase in nutrient digestibility. Isolate 8 feeding increased the digestibility of nutrients, which was identical to the previous studies on other yeast products, such as *Saccharomyces cerevisiae* [34,37]. Therefore, isolate 8 feeding may promote fiber degradation and acetate fermentation by regulating rumen microbiota.

To further explore the effects of isolate 8 on the rumen microbiota, we focused on the bacterial, protozoal and fungal communities, which were the main fermentation microbial populations. Isolate 8 feeding increased bacterial richness, which was consistent with the studies using commercial *S. cerevisiae* on dairy cows [38]. Isolate 8 increased the relative abundance of *Anaerovibrio*, which plays an important role in rumen lipid degradation [39]. By producing extracellular lipase, *Anaerovibrio* decomposes lipids and glycerol into free VFAs [40,41]. This result may explain the increased digestibility of EE by isolate 8. The biomass of *P. kudriavzevii* is rich in glycerol [42], which might be used as a substrate to promote the growth of *Anaerovibrio*. Our data showed that isolate 8 feeding increased the relative abundance of *Pseudomonas*, which can perform functions including hydrocarbon degradation, xenobiotic degradation, nitrification, denitrification, cellulose degradation and lipolysis [43–45]. The higher proportion of *Pseudomonas* may improve feed digestibility in Hu sheep. In addition, isolate 8 feeding also promoted the growth of two probiotics, *Faecalibacterium* and *Bifidobacterium*. *Faecalibacterium* can reduce the severity of inflammation and enhance intestinal barrier function by releasing metabolites [46], while *Bifidobacterium* can stimulate and maintain the intestinal mucosal barrier and immune response and also produce a series of beneficial metabolic substrates to prevent the attachment of pathogens, which exerts positive effects in the intestinal tract of young animals [47]. An improvement in the relative abundance of *Bifidobacterium* can reduce the risk of disease and improve animal performance [48]. Taken together, these findings showed that isolate 8 feeding improved ruminal digestion of lipids and increases the abundance of probiotics.

Tripathi and Karim reported that feeding yeast increased rumen ciliates [49]; although, some studies also have shown the opposite results [50]. In the current study, isolate 8 feeding promoted the growth of the dominant genera of protozoa, *Polyplastron* and unclassified Trichostomatia, and decreased the relative abundance of unclassified Hypotrichia. As common rumen protozoa, *Polyplastron* is capable of producing cellulase and xylanase to decompose structural polysaccharides such as cellulose and hemicellulose [51], contributing to the degradation of fibers in the LPK group. Therefore, our study suggested that isolate 8 enhanced the relative abundance of ruminal-specific protozoa to promote fiber degradation.

Only a few studies, to date, have investigated the effects of yeast on the community of rumen fungi [16]. Chaucheyras-Durand believed that the addition of yeast can promote the fixation of fungi in feed pellets and thus affect the digestive degradation of fiber in the rumen [52]. Isolate 8 increased the relative abundance of Ascomycota and decreased Basidiomycota. Isolate 8 also increased the relative abundance of *Pichia*. The increase in *p. kudriavzevii* was the crucial factor for the increase in *Pichia*. As a fiber-degrading fungus, *P. kudriavzevii* can help the host to degrade cellulose, which was identical to the results of the correlation analysis. *Pichia* was positively correlated with the acetate concentration and acetate-to-propionate ratio. Even though the activity of isolate 8 was not measured in the rumen of Hu sheep, taking all the results into consideration, isolate 8 might have a high potential to survive and exert beneficial effects through interacting with native microbiota in the rumen [4]. In addition, the high biomass of *P. kudriavzevii* might provide rumen microorganisms with rich nutrients such as organic acids, peptides and amino acids [3].

5. Conclusions

A *P. kudriavzevii* isolate was obtained from the rumen of sheep. Live *P. kudriavzevii* feeding promoted the performance of rumen acetate-type fermentation and digestibility of nutrients in Hu sheep. *P. kudriavzevi* enhanced specific ruminal microbial populations, including lipolytic bacteria (e.g., *Anaerovibrio* spp. and *Pseudomonas* spp.), probiotic bacteria (e.g., *Faecalibacterium* spp. and *Bifidobacterium* spp.) and fiber-degrading eukaryotes (e.g., *Pichia* spp. and *Polyplastron* spp.). Therefore, rumen native yeast *P. kudriavzevii* has a high potential for use in sheep production.

Author Contributions: Y.W. conducted experiments and laboratory analysis. Y.W. and Z.L. performed the statistical analysis, and wrote the manuscript. W.J. and S.M. conceived the idea, acquired the funding, supervised the study, and revised the manuscript. All authors have read and agreed to the published version of the manuscript.

Funding: This research was funded by Jiangsu Province Special Project for Carbon Peak & Carbon Neutral Science and Technology Innovation (BE2022309).

Institutional Review Board Statement: Not applicable.

Informed Consent Statement: Not applicable.

Data Availability Statement: The raw reads of the 16S rRNA gene and 18S rRNA gene sequencing obtained by our research are accessible under the Sequence Read Archive of the NCBI with the accession numbers PRJNA896857.

Acknowledgments: The authors thank Limei Lin for her suggestions on this manuscript.

Conflicts of Interest: The authors declare no conflict of interest.

References

1. Weimer, P.J. Redundancy, Resilience, and Host Specificity of the Ruminal Microbiota: Implications for Engineering Improved Ruminal Fermentations. *Front. Microbiol.* **2015**, *6*, 296. [CrossRef] [PubMed]
2. Wang, Y.-L.; Wang, W.-K.; Wu, Q.-C.; Yang, H.-J. The Release and Catabolism of Ferulic Acid in Plant Cell Wall by Rumen Microbes: A Review. *Anim. Nutr.* **2022**, *9*, 335–344. [CrossRef] [PubMed]
3. Fonty, G.; Chaucheyras-Durand, F. Effects and Modes of Action of Live Yeasts in the Rumen. *Biologia* **2006**, *61*, 741–750. [CrossRef]
4. Chaucheyras-Durand, F.; Walker, N.D.; Bach, A. Effects of Active Dry Yeasts on the Rumen Microbial Ecosystem: Past, Present and Future. *Anim. Feed Sci. Technol.* **2008**, *145*, 5–26. [CrossRef]
5. Moya, D.; Ferret, A.; Blanch, M.; Fuentes, M.C.; Fandiño, J.I.; Calsamiglia, S. Effects of Live Yeast (*Saccharomyces cerevisiae*) and Type of Cereal on Rumen Microbial Fermentation in a Dual Flow Continuous Culture Fermentation System. *J. Anim. Physiol. Anim. Nutr.* **2018**, *102*, 1488–1496. [CrossRef] [PubMed]
6. Pancini, S.; Cooke, R.F.; BrandÃO, A.P.; Dias, N.W.; Timlin, C.L.; Fontes, P.L.P.; Sales, A.F.F.; Wicks, J.C.; Murray, A.; Marques, R.S.; et al. Supplementing a Yeast-Derived Product to Feedlot Cattle Consuming Monensin: Impacts on Performance, Physiological Responses, and Carcass Characteristics. *Livest. Sci.* **2020**, *232*, 103907. [CrossRef]
7. Halfen, J.; Carpinelli, N.; Pino, F.A.B.D.; Chapman, J.D.; Sharman, E.D.; Anderson, J.L.; Osorio, J.S. Effects of Yeast Culture Supplementation on Lactation Performance and Rumen Fermentation Profile and Microbial Abundance in Mid-Lactation Holstein Dairy Cows. *J. Dairy Sci.* **2021**, *104*, 11580–11592. [CrossRef]
8. Thrune, M.; Bach, A.; Ruiz-Moreno, M.; Stern, M.D.; Linn, J.G. Effects of *Saccharomyces Cerevisiae* on Ruminal PH and Microbial Fermentation in Dairy Cows. *Livest. Sci.* **2009**, *124*, 261–265. [CrossRef]
9. Ishaq, S.L.; AlZahal, O.; Walker, N.; McBride, B. An Investigation into Rumen Fungal and Protozoal Diversity in Three Rumen Fractions, during High-Fiber or Grain-Induced Sub-Acute Ruminal Acidosis Conditions, with or without Active Dry Yeast Supplementation. *Front. Microbiol.* **2017**, *8*, 1943. [CrossRef]
10. Zoumpopoulou, G.; Kazou, M.; Alexandraki, V.; Angelopoulou, A.; Papadimitriou, K.; Pot, B.; Tsakalidou, E. *Probiotics and Prebiotics in Animal Health and Food Safety*; Gioia, D.D., Biavati, B., Eds.; Springer International Publishing: Berlin, Germany, 2018; pp. 1–34.
11. Lund, A. Yeasts and Moulds in the Bovine Rumen. *Microbiology* **2000**, *81*, 453–462. [CrossRef]
12. Almeida, P.N.M.D.; Duarte, E.R.; AbrÃO, F.O.; Freitas, C.E.S.; Geraseev, L.C.; Rosa, C.A. Aerobic Fungi in the Rumen Fluid from Dairy Cattle Fed Different Sources of Forage. *Rev. Bras. Zootec.* **2012**, *41*, 2336–2342. [CrossRef]
13. Abrão, F.O.; Freitas, C.E.S.; Duarte, E.R.; Geraseev, L.C.; Barreto, S.M.P.; Medeiros, A.O.; Rosa, C.A. Leveduras No Rúmen De Caprinos E Bovinos De Corte Criados Em Pastagem Tropicais. *Arq. Bras. Med. Veterinária E Zootec.* **2011**, *63*, 526–529. [CrossRef]
14. Fernandes, T.; Carvalho, B.F.; Mantovani, H.C.; Schwan, R.F.; Ávila, C.L.S. Identification and Characterization of Yeasts from Bovine Rumen for Potential Use as Probiotics. *J. Appl. Microbiol.* **2019**, *127*, 845–855. [CrossRef]
15. Suntara, C.; Cherdthong, A.; Wanapat, M.; Uriyapongson, S.; Leelavatcharamas, V.; Sawaengkaew, J.; Chanjula, P.; Foiklang, S. Isolation and Characterization of Yeasts from Rumen Fluids for Potential Use as Additives in Ruminant Feeding. *Vet. Sci.* **2021**, *8*, 52. [CrossRef]
16. Ji, Y.; Dong, X.; Liu, Z.; Wang, W.; Yan, H.; Liu, X. Effects of Bovine *Pichia kudriavzevii* T7, *Candida glabrata* B14, and *Lactobacillus plantarum* Y9 on Milk Production, Quality and Digestive Tract Microbiome in Dairy Cows. *Microorganisms* **2022**, *10*, 842. [CrossRef]

17. Valldecabres, A.; Gilmore, S.P.; Embree, J.J.; Zhelev, I.Z.; Gaffney, J.R.; Marotz, C.A.; Yang, F.; Izzo, A.S.; Embree, M.M.; Lago, A. Effects of Rumen-Native Microbial Feed Supplementation on Milk Yield, Composition, and Feed Efficiency in Lactating Dairy Cows. *J. Anim. Sci.* **2022**, *100*, skac275. [CrossRef]

18. Zhang, Y.J.; Zhang, S.; Liu, X.Z.; Wen, H.A.; Wang, M. A Simple Method of Genomic Dna Extraction Suitable for Analysis of Bulk Fungal Strains. *Lett. Appl. Microbiol.* **2010**, *51*, 114–118. [CrossRef]

19. Yang, H. Feeding Standard of Meat-Producing Sheep and Goats. Postdoctoral Thesis, Chinese Academy of Agricultural Sciences (CAAS), Beijing, China, 2003.

20. Shen, J.S.; Chai, Z.; Song, L.J.; Liu, J.X.; Wu, Y.M. Insertion Depth of Oral Stomach Tubes May Affect the Fermentation Parameters of Ruminal Fluid Collected in Dairy Cows. *J. Dairy Sci.* **2012**, *95*, 5978–5984. [CrossRef]

21. AOAC International. *Official Methods of Analysis of AOAC International*, 22nd ed.; Jr, G.W.L., Ed.; Oxford University Press: Oxford, UK, 2007.

22. Van Soest, P.J.; Robertson, J.B.; Lewis, B.A. Methods for Dietary Fiber, Neutral Detergent Fiber, and Nonstarch Polysaccharides in Relation to Animal Nutrition. *J. Dairy Sci.* **1991**, *74*, 3583–3597. [CrossRef]

23. International Organization for Standardization. Animal Feeding Stuffs: Determination of Ash Insoluble in Hydrochloric Acid. *ISO 5985:2002/Cor 1:2005*. Available online: https://www.iso.org/standard/42078.html (accessed on 1 July 2022).

24. Qin, W.L. Determination of Rumen Volatile Fatty Acids by Means of Gas Chromatography. *J. Nanjing Agric. Coll.* **1982**, *4*, 110–116.

25. Baker, S.B.; Summerson, W.H. The Colorimetric Determination of Lactic Acid in Biological Material. *J. Biol. Chem.* **1941**, *138*, 535–554. [CrossRef]

26. Chaney, A.L.; Marbach, E.P. Modified Reagents for Determination of Urea and Ammonia. *Clin. Chem.* **1962**, *8*, 130–132. [CrossRef] [PubMed]

27. Makkar, H.P.S.; Sharma, O.P.; Dawra, R.K.; Negi, S.S. Simple Determination of Microbial Protein in Rumen Liquor. *J. Dairy Sci.* **1982**, *65*, 2170–2173. [CrossRef]

28. Zakrzewski, M.; Goesmann, A.; Jaenicke, S.; Jünemann, S.; Eikmeyer, F.; Szczepanowski, R.; Al-Soud, W.A.; Sørensen, S.; Pühler, A.; Schlüter, A. Profiling of the Metabolically Active Community from a Production-Scale Biogas Plant by Means of High-Throughput Metatranscriptome Sequencing. *J. Biotechnol.* **2012**, *158*, 248–258. [CrossRef] [PubMed]

29. Logares, R.; Audic, S.; Santini, S.; Pernice, M.C.; de Vargas, C.; Massana, R. Diversity Patterns and Activity of Uncultured Marine Heterotrophic Flagellates Unveiled with Pyrosequencing. *ISME J.* **2012**, *6*, 1823–1833. [CrossRef]

30. Magoč, T.; Salzberg, S.L. FLASH: Fast Length Adjustment of Short Reads to Improve Genome Assemblies. *Bioinformatics* **2011**, *27*, 2957–2963. [CrossRef]

31. Schloss, P.D.; Westcott, S.L.; Ryabin, T.; Hall, J.R.; Hartmann, M.; Hollister, E.B.; Lesniewski, R.A.; Oakley, B.B.; Parks, D.H.; Robinson, C.J. Introducing Mothur: Open-Source, Platform-Independent, Community-Supported Software for Describing and Comparing Microbial Communities. *Appl. Environ. Microbiol.* **2009**, *75*, 7537. [CrossRef]

32. García-Ríos, E.; Alonso-Del-Real, J.; Lip, K.Y.F.; Pinheiro, T.; Teixeira, J.; van Gulik, W.; Domingues, L.; Querol, A.; Guillamón, J.M. Genome-Wide Effect of Non-Optimal Temperatures under Anaerobic Conditions on Gene Expression in *Saccharomyces cerevisiae*. *Genomics* **2022**, *114*, 110386. [CrossRef]

33. Marden, J.P.; Julien, C.; Monteils, V.; Auclair, E.; Moncoulon, R.; Bayourthe, C. How Does Live Yeast Differ from Sodium Bicarbonate to Stabilize Ruminal Ph in High-Yielding Dairy Cows? *J. Dairy Sci.* **2008**, *91*, 3528–3535. [CrossRef]

34. Desnoyers, M.; Giger-Reverdin, S.; Bertin, G.; Duvaux-Ponter, C.; Sauvant, D. Meta-Analysis of the Influence of *Saccharomyces cerevisiae* Supplementation on Ruminal Parameters and Milk Production of Ruminants. *J. Dairy Sci.* **2009**, *92*, 1620–1632. [CrossRef]

35. Siciliano-Jones, J.; Murphy, M.R. Production of Volatile Fatty Acids in the Rumen and Cecum-Colon of Steers as Affected by Forage: Concentrate and Forage Physical Form. *J. Dairy Sci.* **1989**, *72*, 485–492. [CrossRef] [PubMed]

36. Opsi, F.; Fortina, R.; Tassone, S.; Bodas, R.; López, S. Effects of Inactivated and Live Cells of *Saccharomyces cerevisiae* on *In Vitro* Ruminal Fermentation of Diets with Different Forage: Concentrate Ratio. *J. Agric. Sci.* **2012**, *150*, 271–283. [CrossRef]

37. Perdomo, M.C.; Marsola, R.S.; Favoreto, M.G.; Adesogan, A.; Staples, C.R.; Santos, J.E.P. Effects of feeding live yeast at 2 dosages on performance and feeding behavior of dairy cows under heat stress. *J. Dairy Sci.* **2019**, *103*, 325–339. [CrossRef] [PubMed]

38. Xiao, J.X.; Alugongo, G.M.; Chung, R.; Dong, S.Z.; Li, S.L.; Yoon, I.; Wu, Z.H.; Cao, Z.J. Effects of *Saccharomyces cerevisiae* Fermentation Products on Dairy Calves: Ruminal Fermentation, Gastrointestinal Morphology, and Microbial Community. *J. Dairy Sci.* **2016**, *99*, 5401–5412. [CrossRef] [PubMed]

39. Strömpl, C.; Tindall, B.J.; Jarvis, G.N.; Lünsdorf, H.; Moore, E.R.B.; Hippe, H. A Re-Evaluation of the Taxonomy of the Genus *Anaerovibrio*, with the Reclassification of *Anaerovibrio glycerini* as *Anaerosinus glycerini* Gen. Nov., Comb. Nov., and *Anaerovibrio burkinabensis* as *Anaeroarcus burkinensis*. Gen. Nov., Comb. Nov. *Int. J. Syst. Evol. Microbiol.* **1999**, *49*, 1861–1872. [CrossRef]

40. Hobson, P.N.; Mann, S.O. The Isolation of Glycerol-Fermenting and Lipolytic Bacteria from the Rumen of the Sheep. *J. Gen. Microbiol.* **1961**, *25*, 227–240. [CrossRef]

41. Schauder, R.; Schink, B. *Anaerovibrio Glycerini* Sp. Nov., an Anaerobic Bacterium Fermenting Glycerol to Propionate, Cell Matter, and Hydrogen. *Arch. Microbiol.* **1989**, *152*, 473–478. [CrossRef]

42. Jennings, D.H. Polyol Metabolism in Fungi. In *Advances in Microbial Physiology*; Rose, A.H., Tempest, D.W., Eds.; Academic Press: Cambridge, MA, USA, 1985; Volume 25, pp. 149–193. [CrossRef]

43. Lalucat, J.; Bennasar, A.; Bosch, R.; García-Valdés, E.; Palleroni, N.J. Biology of *Pseudomonas stutzeri*. *Microbiol. Mol. Biol. Rev.* **2006**, *70*, 510–547. [CrossRef]

44. Priji, P.; Sajith, S.; Unni, K.N.; Anderson, R.C.; Benjamin, S. *Pseudomonas* sp. BUP6, a Novel Isolate from Malabari Goat Produces an Efficient Rhamnolipid Type Biosurfactant. *J. Basic Microbiol.* **2017**, *57*, 21–33. [CrossRef]

45. Patel, M.; Patel, H.M.; Vohra, N.; Dave, S. Complete Genome Sequencing and Comparative Genome Characterization of the Lignocellulosic Biomass Degrading Bacterium *Pseudomonas stutzeri* Mp4687 from Cattle Rumen. *Biotechnol. Rep.* **2020**, *28*, e00530. [CrossRef]

46. Lopez-Siles, M.; Duncan, S.H.; Garcia-Gil, L.J.; Martinez-Medina, M. *Faecalibacterium prausnitzii*: From Microbiology to Diagnostics and Prognostics. *J. ISME* **2017**, *11*, 841–852. [CrossRef]

47. Leahy, S.C.; Higgins, D.G.; Fitzgerald, G.F.; Sinderen, D. Getting Better with Bifidobacteria. *J. Appl. Microbiol.* **2005**, *98*, 1303–1315. [CrossRef]

48. Kelly, W.J.; Cookson, A.L.; Altermann, E.; Lambie, S.C.; Perry, R.; Teh, K.H.; Otter, D.E.; Shapiro, N.; Woyke, T.; Leahy, S.C. Genomic Analysis of Three *Bifidobacterium* Species Isolated from the Calf Gastrointestinal Tract. *Sci. Rep.* **2016**, *6*, 30768. [CrossRef]

49. Tripathi, M.K.; Karim, S.A. Effect of Yeast Cultures Supplementation on Live Weight Change, Rumen Fermentation, Ciliate Protozoa Population, Microbial Hydrolytic Enzymes Status and Slaughtering Performance of Growing Lamb. *Livest. Sci.* **2011**, *135*, 17–25. [CrossRef]

50. Majewska, M.P.; Miltko, R.; Bełżecki, G.; Kowalik, B. Population of Protozoa and Carbohydrate-Digesting Enzymes in the Rumen of Sheep Fed a Diet Supplemented with Yeast *Saccharomyces cerevisiae*. *Small Rumin. Res.* **2021**, *205*, 106544. [CrossRef]

51. Akkada, A.; Eadie, J.; Howard, B. The Biochemistry of Rumen Protozoa. 7. the Carbohydrases of *Polyplastron multivesiculatum* (Dogiel & Fedorowa). *Biochem. J.* **1963**, *89*, 268–272. [CrossRef]

52. Chaucheyras-Durand, F.; Ameilbonne, A.; Bichat, A.; Mosoni, P.; Ossa, F.; Forano, E. Live Yeasts Enhance Fibre Degradation in the Cow Rumen through an Increase in Plant Substrate Colonization by Fibrolytic Bacteria and Fungi. *J. Appl. Microbiol.* **2016**, *120*, 560–570. [CrossRef]

MDPI

St. Alban-Anlage 66

4052 Basel

Switzerland

www.mdpi.com

Journal of Fungi Editorial Office

E-mail: jof@mdpi.com

www.mdpi.com/journal/jof